Solidification and Solid-State Transformations of Metals and Alloys

Solidification and Solid-State Transformations of Metals and Alloys

José Antonio Pero-Sanz Elorz

María José Quintana Hernández

Luis Felipe Verdeja González

ELSEVIER

elsevier.com

Elsevier
Radarweg 29, PO Box 211, 1000 AE Amsterdam, Netherlands
The Boulevard, Langford Lane, Kidlington, Oxford OX5 1GB, United Kingdom
50 Hampshire Street, 5th Floor, Cambridge, MA 02139, United States

Library of Congress Cataloging-in-Publication Data
A catalog record for this book is available from the Library of Congress

British Library Cataloguing-in-Publication Data
A catalogue record for this book is available from the British Library

ISBN: 978-0-12-812607-3

Publisher: Matthew Deans
Acquisition Editor: Christina Gifford
Editorial Project Manager: Mariana L. Kuhl
Production Project Manager: Anusha Sambamoorthy
Cover Designer: Christian J. Bilbow

Typeset by SPi Global, India

Dedication

This book is dedicated to the memory of
José Antonio Pero-Sanz Elorz

Endorsement

Steel is a robust and enduring word, because the level of complexity involved in metallurgy and material science, being multiparametric and combinational, has something in common with that of biology. If steel is invented today, it would most certainly be called a nanotech material.

Jean-Pierre Birat, 2004

Contents

About the Authors xiii
Preface xv
Acknowledgments xvii

1. Solidification of Metals

1.1 Metals 1
1.2 From the Gaseous State to the Crystalline State 4
1.3 Crystalline Systems for Metals 6
1.4 Solidification Temperature 18
References 28
Bibliography 28

2. Phase Transformation Kinetics

2.1 Surface Free Energy 31
2.2 Homogeneous Nucleation and Critical Nucleus Size 39
2.3 Discontinuous Nature of Solidification 42
2.3.1 Nucleation Rate 45
2.3.2 Growth Rate 50
2.4 Solidification of a Metal by Continuous Cooling 56
2.4.1 Solidification Curves 56
2.4.2 Considerations on the Nuclei Growth With Unidirectional Heat Flow (Chvorinov Law) 59
Bibliography 65

3. Total Insolubility and Solubility in Alloys

3.1 Total Insolubility Between Two Metals (A and B) 67
3.1.1 A and B do not Have Affinity Between Them 68
3.1.2 A and B Form an Intermetallic Compound With Stoichiometric Composition A_xB_y 71
3.1.3 A and B, are Insoluble in Solid State, but Present Eutectic Affinity 74
3.2 Solubility 78
3.2.1 Substitutional Solid Solutions 78
3.2.2 Interstitial Solid Solutions 79

3.2.3 Total and Partial Solubility 83
3.2.4 Total Solubility Binary Phase Diagram 85
References 91
Bibliography 91

4. Invariant Solidification

4.1 Introduction 93
4.2 System at Equilibrium and the Phase Rule 94
4.3 Binary Eutectic Reaction 96
4.3.1 Abnormal Eutectics. Al-Si System 104
4.3.2 Eutectics With S 108
4.3.3 Cu-O Eutectic. Types of Commercial Copper 109
4.4 Binary Peritectic Reaction 112
4.5 Binary Monotectic Reaction 119
4.6 Binary Sintectic Reaction 123
4.7 Other Invariant Reactions 124
4.8 Computational Calculation Methods (CALPHAD) 126
4.8.1 Fe-C System 127
4.8.2 Al-Si Pressure Variation 130
References 131
Bibliography 131

5. Nonequilibrium Solidification and Chemical Heterogeneities

5.1 Unidirectional Solidification and Zone Melting Solidification 133
5.1.1 Unidirectional Solidification 135
5.1.2 Zone Melting Solidification 139
5.2 Macrosegregation 142
5.2.1 Normal Segregation 143
5.2.2 Inverse Segregation 153
5.2.3 Gravity Segregation 154
5.2.4 Local Segregation 156
5.2.5 Macrosegregation Indexes 158
5.3 Microsegregation 161
5.3.1 Homogenization Heat Treatment 162
5.4 Nonequilibrium by Eutectic and Peritectic Reactions 164
References 169
Bibliography 170

6. Physical Heterogeneities in Solidification

6.1 Factors That Influence the Columnar Structure 171
6.2 Contraction Caused by Solidification 176
6.2.1 Shrinkage and Solid Contraction 176
6.3 Evolution of Gasses During Solidification 187

6.3.1 Boiling Produced by Carbon Monoxide in Steels 194
6.3.2 Macrostructure of Steel Ingots According to Their Degree of Deoxidation 199
6.4 Moldability 202
6.4.1 Castability 202
6.4.2 Soundness 204
6.4.3 Hot Tearing 206
Reference 208
Bibliography 208

7. Equilibrium Transformations

7.1 Transformations Caused by Solubility Variations 210
7.1.1 Precipitation Process 210
7.1.2 Heat Treatments and Precipitation in Age-Hardening Alloys 221
7.1.3 Final and Intermediate Phases (Intermetallic Compounds) 230
7.2 Transformations Caused by Allotropic Changes 238
7.3 Transformations Caused by Allotropy and Solubility Changes 244
7.3.1 Cu-Zn System 244
7.3.2 Cu-Sn System 245
7.3.3 Cu-Al System 247
7.4 Order/Disorder Transformations 248
Reference 254
Bibliography 254

8. Solid-State Transformations in the Fe-C System

8.1 Equilibrium Transformations in the Metastable Fe-C System 255
8.1.1 Constituents of the Metastable Fe-C System 255
8.1.2 Austenite-Ferrite Transformation Kinetics 258
8.1.3 Fe-Fe_3C Metastable Diagram 262
8.1.4 Equilibrium Cooling Microstructures of Steels and White Cast Irons 268
8.1.5 Ultimate Stress Estimation in Binary Ferritic-Pearlitic Steels 273
8.1.6 Cooling at a Rate Faster Than Equilibrium 275
8.1.7 Influence of Alloying Elements in the Metastable Fe-Fe_3C Diagram 278
8.2 Nonequilibrium Transformations of Austenite During Cooling 289
8.2.1 Pearlitic Transformation 289
8.2.2 Bainitic Transformation 293
8.2.3 Martensitic Transformation 296
8.2.4 TTT (Time-Temperature-Transformation) Curves 303

8.3 Heat Treatments 314
8.3.1 Austenite Formation by Heating at $T > A_3$ 314
8.3.2 Tempering of Martensite 319
8.3.3 Tempering of Bainite and Pearlite 323
References 323
Bibliography 324

9. Ternary Systems

9.1 Interpretation of Ternary Diagrams 325
9.1.1 Graphic Representation 325
9.1.2 Equilibrium of Two Phases 329
9.1.3 Equilibrium Between Three Phases 330
9.1.4 Equilibrium Between Four Phases 334
9.2 Ternary Diagrams in Metallography 338
9.2.1 Pb-Sb-Sn System 341
Reference 358
Bibliography 358

Index 359

About the Authors

The late *José Antonio Pero-Sanz* (2012) received his PhD in Engineering from the University of Barcelona (Spain), was a founding member of the International Metallographic Society (USA), a fellow of the Institute of Materials, Minerals and Mining (UK), and *Membre d'Honneur* of the *Société Française de Métallurgie et des Matériaux*. He was an advisor on Physical Metallurgy topics for the United Nations Industrial Development Organization (UNIDO) as well as for Arcelor Mittal - Europe and was a member of the *Conseil Scientifique des Usines Renault*. For more than 30 years, he was the head of research of the Materials groups in the Oviedo and Madrid Schools of Mines.

María José Quintana (mquintana@up.edu.mx) has a European PhD in Science and Technology of Materials from the University of Oviedo (Spain), and is a professor and researcher at the School of Engineering of Universidad Panamericana (campus Mexico). Her research activities include the thermomechanical treatment and characterization of steels and other metallic materials, superplasticity, microscopy, and mechanical testing of manufactured products, as well as the application of design theory and simulation from a macro and micro scales.

Luis Felipe Verdeja (lfv@uniovi.es) has a PhD in Chemical Sciences from the University of Oviedo (Spain), where he is a professor of Materials Science and head of the Siderurgy, Metals and Materials Group (Sid-Met-Mat). His work includes books such as *Metalurgia Extractiva* and Refractory and Ceramic Materials, the last one having Spanish and English versions. His research focuses in the application, maintenance, and wear of refractory linings in blast furnaces and other metal and steels production processes.

Preface

This book is the result of the activity undertaken by Dr. Pero-Sanz between 1971 and 2008, which included various lectures at the Oviedo and Madrid Schools of Mines and the publishing of books in Spanish, such as *Materiales Metálicos*, *Ciencia e Ingeniería de Materiales*, *Fundiciones*, and *Aceros*. The work also includes concepts and solved examples by Dr. José Ignacio Verdeja, Dr. Luis Felipe Verdeja, and José Ovidio García, developed for the Materials Science and Engineering courses in the last decades. The book also includes more than 80 exercises with their detailed solution, in order to explain and clarify concepts and theoretical models.

Through this book, the former students of Dr. Pero-Sanz pay tribute to his memory and present some of his teachings where they can be helpful.

This book is intended for understanding the subject of Metallic Materials and Physical Metallurgy, by undergraduate students, physicists, chemists, and engineers, as its content addresses the fundamentals of manufacturing, treatment, and properties of metals and alloys, not including advanced theoretical concepts: the close relationship between structure and properties is considered through the nine chapters and would help to determine specific applications of metallic alloys.

The solidification phenomenon during industrial operations is analyzed when manufacturing structural parts by casting, or semiproducts for forging, in order to obtain large, flat or specifically shaped cross-sections. Nucleation and growth models are used to describe solidification and solid-state transformations, such as those taking place because of changes in solubility and allotropy or changes produced by recrystallization.

Furthermore, heat treatments involving controlled heating, holding, and cooling, are related to specific structures and properties of metals and alloys. The interpretation of phase diagrams, both binary and ternary, is explained in detail to provide readers a better understanding of iron, aluminum, copper, lead, tin, nickel, titanium, etc., alloys and the effect of other metallic or metalloid elements.

Experimental data along with optical and scanning electron micrographs are presented to distinguish between theoretical calculations and the effect of industrial processes on the properties of solidified or heat treated products. The combination of simple mathematical models, statistical calculations, and actual metallic alloys is not usually found in basic texts, as nucleation and growth models are commonly analyzed from a mathematical point of view and microstructures are described in handbooks aimed for industrial practice technicians. Consequently, the book may be a very important tool to clarify quality control parameters for engineers and technical staff involved in the manufacture of metallic parts or raw material products.

Acknowledgments

This work would not have been possible without the help of graduate student Daniel Fernández González of the School of Mines, Energy and Materials of University of Oviedo and Elena Gómez Lovera, Fabián Gómez López, Melissa Amaris Muñoz Gómez, Juan Pablo Terrazas Jiménez, and Carlos Alfonso Ponce Ramírez, all undergraduate students at Universidad Panamericana, campus Mexico who helped with the revision of text, exercises, figures, and diagrams.

The authors acknowledge the invaluable help of Dr. Roberto Gonzalez at Universidad Panamericana campus Mexico in the revision of this text.

Chapter 1

Solidification of Metals

1.1 METALS

The main difference between metals and nonmetals lies in the number of electrons in the external orbit of the atoms: metals have a lower number of electrons which are easily released in order to form complete and stable orbits.

Table 1.1 shows the electronegativity of metals, calculated as the energy necessary for an atom to attract an electron using 3.98 as the base value (Pauling criterion) assigned to Fluorine (the most electronegative element). Corresponding values for some nonmetals are: Boron (2.04), Phosphorous (2.19), Hydrogen (2.20), Carbon (2.55), Sulfur (2.58), Iodine (2.66), Bromine (2.96), Nitrogen (3.04), Chlorine (3.16), and Oxygen (3.44).

When the difference in electronegativity between two metals is considerable, the bond between them will be of the ionic type. Other possible atomic or molecular bonds are covalent, coordinate covalent, polar covalent, and metallic.

In the case of a metallic solid, each atom loses peripheral electrons to an electron cloud and this, due to its electronegativity, brings positively charged atoms together. This bond between atoms of the metallic crystal or grain is only observed in metals and therefore called *metallic bond*. A characteristic of this structure is the anonymity of the bond between atoms where each atom is not specifically connected to any other atom, which is in contrast from other types of chemical bonds. Another difference is the mobility of the cloud formed by valence electrons; this easiness in their displacement results in high *thermal* and *electric conductivities*.

Another property of metals related to their bond is the amount of *deformation* before rupture, which compared to nonmetallic materials, such as ceramics, glasses, ionic solids, etc., is considerably larger. Once deformation reaches the yield stress value, the metallic bond does not break as the atoms can slide over each other, which is translated into plastic deformation (at atomic, microscopic, and macroscopic levels). On the other hand when nonmetallic materials reach elastic separation energy between the atoms of a molecule, the bond breaks and the material fractures.

The metallic bond also results in the ability of metals to form *alloys* either by substitution or insertion of foreign atoms: due to the anonymity of the bond in the crystalline lattice (solvent) some of its atoms can be substituted by other

Solidification and Solid-State Transformations of Metals and Alloys. http://dx.doi.org/10.1016/B978-0-12-812607-3.00001-2

TABLE 1.1 **Electronegativity (in Increasing Order) of Metals According to the Pauling Criteria**

Element	Electronegativity (eV)
Cs	0.79
Rb	0.82
K	0.82
Ba	0.89
Na	0.93
Sr	0.95
Li	0.98
Ca	1.00
La	1.10
Ce	1.12
Pr	1.13
Nd	1.14
Sn	1.17
Gd	1.20
Dy	1.22
Y	1.22
Ho	1.23
Er	1.24
Lu	1.27
Pu	1.28
Mg	1.31
Zr	1.33
Sc	1.36
Np	1.36
U	1.38
Ti	1.54
Be	1.57
Mn	1.55
Al	1.16

TABLE 1.1 Electronegativity (in Increasing Order) of Metals According to the Pauling Criteria—cont'd

Element	Electronegativity (eV)
V	1.63
Zn	1.65
Cr	1.66
Cd	1.69
In	1.78
Ga	1.81
Fe	1.83
Co	1.88
Cu	1.90
Si	1.90
Ni	1.91
Ag	1.93
Sn	1.96
Hg	2.00
Ge	2.01
Bi	2.02
Tl	2.04
Sb	2.05
Mo	2.16
As	2.18
Pd	2.20
Ir	2.20
Rh	2.28
Pt	2.28
Pb	2.33
W	2.36
Au	2.54
Se	2.55

metals (solute). The substitution of one kind of atom by another can be complete if atomic radius, electronegativity, valence (number of electrons offered to the electronic cloud), and crystalline lattice are similar to the ones of the solvent; for example, copper and nickel can form alloys in any proportion, such as Monel (75% Ni and 25% Cu), Constantan (45% Ni and 55% Cu), or Cupronickels (75% Cu and 25% Ni).

1.2 FROM THE GASEOUS STATE TO THE CRYSTALLINE STATE

Metals, like all other elements, will be in a solid, liquid, or gaseous state depending on the combination of temperature and pressure. As an example, Fig. 1.1 shows the equilibrium temperature-time curve during cooling for aluminum (Al) at a pressure of 1 atm: pure Al is solid below 660°C, liquid between 660°C and 2450°C, and gas above 2450°C. The gasified metal inside a closed environment behaves according to the kinetic model of a group of atoms with Brownian movement: atoms move in a disordered fashion, gravity has no influence in these movements and collisions follow, almost exactly, the laws for conservation of momentum and energy. Due to this high kinetic energy, it is practically impossible for the formation of any group of atoms.

The total kinetic energy of the atoms can be evaluated by the *pressure* of the gas against the walls, while the *temperature* of the gas (in Kelvin degrees)

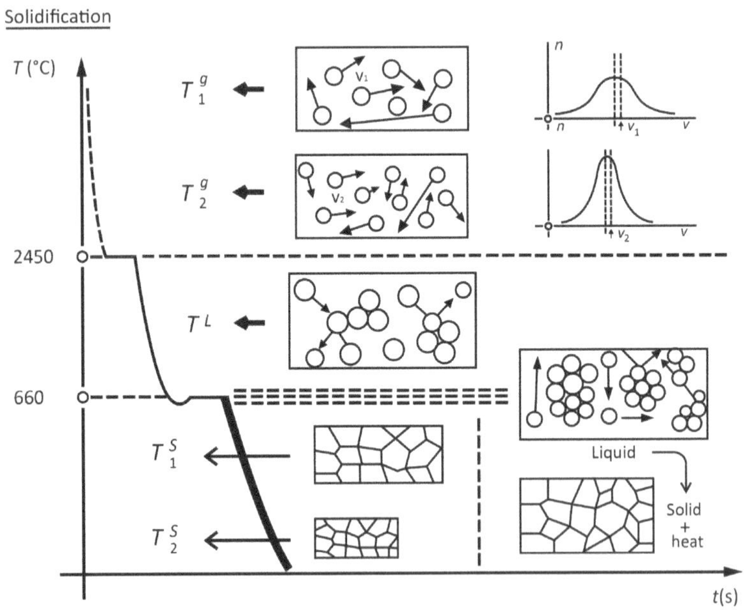

FIG. 1.1 Cooling curve of a metal from gas to solid.

measures the mean kinetic energy, though not all of them have the same kinetic value: some will have higher or lower values compared to the mean one and the true value of the kinetic energy for each atom, and therefore its speed, follows at any instant in time the Maxwell-Boltzmann law (mean speed and its standard deviation are directly proportional to the temperature of the gas). The most probable speed, according to the Maxwell-Boltzmann statistic distribution, at a temperature T is determined by:

$$v_0 = \sqrt{\frac{2kT}{m}} \tag{1.1}$$

where m is the mass of the atom and k is the Boltzmann constant ($k = 1.38 \times 10^{-23}\,\text{J/K}$). Fig. 1.1 also shows the statistical distribution and the dn/N ratio of atoms with a certain speed v:

$$\frac{dn}{N} = \frac{4}{\sqrt{\pi}} \left(\frac{v}{v_0}\right)^2 e^{-\left(\frac{v}{v_0}\right)^2} d\left(\frac{v}{v_0}\right) \tag{1.2}$$

EXERCISE 1.1

For Al gas at a temperature of 3000 K, calculate the most probable speed for one of its atoms.

Solution

As Eq. (1.1) requires the mass of the atom, it can be obtained by dividing the atomic weight by the Avogadro number (N_A):

$$m = \frac{\text{atomic weight}}{N_A} = \frac{26.981\,\text{u}}{6.023 \times 10^{23}\,\frac{\text{u}}{\text{g}}} = 4.48 \times 10^{-23}\,\text{g} = 4.48 \times 10^{-26}\,\text{kg}$$

And using this value in Eq. (1.1):

$$v_0 = \sqrt{\frac{2\left(1.38 \times 10^{-23}\,\frac{\text{J}}{\text{K}}\right)(3000\,\text{K})}{4.48 \times 10^{-26}\,\text{kg}}} = 1359.54\,\frac{\text{m}}{\text{s}} = 4894.35\,\frac{\text{km}}{\text{h}}$$

As an example, at T_1^g, the curve follows the shape of the Maxwell-Boltzmann distribution with the most probable speed value v_1. When temperature decreases to T_2^g (gaseous state), the new curve shows a smaller dispersion in the speed of the atoms and the mean speed v_2 is lower than v_1: a drop in speed produces the mean displacement of the atoms at T_2^g, lower than at T_1^g. As temperature decreases, a threshold temperature is reached where the attraction forces between the atoms of the gas start to balance the repulsion effect of kinetic energy. These attraction forces (caused by the metallic bond) are the reason why atoms tend to form groups and eventually the characteristic lattice of their crystalline state. Furthermore, metallic atoms are subjected to two types of interatomic forces:

- *Repulsion* forces: caused by the kinetic energy of each atom which promotes dispersion (influenced by the collisions with the walls of the container and with each other).
- *Attraction* forces: caused by the tendency to form metallic bonds.

If temperature is lowered, attraction forces start to manifest and, therefore, begin to neutralize the repulsion ones (e.g., for Al at 2450°C). The metal changes from the gaseous state to the liquid one, by a mechanism of nonstop formation and breaking of bonds. In the liquid state, at temperatures close to evaporation (or overheated liquids), atoms have a freedom of movement similar to the one they have at the gaseous state. Each atom is surrounded by a defined number of anonymous neighbor atoms, without the tendency to adopt a crystalline structure, and with a behavior similar to a compressed gas. Yet, the difference with a gaseous state is that groups of atoms are formed, even though their lifetimes are short and their bonds easily disappear.

If the temperature of the liquid metal drops even further, the groups last longer until they become permanent, and solid state characteristics are reached: the liquid presents a certain crystalline structure (i.e., the vibration of particles around certain equilibrium positions) even though it is a transitory state and can easily be dissolved. The atoms can move short distances and the thermal movement of each of them is constituted by vibrations with a defined mean frequency but with continuously changing direction and amplitude. The true values for the kinetic energy at any moment are also distributed according to the Maxwell-Boltzmann law. As a specific temperature is reached (660°C for Al), the liquid metal turns to solid, in the form of crystalline aggregates; each of them formed by atoms that vibrate and are positioned at locations of the typical crystalline lattice of each metal.

If a crystalline solid is cooled even further, the amount of vibration also diminishes and the atoms become closer to one another until reaching a new equilibrium between attraction forces (metallic bond) and repulsion ones (addition of vibration energy and electrostatic rejection between the positive charges of the nuclei). The equilibrium distance between the atoms at the interior of each grain in solid state can increase if energy is added to the metal by heat, or can decrease if energy is removed by cooling: the dimensions of the crystalline lattice are reduced, which at a macroscopic scale results in contraction of the material.

1.3 CRYSTALLINE SYSTEMS FOR METALS

When metallic sample is properly polished, etched with an adequate chemical reagent, and observed through an optical microscope, a cellular structure similar to the one presented in Fig. 1.2 is revealed, where each cell is a *crystalline grain*

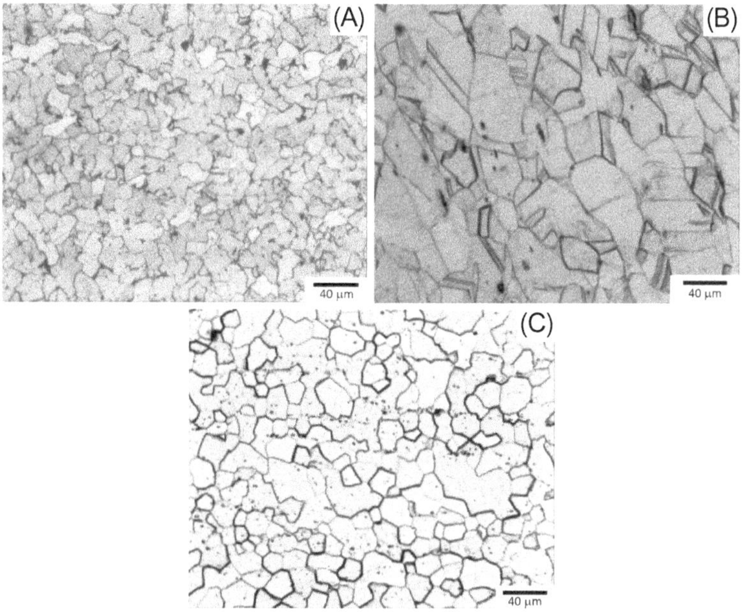

FIG. 1.2 Optical micrographs of equiaxed grain structure in a metal: (A) hot-rolled and tensile deformed at high temperatures for shipbuilding steel, (B) as-extruded tellurium-copper alloy for electrical applications, and (C) hot-rolled and recrystallized titanium alloy for heat exchanger applications.

or *crystal*. If X-ray diffraction techniques are used, it can be proved that inside each grain, the metallic atoms are positioned in a regular manner; repeated systematically in the three spatial directions. The fundamental principle of crystal structures is that the atoms are set in space either on the points of a Bravais lattice or in some fixed relation to those points. Therefore, the fundamental stacking unit is known as *unit cell*. Though it has an ideal or abstract character, crystals are formed by groups of these cells where an atom may be shared by several of these ideally sectioned units.

Most materials used in industrial applications have, at room temperature, the following crystalline systems: face-centered cubic (e.g., γ-Fe, Al, Cu, Pb, Ni, Ag, Pt, and Au), body-centered cubic (e.g., α-Fe, V, Cr, Nb, Mo, Ta, and W), or hexagonal (e.g., Mg, Zn, Zr, Ti, Be, and Co). Fig. 1.3 shows the unit cell of the face-centered cubic system, where the centers of the atoms are located in the corners of the cube and at the center of the faces (each cell contains the mass of four atoms). Fig. 1.4 shows the unit cell for the body-centered cubic (bcc) system (with a total of two atoms per unit cell), with atoms located at the corners of the cube and at the center of the cell.

Furthermore, the compact hexagonal system has as unit cell a prism with a rhombus as a base (of sides a and a) and height c; each corner is associated with two atoms: one occupies the (0,0,0) position and the other occupies the

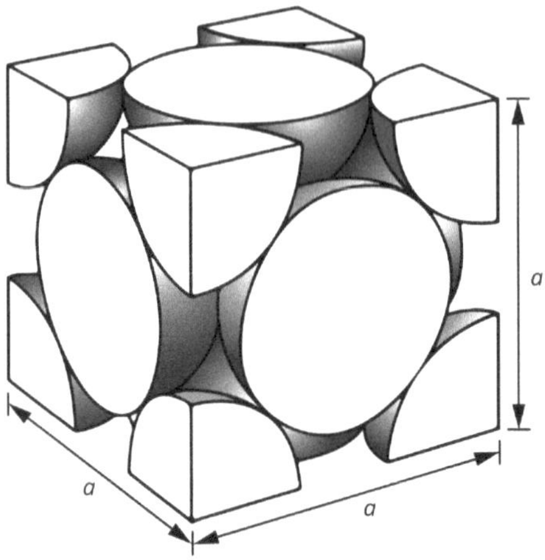

FIG. 1.3 Face-centered cubic unit cell.

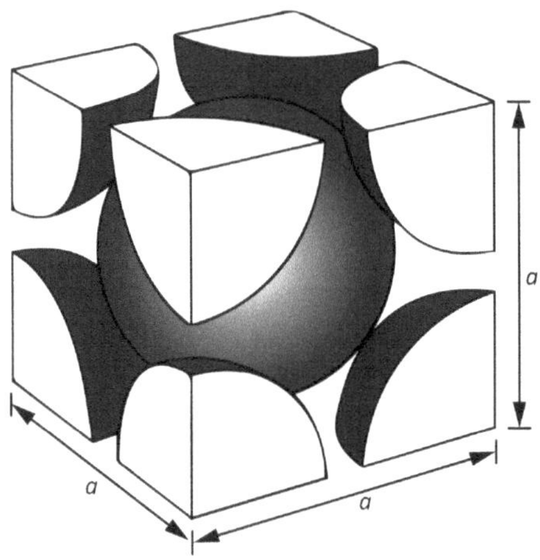

FIG. 1.4 Body-centered cubic unit cell.

$(2/3a, 1/3a, c/2)$ position as indicated in Fig. 1.5. Joining two rhombic prisms and two semiprisms together results in what is erroneously, but commonly, known as a hexagonal unit cell. Fig. 1.5 also shows the base of the "hexagonal cell": a hexagon formed by three rhombs. As the height of the cell is c, the centers of the atoms are located at the corners of the hexagonal prism, at the centers of the hexagonal bases, and at the middle plane of the prism (in total, there are six atoms per "unit cell").

The hexagonal system behaves as a compact one when the atoms at the middle plane are tangent to the base atoms. In this case, each atom is tangent to 12 others: 6 of its own plane, 3 of the upper plane, and 3 of the lower plane. If c is

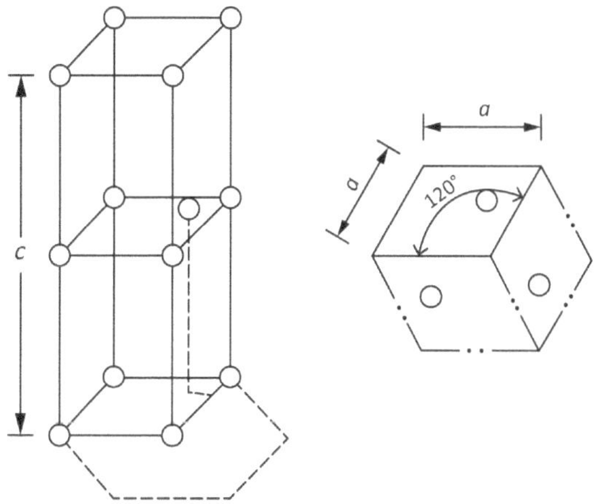

FIG. 1.5 Hexagonal compact packing unit cell.

the height of the prism and a is the side of the hexagon (or the rhomb for the unit cell), the c/a ratio in the hexagonal compact system equals 1.633, which may be confirmed by the geometric considerations shown in Fig. 1.6, where

$$MH = \frac{a}{\sqrt{3}} \tag{1.3}$$

$$MH^2 = MB^2 - HB^2 = a^2 - \left(\frac{c}{2}\right)^2 \tag{1.4}$$

$$\frac{c}{a} = \sqrt{\frac{8}{3}} = 1.633 \tag{1.5}$$

Hexagonal compact packing (hcp) and face-centered cubic (fcc) systems are the only crystalline arrangements in which maximum tangency of an atom to the surrounding ones is achieved. Each atom is in contact with 12 others, so there is an analogy between both systems; in fcc, the compact packing of atoms

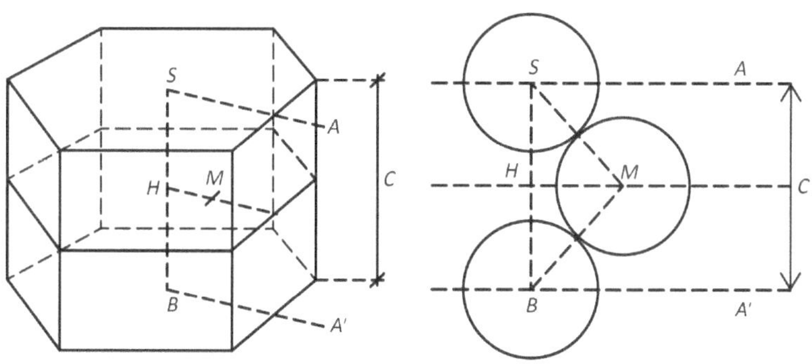

FIG. 1.6 Compact condition in the hexagonal system.

corresponds to diagonal planes that are crystallographically known as (111); while in hcp, the compact packing of atoms corresponds to basal planes that are crystallographically known as (0001).

Other metals adopt more complex structures. For example, tin crystallizes in the body-centered tetragonal system and uranium in the orthorhombic one. Table 1.2 indicates the system in which metals crystallize, ordered by atomic number, along with their atomic weight (the weight in grams of 6.023×10^{23} atoms).

TABLE 1.2 Crystalline System and Density of Metals

Atomic Number	Metal	Symbol	Atomic Weight	Crystalline System	Density at 20°C (g/cm^3)
3	Lithium	Li	6.939	bcc	0.534
4	Beryllium	Be	9.012	hcp ($T < 1260°$) bcc ($T > 1260°$)	1.840
11	Sodium	Na	22.989	bcc	0.9712
12	Magnesium	Mg	24.305	hcp	1.74
13	Aluminum	Al	26.981	fcc	2.669
14	Silicon	Si	28.085	Cubic diamond	2.33
19	Potassium	K	39.098	bcc	0.85
20	Calcium	Ca	40.08	fcc ($T < 300°$) hcp ($T > 450°$)	1.55
21	Scandium	Sc	44.955	hcp, fcc	2.99
22	Titanium	Ti	47.90	α-hcp ($T < 882.5°$) β-bcc ($T > 882.5°$)	4.507
23	Vanadium	V	50.941	bcc	6.1
24	Chromium	Cr	51.996	bcc (under 20° in electroplating is hcp)	7.19
25	Manganese	Mn	54.938	Cubic complex	7.43
26	Iron	Fe	55.847	bcc ($T < 910°$) fcc ($910 < T < 1400°$) bcc ($T > 1400°$)	7.97

TABLE 1.2 Crystalline System and Density of Metals—cont'd

Atomic Number	Metal	Symbol	Atomic Weight	Crystalline System	Density at 20°C (g/cm^3)
27	Cobalt	Co	58.933	hcp ($T<417°$) fcc ($T>417°$)	8.85
28	Nickel	Ni	58.71	fcc	8.902
29	Copper	Cu	63.546	fcc	8.96
30	Zinc	Zn	65.38	hcp	7.133
31	Gallium	Ga	69.72	Orthorhombic	5.907
32	Germanium	Ge	72.59	Cubic diamond	5.323
33	Arsenic	As	74.921	Rhombohedral	5.72
34	Selenium	Se	78.96	*h* (other metastable monoclinic varieties at low *T*)	4.79
37	Rubidium	Rb	85.467	bcc	1.53
38	Strontium	Sr	87.62	fcc	2.60
39	Yttrium	Y	88.905	hcp ($T<1460°$) bcc ($T>1460°$)	4.47
40	Zirconium	Zr	91.22	hcp ($T<862°$) fcc ($T>862°$)	6.439
41	Niobium (Columbium)	Nb (Cb)	92.906	bcc	8.57
42	Molybdenum	Mo	95.94	bcc	10.22
43	Technetium	Tc	98.906	–	–
44	Ruthenium	Ru	101.07	hcp	12.2
45	Rhodium	Rh	102.905	fcc	12.44
46	Palladium	Pd	106.4	fcc	12.02
47	Silver	Ag	107.868	fcc	10.49
48	Cadmium	Cd	112.41	hcp	8.65
49	Indium	In	114.82	Tetragonal	7.31
50	Tin	Sn	118.69	Tetragonal ($T>13.2°$) Cubic ($T>13.2°$)	7.2994

Continued

TABLE 1.2 Crystalline System and Density of Metals—cont'd

Atomic Number	Metal	Symbol	Atomic Weight	Crystalline System	Density at 20°C (g/cm^3)
51	Antimony	Sb	121.75	Rhombohedral	6.62
52	Tellurium	Te	127.60	*h*	6.24
55	Cesium	Cs	132.905	bcc	1.903
56	Barium	Ba	137.33	bcc	3.5
57	Lanthanum	La	138.905	*h*; bcc; fcc	6.19
58	Cerium	Ce	140.12	fcc; *h*; bcc	6.77
59	Praseodymium	Pr	140.907	*h*, bcc	5.77
60	Neodymium	Nd	140.24	*h*; bcc	7
61	Promethium	Pm	145.00	–	–
62	Samarium	Sm	150.4	Rhombohedral	7.49
63	Europium	Eu	151.96	bcc	5.245
64	Gadolinium	Gd	157.25	hcp	7.86
65	Terbium	Tb	158.925	hcp	8.25
66	Dysprosium	Dy	162.50	hcp	8.55
67	Holmium	Ho	164.93	hcp	6.79
68	Erbium	Er	167.26	hcp	9.15
69	Thulium	Tm	168.934	hcp	9.31
70	Ytterbium	Yb	173.04	fcc	6.96
71	Lutetium	Lu	174.97	hcp	9.85
72	Hafnium	Hf	178.49	hcp; fcc ($T > 1760°$)	13.09
73	Tantalum	Ta	180.947	bcc	16.6
74	Tungsten	W	183.85	bcc	19.3
75	Rhenium	Re	186.20	hcp	21.04
76	Osmium	Os	190.20	hcp	22.57
77	Iridium	Ir	192.22	fcc	22.5
78	Platinum	Pt	195.09	fcc	21.45
79	Gold	Au	196.966	fcc	19.32
80	Mercury	Hg	200.59	Rhombohedral	13.546

TABLE 1.2 Crystalline System and Density of Metals—cont'd

Atomic Number	Metal	Symbol	Atomic Weight	Crystalline System	Density at 20°C (g/cm^3)
81	Thallium	Tl	204.37	hcp; bcc	11.85
82	Lead	Pb	207.20	fcc	11.36
83	Bismuth	Bi	208.98	Rhombohedral	9.8
84	Polonium	Po	209.00	hcp; bcc	–
87	Francium	Fr	223.00	fcc	–
88	Radium	Ra	226.00	Rhombohedral	5
89	Actinium	Ac	227.00	Monoclinic	–
90	Thorium	Th	232.038	–	11.66
91	Protactinium	Pa	231.035	–	15.4
92	Uranium	U	238.029	–	19.07
93	Neptunium	Np	237.048	fcc	–
94	Plutonium	Pu	244.00	–	19–19.72
95	Americium	Am	243.00	Orthorhombic	11.7
96	Curium	Cm	247.00	–	–
97	Berkelium	Bk	247.00	Monoclinic	–
98	Californium	Cf	251.00	–	–
99	Einsteinium	Es	254.00	–	–
100	Fermium	Fm	257.00	–	–
101	Mendelevium	Md	258.00	–	–
102	Nobelium	No	259.00	–	–
103	Lawrencium	Lr	262.00	–	–

bcc, body-centered cubic; *fcc*, face-centered cubic; *h*, hexagonal; *hcp*, hexagonal compact packing.

The crystalline system in which metallic elements arrange themselves is not related to their position in the periodic table, as this is done according to their atomic number. While atomic weight is directly proportional to the atomic number, density is instead a function of both the Bravais lattice (four atoms per cell in the face-centered cubic system; two atoms in the body-centered cubic system; two atoms—and six per hexagonal prism—in hexagonal compact packing system) and the volume of the cell. If a is the side of the cubic unit cell (or lattice parameter), the densities of the main Bravais lattices can be calculated through:

- *Body-centered cubic*: $2 \cdot \text{atomic weight}/N_A \cdot a^3$
- *Face-centered cubic metals*: $4 \cdot \text{atomic weight}/N_A \cdot a^3$
- *Hexagonal compact packing*: $\sqrt{2} \cdot \text{atomic weight}/N_A \cdot a^3$

Unit cell parameters can be determined using X-ray diffractometry techniques (Debye-Scherrer method).

If atomic radius is considered, the crystalline system is an important factor to be analyzed. When the same element crystallizes in more than one system (allotropy), atomic radius changes. The atomic diameter is therefore, the distance between centers or nuclei of two tangent atoms, and its value equals $\sqrt{2}a/2$ for bcc, $\sqrt{3}a/2$ for fcc, and $a/2$ for hcp systems, respectively. The atomic radius also depends on temperature, which affects expansion or contraction of the unit cell, resulting in density variations.

In order to compare the diameters of different elements, Goldschmidt (1929) calculated them assuming a coordination (number of atoms tangent to each other) equal to 12 for all metals. As previously mentioned, only face-centered cubic and hexagonal compact packing systems have such a high coordination. Table 1.3 shows the atomic diameters according to both Goldschmidt (1929) and Barret (1986).

TABLE 1.3 Atomic Diameters of Metals

Element	Goldschmidt Diameter	Barrett Diameter	Crystalline System	Cell Parameters (Å) a	c
Be	2.25	2.225	hcp ($\theta < 1260$)	2.2854	3.5841
Ni	2.487	2.491	fcc	3.5238	
Co	2.50	2.506	hcp	2.507	4.069
		2.511	fcc	3.552	
Fe	2.52	2.431	α-bcc	2.8664	
		2.585	γ-fcc	3.656 (at 950°C)	
		2.54	δ-bcc	2.94 (at 1425°C)	
Cu	2.551	2.556	fcc	3.6153	

TABLE 1.3 Atomic Diameters of Metals—cont'd

Element	Goldschmidt Diameter	Barrett Diameter	Crystalline System	Cell Parameters (Å)	
				a	c
Cr	2.57	2.498	bcc	2.8845	
Ru	2.67	2.649	hcp	2.7038	4.2816
Rh	2.684	2.689	fcc	3.8034	
Os	2.70	2.675	hcp	2.733	4.3191
Ir	2.709	2.714	fcc	3.8389	
V	2.71	2.632	bcc	3.039	
Pd	2.745	2.750	fcc	3.8902	
Zn	2.748	3.60	hcp	2.664	4.945
Mn	2.75	2.24	Cubic complex	8.912	
Pt	2.769	2.775	fcc	3.9237	
Ge	2.788	2.450	Cubic diamond	5.658	
Mo	2.80	2.725	bcc	3.1466	
W	2.82	2.739	bcc	3.1648	
Al	2.85	2.862	fcc	4.0490	
Te		2.87	h	4.4559	5.9268
Au	2.878	2.884	fcc	4.0783	
Ag	2.883	2.888	fcc	4.0855	
Ti	2.93	2.89	α-hcp	2.9504	4.6833
		2.89	β-bcc	3.33 (at 900°C)	
Nb	2.94	2.859	bcc	3.3007	
Ta	2.94	2.86	bcc	3.3026	
Cd	3.042	2.979	hcp	2.9787	
Hg	3.10	3.006	Rhombohedral		
Li	3.13	3.039	bcc	3.5089	
In	3.14	3.25	Tetragonal centered	4.594	4.951

Continued

TABLE 1.3 Atomic Diameters of Metals—cont'd

Element	Goldschmidt Diameter	Barrett Diameter	Crystalline System	Cell Parameters (Å) a	c
Sn	3.164	3.022	β-white tetragonal	5.8311	3.1817 (at 20° C)
		2.81	α-gray cubic diamond	6.47 (at 18°C)	
Ht	3.17	3.15	hcp	3.206	5.087
Zr	3.19	3.17	α-hcp	3.230	5.133
		3.13	β-bcc	3.62 (at 867°C)	
Mg	3.20	3.196	hcp	3.2092	5.2103
Sc	3.2	3.211	α-fcc	4.541	5.24 (at 20°C)
		3.24	β-fcc	3.31	
Sb	3.228	2.903	Rhombohedral		
Tl	3.42	3.407	α-hcp	3.4564	5.531
		3.362	β-bcc	3.882 (at 262°C)	
Pb	3.49	3.499	fcc	4.9495	
Y	3.62	3.60	hcp	3.670	5.826
Bi	3.64	3.111	Rhombohedral		
La	3.741	3.74	h	3.762	6.075 (at 20°C)
Na	3.83	3.715	bcc	4.2906	
Ca	3.93	3.94	fcc	5.57	6.53 (at 460°C)
		3.95	hcp	3.99	
Sr	4.29	4.31	fcc	6.087	
Ba	4.48	4.35	bcc	5.025	
K	4.76	4.627	bcc	5.344	
Rb	5.02	4.88	bcc	5.63 (at −173°C)	
Cs	5.4	5.25	bcc	6.06 (at −173°C)	

bcc, body-centered cubic; *fcc*, face-centered cubic; *h*, hexagonal; *hcp*, hexagonal compact packing.

EXERCISE 1.2

Iron has an atomic weight of 55.847 u and a lattice parameter of 2.86 Å at room temperature. Calculate: (a) its density, (b) the volume change associated with the allotropic transformation of α-Fe into γ-Fe at 912°C, and (c) the linear expansion coefficient of α-Fe.

Data: at 912°C, $a_{\alpha\text{-Fe}} = 2.898$ Å, and $a_{\gamma\text{-Fe}} = 3.639$ Å (Tables 1.2 and 1.3).

Solution

(a) Density of α-Fe

α-Fe has a bcc crystalline system, thus the density can be calculated using the equation mentioned previously, making the appropriate unit transformations as well:

$$\rho_{bcc} = \frac{2 \cdot \text{atomic weight}}{N_A \cdot a^3} = \frac{2(55.847\,\text{u})}{6.023 \times 10^{23}\frac{\text{u}}{\text{g}}\left(2.86\text{Å}\right)^3} = 7927.18\frac{\text{kg}}{\text{m}^3}$$

(b) Volume change

A bcc unit cell has two atoms while an fcc unit cell has four, meaning that two bcc crystals will form one fcc crystal. Considering the number of crystals of each type, the volume change associated with the allotropic transformation of ferrite (α-Fe) into austenite (γ-Fe) can be calculated through:

$$\Delta V = \frac{a^3_{\gamma-\text{Fe}} - 2a^3_{\alpha-\text{Fe}}}{2a^3_{\alpha-\text{Fe}}} = \frac{\left(3.639\text{Å}\right)^3 - 2\left(2.898\text{Å}\right)^3}{2\left(2.898\text{Å}\right)^3} = -0.01 = -1\%$$

Comment: It is important to consider isothermal density changes caused by allotropic transformations during heat treatments.

(c) Linear expansion coefficient

Adjusting the linear expansion coefficient to this exercise:

$$a_{912°\text{C}} = a_{23°\text{C}}(1 + \alpha\Delta T)$$

and solving for α:

$$\alpha = \frac{1}{\Delta T}\left(\frac{a_{912°\text{C}}}{a_{23°\text{C}}} - 1\right) = \frac{1}{(912 - 23)}\left(\frac{2.898}{2.866} - 1\right) = 12.56 \times 10^{-6}\,°\text{C}^{-1}$$

which can be compared to the value in Table 1.6.

EXERCISE 1.3

Magnesium has a hexagonal compact structure with an atomic weight of 24.305 u and a density of 1.74 g/cm^3 (Table 1.2). Calculate (a) the lattice parameters and (b) its atomic packing factor.

Solution

(a) Lattice parameters

Using the equation for density of hcp materials, the lattice parameter *a* can be obtained by:

$$\rho_{hcp} = \frac{\sqrt{2} \cdot \text{atomic weight}}{N_A \cdot a^3}$$

$$a = \sqrt[3]{\frac{\sqrt{2} \cdot \text{atomic weight}}{N_A \cdot \rho_{hcp}}} = 3.2 \times 10^{-8}\,\text{cm} = 3.20\,\text{Å}$$

and knowing that the *c/a* ratio is 1.633:

$$c = 1.633a = 5.22\,\text{Å}$$

(b) Atomic packing factor (*APF*)
This factor can be calculated by dividing the volume occupied by atoms in the cell by the volume of the cell:

$$APF = \frac{V_{atoms}}{V_{cell}}$$

The volume of the hcp cell is:

$$V_{cell} = a \cdot \frac{\sqrt{3}}{2} a \cdot c = \frac{\sqrt{3}}{2} a^2 \cdot \sqrt{\frac{8}{3}} a = \sqrt{2} a^3$$

Knowing that the hcp crystals have two atoms per cell, the *APF* can be obtained as:

$$APF = \frac{2\left(\frac{4}{3}\pi r^3\right)}{\sqrt{2}a^3}$$

As previously mentioned, the atomic radius corresponds to *a*/2 in hcp crystals, thus:

$$APF = \frac{2\left(\frac{4}{3}\pi\right)\left(\frac{a}{2}\right)^3}{\sqrt{2}a^3} = 0.74$$

Comment: The *APF* for both hcp and fcc systems is the same (0.74), while the one for bcc is 0.68.

1.4 SOLIDIFICATION TEMPERATURE

Fig. 1.7 shows the empiric variation of free energy for a pure metal, both in solid and liquid states, as a function of temperature; the horizontal axis presents temperature in K, while the vertical axis presents free energy per volume unit.

The *free energy* G is the extensive magnitude with value $H - TS$, where G, H (enthalpy), and S (entropy) are scalar functions between 0 K (−273°C) and T K (as long as a change in state is not produced), and can be calculated through:

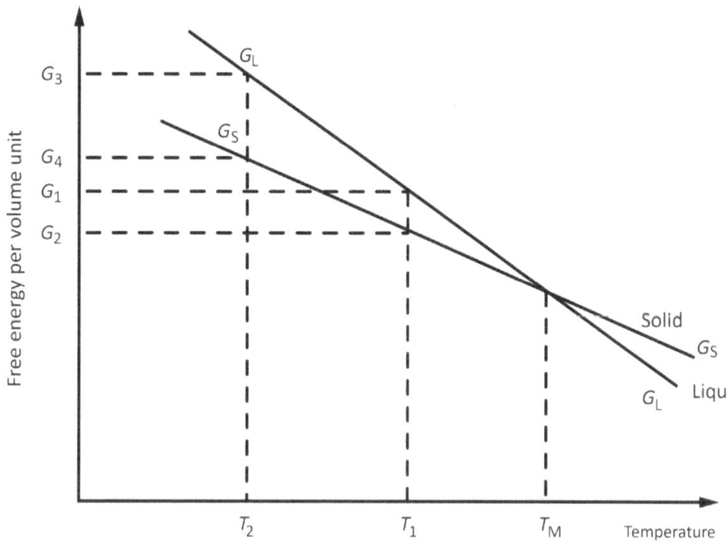

FIG. 1.7 Difference in free energy between liquid and solid, for a pure metal (the curvature of the G_S and G_L lines has been ignored).

$$H = \int_0^T c_p dT \tag{1.6}$$

$$S = \int_0^T \frac{c_p dT}{T} \tag{1.7}$$

where c_p is the specific heat at a constant pressure (i.e., the amount of heat required to raise the temperature of 1 mol of substance by 1 degree Kelvin). Therefore, the pure metal will adopt the lower energy condition and will be in the solid state at temperatures below T_M, which is the equilibrium melting temperature where liquid and solid coexist, known as solidification temperature. If H_L and H_S are the enthalpies for liquid and solid, respectively, and S_L and S_S are the respective entropies, the value of the solidification temperature T_M in K is obtained from:

$$G_L - G_S = 0 = (H_L - H_S) - T_E(S_L - S_S) = \Delta H - T_M \cdot \Delta S_{LS} \tag{1.8}$$

$$T_M = \frac{\Delta H_f}{\Delta S_{LS}} \tag{1.9}$$

where ΔH_f is the change in enthalpy or *latent heat of melting*, defined by the amount of heat (in Joules) per volume unit released (or per mol of mass if

ΔS_{LS} is expressed in Joules per K and per mol) by the metal when it changes from liquid to solid. This heat release results in the following solidification process; when an atom of liquid is integrated to the solid, the loss of both momentum and energy of this atom will be transformed into an emission of heat. The latent heat of melting is higher in metals whose bonds have higher energy.

Bond energy determines separation between atoms and depends on the number of valence electrons transferred by each atom to the electronic cloud. The higher the number of electrons transferred and the smaller the separation between them, the higher the bond energy will be. In other words, the melting temperature T_M is proportional to the latent heat of fusion, with a proportionality coefficient between 0.3 and 0.6 depending on the metal (0.5 as an average). Table 1.4 shows crystalline system, latent heat of fusion ΔH in Joules per mol, melting temperature T_M in K, and ΔS_{LS} ($\Delta H_f/T_M$ ratio) for various metals, which determines the variation of entropy during the melting of 1 mol of metal (difference between liquid and solid entropies). Most metals have $\Delta S_{LS} \cong R = 8.314 \text{J mol}^{-1} \text{K}^{-1}$, as seen in Table 1.4.

Table 1.5 presents, in an increasing order, the melting temperatures of metals; those with values higher than the one for Nb are industrially known as *refractory metals*. Though the temperatures shown in this table correspond to pure metals, solid solution atoms acting as impurities generally diminish the melting temperature of the base material.

The mechanical behavior of metals depends on the working temperature of the subjected part, though industrial use at temperatures higher than $0.5T_M$ (being T_M the absolute melting temperature) is almost always unviable. Even at loads below their yield stress value, plastic deformation under constant stress (a phenomenon known as *creep*) is significant. The effective temperature threshold for industrial use of a metal in order to avoid creep is considerably lower than $0.5T_M$ as other factors, such as thermal stresses, recrystallization of microstructure, oxidation, and corrosion, must also be taken into account.

If a metal has low thermal conductivity, then any local decrease in temperature will produce thermal stresses, which strongly depend on the expansion coefficient. This last property is the inverse function of the melting temperatures (Table 1.6), and a low T_M is related to small bond energies, thus, the rate of expansion when temperature is increased ($< T_M$) is inversely proportional to the melting temperature.

On the other hand, solidification temperature influences *static recrystallization* temperatures. When a metal is cold worked ($0.25T_M$) by a mechanical process such as rolling or wire drawing, the crystalline grains elongate in the forming direction. The metal acquires, as a consequence of the interaction between the crystalline defects, or dislocations, a *cold worked structure* that provides, among other structural variations, an increase in both hardness and maximum stress in the tensile test, and a decrease in elongation.

If a metal with enough cold work deformation is subsequently heat treated ($0.5T_M$), a gradual recovery of properties corresponding to those of a

TABLE 1.4 Differences in Entropy per mol During the Melting of Various Metals and Their Crystalline System

Metal	Latent Heat of Melting ΔH(J/mol)	Melting Temperature T_M(K)	$\Delta S_{LS} = \frac{\Delta H_f}{T_M}\left[\frac{J}{molK}\right]$	Crystalline System
Al	10,659.0	933	11.42	fcc
Ag	11,286.0	1233.8	9.15	fcc
Au	12,665.4	1336	9.48	fcc
Ca	8711.1	1111	7.84	fcc
Cu	12,999.8	1356	9.59	fcc
Cr	20,866.6	2148	9.71	bcc
Fe−δ	15,290.4	1809.5	8.45	bcc
K	2399.3	336.7	7.13	bcc
Mg	9028.8	923	9.78	hcp
Mn	14,421.0	1518	9.50	Cubic complex
Na	2633.4	370.8	7.10	bcc
Ni	17,556.0	1726	10.17	fcc
Pb	5116.3	600	8.53	fcc
Pt	19,646.0	2042	9.62	fcc
Sn	7189.6	504.9	14.24	Tetragonal
Ti	20,900.0	1941	10.77	hcp
W	35,195.6	3683	9.56	bcc
Zr	22,990.0	2125	10.82	hcp
Zn	6667.1	692.5	9.63	hcp

fcc, face-centered cubic; *bcc*, body-centered cubic; *hcp*, hexagonal compact packing.

nondeformed state may be achieved. When the treatment is performed at a temperature above the *static recrystallization* limit (which depends on the amount of previous cold work deformation), the recovery of properties is accompanied with the formation of a regular grained microstructure known as recrystallized state. If temperature is high enough ($0.75T_M$), both, *work hardening* and *dynamic recrystallization*, occur simultaneously, thus hot work deformations (forging, rolling, extrusion, etc.) can result in recrystallization.

TABLE 1.5 Melting Temperature of Metals

Metal	°C
Mercury	−38.36
Cesium	28.7
Gallium	29.5
Rubidium	38.9
Potassium	63.7
Sodium	97.8
Indium	156
Lithium	180.5
Selenium	217
Tin	232
Polonium	254
Bismuth	271.3
Thallium	303
Cadmium	320.9
Lead	327.4
Zinc	419.5
Tellurium	449.5
Antimony	630.5
Plutonium	640
Magnesium	650
Aluminum	660
Radium	700
Barium	714
Strontium	768
Cerium	804
Arsenic	810
Europium	826
Calcium	838
Praseodymium	919

TABLE 1.5 Melting Temperature of Metals—cont'd

Metal	°C
Lanthanum	920
Germanium	936
Silver	960.8
Neodymium	1019
Gold	1063
Samarium	1072
Copper	1083
Uranium	1132.3
Manganese	1245
Beryllium	1277
Gadolinium	1312
Dysprosium	1407
Silicon	1404
Nickel	1453
Holmium	1461
Cobalt	1495
Erbium	1497
Iron	1536.5
Scandium	1539
Thulium	1545
Palladium	1552
Titanium	1668
Thorium	1750
Platinum	1769
Zirconium	1852
Chromium	1875
Vanadium	1900
Rhodium	1966
Hafnium	2222

Continued

TABLE 1.5 Melting Temperature of Metals—cont'd

Metal	°C
Iridium	2454
Niobium	2468
Ruthenium	2500
Molybdenum	2610
Tantalum	2996
Rhenium	3180
Tungsten	3410

TABLE 1.6 Linear Expansion Coefficient Along the Lattice Parameter (*a* or *c* in the Case of Hexagonal) of the Cell and Melting Temperatures of Common Metals

Metal	Linear Expansion Coefficient α[°C^{-1}] Between 0°C and 100°C	Melting Temperature T_M(°C)
Sodium	71×10^{-6}	97.8
Lead	29.3×10^{-6}	327.4
Zinc	39.7×10^{-6}	419.5
Magnesium	26.1×10^{-6}	650
Aluminum	23.6×10^{-6}	660
Silver	19.6×10^{-6}	960.8
Copper	16.4×10^{-6}	1083
Nickel	13.1×10^{-6}	1453
Iron	12.2×10^{-6}	1536.5
Titanium	8.41×10^{-6}	1668
Platinum	8.9×10^{-6}	1769
Chromium	6.2×10^{-6}	1875
Tungsten	4.6×10^{-6}	3410

As an example, if the cold work ($0.25T_M$), static recrystallization ($\sim 0.50T_M$), and hot work temperatures ($0.75T_M$) for Fe, Cu, and Zn are to be calculated, it is interesting to note that at room temperature both iron and copper are *cold*, while zinc is *warm*.

Fig. 1.8A presents the microstructure of a commercial steel alloy in a deformed (cold work) state showing elongated grains, while Fig. 1.8B shows the same alloy when recrystallized, where grains are now equiaxed. Furthermore, Fig. 1.9 presents a commercial zinc alloy with a *banded* microstructure formed by *recrystallization* and *recovery* (nonrecrystallized grains) and when the material is kept for long periods of time at the recrystallization temperature (or higher), a *strain-free* structure with equiaxed regular grain growth is obtained. For recrystallization to take place, the temperatures must be higher than $T_M/2$ for small amounts of cold work.

The *stiffness* of metals is closely related to their melting temperature, which is defined as the ability of a material to withstand stresses without suffering irreversible, permanent, or plastic deformations.

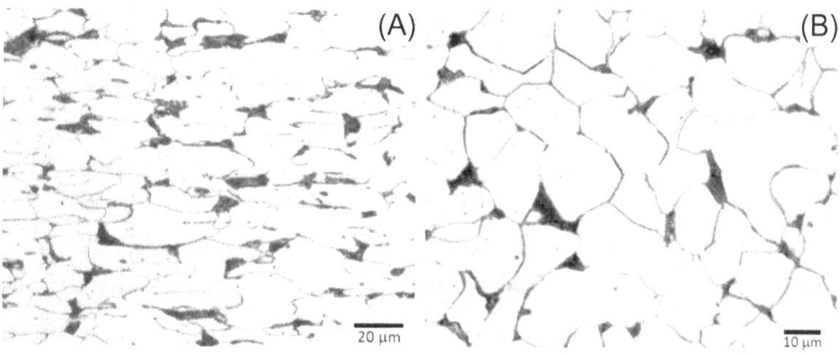

FIG. 1.8 (A) Cold work structure and (B) recrystallized structure of a commercial low-carbon steel.

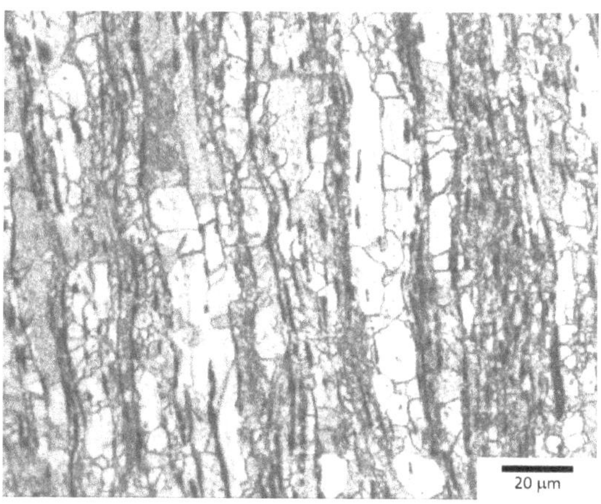

FIG. 1.9 Partially recrystallized structure in a commercial Zn-Cu-Ti alloy.

Macroscopically, a measure of stiffness is the Young's modulus E and can be determined by the propagation speed of longitudinal waves in the material as:

$$V_1 = (E/\rho)^{\frac{1}{2}} \tag{1.10}$$

where ρ is the density of the material. Another option to measure the stiffness is through extensometry, by detecting the deformations (ε) caused by the application of small stresses (σ):

$$\sigma = E\varepsilon \tag{1.11}$$

is also known as *Hooke's law*. However, a more precise method is the use of ultrasound to produce longitudinal waves with a certain speed (V_l).

At atomic scale, the higher the bond energy between atoms, the higher the value of the stiffness. The *stiffness of the bond* (dF/dr) is an indication of the difficulty to separate, by an external force (traction), two atoms whose initial equilibrium interatomic distance is r_0:

$$\frac{dF}{dr} = S \tag{1.12}$$

$$F = S(r - r_0) \tag{1.13}$$

$$F = \sigma A = S(r - r_0) \tag{1.14}$$

where S is the stiffness and r is the distance between atoms. Considering that $A \approx r_0^2$ and Eq. (1.14), the stress becomes:

$$\sigma = \frac{S}{r_0}\left(\frac{r - r_0}{r_0}\right) = E\varepsilon \tag{1.15}$$

and since deformation is defined as:

$$\varepsilon = \left(\frac{r - r_0}{r_0}\right) \tag{1.16}$$

then:

$$E = \frac{S}{r_0} \tag{1.17}$$

This stiffness of the bond is constant as long as the separation between atoms does not exceed r_1, since the atoms are unable to go back to their original equilibrium distance (r_0) by elastic recovery. It is evident that the Young's modulus and the stiffness of a material are proportional (Eq. 1.17), and both depend on the melting temperature of the material (Table 1.7). The value of the stiffness of the bond of pure metals is 15–40 N/m.

TABLE 1.7 Young's Modulus for Some Metals and Their Melting Temperature

Metal	Young's Modulus E (GPa)	Melting Temperature T_M (°C)
Tungsten	405	3410
Molybdenum	325	2610
Beryllium	250	1277
Nickel	210	1453
Steel and castings	200–170	1538–1150
Uranium	175	1132
Copper	125	1083
Titanium	115	1668
Zinc	80	419.5
Aluminum	70	660
Magnesium	40	650
Tin	40	232
Lead	15	327

EXERCISE 1.4

Calculate the lattice parameter, the Young's modulus, and the speed of propagation of longitudinal waves in Cu.

Data: The stiffness of the bond is 42 N/m, density is 8.96 g/cm^3, and atomic weight is 63.55 u (Table 1.2).

Solution

(a) Lattice parameter

$$\rho_{Cu} = \frac{4(\text{atomic weight})}{N_A a^3} \Rightarrow a = \sqrt[3]{\frac{4(\text{atomic weight})}{N_A \rho_{Cu}}} = \sqrt[3]{\frac{4(63.55\,\text{u})}{\left(6.023 \times 10^{23}\,\frac{\text{u}}{\text{g}}\right)\left(8.96\,\frac{\text{g}}{\text{cm}^3}\right)}}$$

$$a = 3.61 \times 10^{-8}\,\text{cm} = 3.61\,\text{Å}$$

(b) Young's modulus

Using Eq. (1.17) and the lattice parameter just obtained, the elastic modulus can be calculated as:

$$E = \frac{S}{r_0} = \frac{42\,\frac{\text{N}}{\text{m}}}{3.61 \times 10^{-10}\,\text{m}} = 1.16 \times 10^{11}\,\frac{\text{N}}{\text{m}^2} = 116\,\text{GPa}$$

(c) Speed of longitudinal waves
And finally, using Eq. (1.10), the propagation speed is:

$$V_l = \sqrt{\frac{E}{\rho}} = \sqrt{\frac{1.16 \times 10^{11}\,\frac{\text{N}}{\text{m}^2}}{8960\,\frac{\text{kg}}{\text{m}^3}}} = 3602.74\,\frac{\text{m}}{\text{s}}$$

EXERCISE 1.5

When heat treating a ferritic steel, there can be a temperature gradient of 10°C in the subjected parts with thickness of 10 cm. Using data from Tables 1.6 and 1.7, calculate the stress generated by this temperature gradient.

Solution

The Hooke's law relating stress and deformation is:

$$\sigma = E\varepsilon$$

In this case, deformation is caused by a thermal gradient, therefore:

$$\varepsilon = \alpha\Delta T$$

When combining both expressions:

$$\sigma = E\alpha\Delta T = (2 \times 10^5\,\text{MPa})\left(12.2 \times 10^{-6}\,\frac{1}{°\text{C}}\right)(10°\text{C}) = 24.4\,\text{MPa}$$

Comment: The material can easily withstand this value of stress, since it is considerably lower than its maximum one (~ 300 MPa).

REFERENCES

Barret, C., 1986. Structure of Metals. Crystallographic Methods. Principles and Data, third ed. McGraw-Hill, New York.

Goldschmidt, V., 1929. Crystal structure and chemical constitution. Trans. Faraday Soc. 25, 253–283.

BIBLIOGRAPHY

A.S.M., 1978. Properties and Selection. American Society for Metals, Metals Park, OH.

Cullity, B., 2001. Elements of X-Ray Diffraction, third ed. Prentice-Hall, London.

Frenkel, J., 1955. Kinetic Theory of Liquids. Dover Publications, New York.

Jeans, J., 2009. An Introduction to the Kinetic Theory of Gasses. Cambridge University Press, Cambridge.

Pero-Sanz, J., 1971. Restauración-recristalización en latón 70/30 reducido 90% en frío. Dyna 4, 197–202.

Pero-Sanz, J., 1977. Comentario a la cinética de Recristalización expresada por la ecuación de Avrami. Revista de Metalurgia del Centro Nacional de Investigaciones Metalúrgicas 3, 217–220.

Pero-Sanz, J., 2006. Ciencia e ingeniería de materiales, fifth ed. CIE Dossat, Madrid.

Pero-Sanz, J., Verdeja, J., 1977. Heat treatments on hot rolled Fe-17% Cr sheet subjected to ridging phenomena. Microstruct. Sci. 15, 177–200.

Porter, D., Easterling, K., 2009. Phase Transformations in Metals and Alloys, third ed. CRC Press, Boca Raton, FL.

Reed-Hill, R., Abbaschian, R., 2009. Physical Metallurgy Principles, fourth ed. Cengage Learning, Fairford.

Van Vlack, L., 1982. Materials for Engineering. Concepts and Its Applications. Addison-Wesley, Menlo Park, CA.

Vander Voort, G., 1999. Metallography. Principles and Practice. ASM International, Materials Park, OH.

Weast, R., 2014. Handbook of Chemistry and Physics, 95th ed. The Chemical Rubber Co., Cleveland, OH.

Winegard, W., 1964. An Introduction to the Solidification of Metals. The Institute of Metals, London.

Chapter 2

Phase Transformation Kinetics

2.1 SURFACE FREE ENERGY

The solidification temperature T_M for a pure metal is constant and evident at the intersection of the free energy curves for solid and liquid (Fig. 1.7). For solidification to fully take place, a certain degree of *undercooling* from the equilibrium temperature is required, since liquid and solid coexist in equilibrium at the solidification temperature $\left(\text{liquid} \leftrightarrow \text{solid} + \Delta H_f\right)$; in order to solidify, latent heat ΔH_f must be released. Otherwise, solidification would stop because, at T_M, some atoms would reach solid state while others would abandon it exactly at the same rate (i.e., the activity would not completely cease at the solid-liquid interface).

When reaching a certain level of undercooling, the metal solidifies, though this does not happen immediately since a certain time is necessary to fully reach solid phase. During a period of coexistence of both liquid and solid phases, the amount of liquid diminishes while the amount of solid formed by nucleation and growth, increases.

The physical processes that rule the relative stabilities of two phases (i.e., solidification, evaporation, solid-state transformations, etc.) always take place either through the contact surface or the *interface* between phases (Fig. 2.1). Between two phases or atomic groups A (solid) and B (liquid) formed by atoms of the same metal in contact through the interface (ab), phase A will grow at the expense of phase B if more atoms of B cross to phase A than atoms of A crossing to phase B, per time unit. The only atoms that take part in the flux, in both directions, are the ones, of phase A or phase B, close to the interface. The rest of the atoms of A or B can be considered disconnected from the phase change happening at the interface. During solidification, and due to the rapid changes occurring at the interface, the behavior at the surface of the phase differs from the one of the interior of the material; the atoms of the interface are not in equilibrium state, as they are definitely not in neither phase: their bonds are rapidly formed and broken. Thus, there are many atomic positions of the solid in the periphery not occupied by atoms.

The coordination of an atom located at the periphery of the solid is considerably smaller than that for an atom located at the interior of the solid mass. For example, in the face-centered cubic system, an atom at the surface does not have 12 tangent atoms, contrary to the atoms at the interior of the solid. Moving one

Solidification and Solid-State Transformations of Metals and Alloys. http://dx.doi.org/10.1016/B978-0-12-812607-3.00002-4

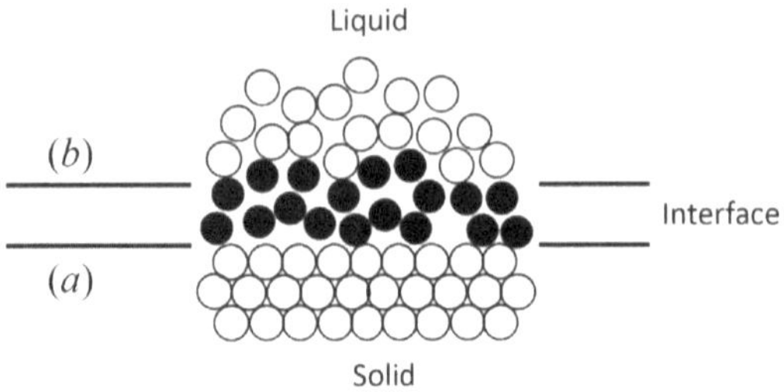

FIG. 2.1 Solid (*a*)-liquid (*b*) interface.

atom from the interior towards the liquid requires an amount of heat equal to the latent heat per atom; while the amount of heat to take an atom of the surface to the liquid will be only a fraction of that latent heat and a function of the coordination index.

Melting phenomena (variation of enthalpy) can also be applied to other variations of free energy: when solidifying a monocrystal of a face-centered cubic crystalline system metal, the variation of total free energy will be the algebraic addition of the free energy of the solid (considering that there are 12 atoms in coordination) plus the excess of free energy of the atoms located at the periphery of the monocrystal (surface free energy). If the total free energy of the solid is to be calculated, the value of the surface free energy will have special importance when the relationship between volume and area of the solid is very small.

The *surface free energy* is proportional to the difference between the free energies of an atom located at the interior of the solid (with a coordination defined by its crystalline system, i.e., 12 for fcc, 12 for hcp, and 8 for bcc) and of another atom located at the surface. Furthermore, it depends on the metallic bond energy, which is directly proportional to the melting temperature, meaning the surface free energy of metals generally increases with T_M (Tables 1.5 and 2.1).

TABLE 2.1 Surface Free Energies for Some Metals

Metal	γ_{LS}(mJ/m^2)	Metal	γ_{LS}(mJ/m^2)	Metal	γ_{LS}(mJ/m^2)
Al	121	Pd	209	Ge	181
Mn	206	Ag	126	Sn	59
Fe	204	Pt	240	Sb	101
Co	234	Au	132	Hg	28
Ni	255	Pb	33	Bi	54
Cu	177	Ga	56		

EXERCISE 2.1

Calculate the liquid-solid surface free energy of Fe.

Solution

The surface free energy can be calculated by:

$$\gamma_{LS} = \frac{\text{energy}}{\text{area}}$$

being the energy equal to $\Delta H_f/2$ since it only has to break half of the bonds (surface of a solid). The latent heat for Fe from Table 1.4 is 15,290.4 J/mol.

On the other hand, Fe is a bcc metal, and its most closed-packed plane is the one joining two opposite edges of the cube (unit cell), i.e., the {1 1 0} plane: the height of this rectangle is a while its base is $\sqrt{2}a$, making the area of this plane $\sqrt{2}a^2$. Furthermore this plane has two atoms in total (a quarter of circle at each corner and one circle in the center).

The latent heat of Fe must be divided by the Avogadro number to have units of energy per atom:

$$\Delta H_f = 15,290.4\ \text{J/mol}\left(\frac{1}{6.023\times 10^{23}\ \text{atoms/mol}}\right) = 2.54\times 10^{-20}\ \text{J/atom}$$

and multiplied by the number of atoms in the plane (2):

$$\Delta H_f = \left(2.54\times 10^{-20}\ \text{J/atom}\right)(2\ \text{atoms}) = 5.08\times 10^{-20}\ \text{J}$$

Finally, considering the lattice parameter for Fe ($a = 2.94$Å, Table 1.3), the surface free energy is:

$$\gamma_{LS} = \frac{\frac{5.08\times 10^{-20}\ \text{J}}{2}}{\sqrt{2}\left(2.94\times 10^{-10}\ \text{m}\right)^2} = 0.21\ \text{J/m}^2 = 210\ \text{mJ/m}^2$$

which is very similar to the value in Table 2.1 (204 mJ/m^2).

EXERCISE 2.2

Calculate the liquid-solid surface free energy of Al.

Solution

From Table 1.4, the latent heat for Al is10,659 J/mol.

Al is an fcc metal, and (1 1 1) is its most atomically dense plane, which is an equilateral triangle: $\sqrt{2}a$ sides and area of $\sqrt{3}a^2/2$. Analyzing the atoms in the plane, each corner has a sixth of circle (multiplied by three corners equals half a circle) and the sides have half circle each (multiplied by three sides equals one and a half circles) which totals two circles inside this plane.

The latent heat of Al must be divided by the Avogadro's number to get units of energy per atom:

$$\Delta H_f = 10,659\text{J/mol}\left(\frac{1}{6.023\times 10^{23}\ \text{atoms/mol}}\right) = 1.77\times 10^{-20}\ \text{J/atom}$$

and multiplied by the number of atoms in the plane:

$$\Delta H_f = (1.77 \times 10^{-20}\,\mathrm{J/atom})(2\ \mathrm{atoms}) = 3.54 \times 10^{-20}\,\mathrm{J}$$

Finally, considering the lattice parameter of Al ($a = 4.05$Å, Table 1.3), the surface free energy is:

$$\gamma_{LS} = \frac{\dfrac{3.54 \times 10^{-20}\,\mathrm{J}}{2}}{\dfrac{\sqrt{3}}{2}(4.05 \times 10^{-10}\,\mathrm{m})^2} = 0.12\,\mathrm{J/m^2} = 120\,\mathrm{mJ/m^2}$$

And once again, the value is very similar to that of Table 2.1 (121 mJ/m^2).

During solidification, as a result of the excess free energy that the solid has because of its solid/liquid interface, the material has a tendency to decrease the area of that interface. This tendency (also observed as surface tension) promotes the formation of rounded profiles in the crystals in order to reduce their surface to volume ratio. This phenomenon is more common for metals with high values of γ_{LS}, and will form *dendrites* with rounded tips, while metals with low γ_{LS} will form *dendrites* with planar tips (the dendritic shape of the crystals is caused by their tendency to grow in certain preferable crystallographic directions, Section 2.3.2).

Other processes analog to solidification are those related to the binary liquid-gas equilibrium (evaporation or condensation): atoms in the surface of the liquid are less bonded than the ones in the interior and therefore have a surface free energy (γ_{LG}) with a tendency to decrease the surface/volume ratio (which explains why rain drops would be spherical if gravity did not deform them).

When analyzing the equilibrium between the atoms in the interior of the grain and those in their boundaries in a polycrystalline metal (Fig. 1.2), the excess free energy between them (*grain boundary free energy*) is known as γ_{GB}. Grain boundaries play an important role in recrystallization processes (the transformation of highly deformed grained structures into new nondeformed grains), and though no new phases are involved in recrystallization, it does have many features in common with solidification.

When there is an interaction between three phases (liquid L, solid S, and gas G), an interrelation with a tendency to minimize the surface free energy of the group exists: consider a drop of liquid L over a solid S, the free energy per surface unit between liquid and solid (γ_{LS}), the free energy per surface unit between solid and gas (γ_{SG}), and the free energy per surface unit between liquid and gas (γ_{LG}) will tend to the equilibrium and the drop (Fig. 2.2) will extend over the surface of the solid, wetting it. When increasing the contact area by dA, ΔG will have values lower than zero:

$$\Delta G = \gamma_{LS} \cdot dA + \gamma_{LG} \cdot dA \cdot \cos\theta - \gamma_{SG} \cdot dA < 0 \tag{2.1}$$

Wetting will reach equilibrium when $\Delta G = 0$, which means that the contact meniscus angle will be constant and determined by:

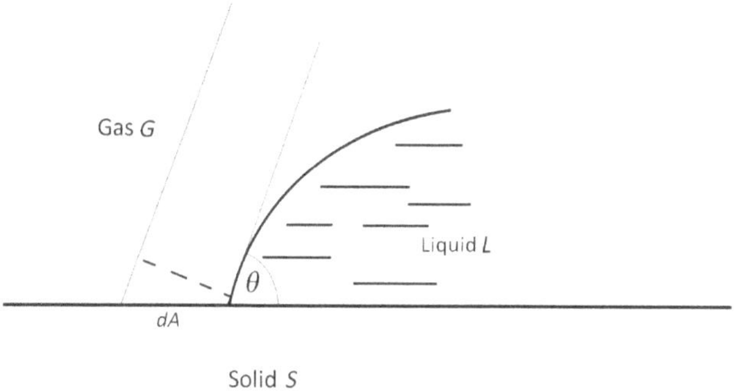

FIG. 2.2 Contact angle between gas-liquid-solid interfaces.

$$\cos\theta = (\gamma_{SG} - \gamma_{LS})/\gamma_{LG} \tag{2.2}$$

The liquid will not wet the surface and will adopt, as a consequence, a spherical shape with $\theta = 180$degrees if $\Delta G > 0$, indicating that $\gamma_{LS} > \gamma_{LG} + \gamma_{SG}$. On the other hand, wetting will be total ($\theta = 0$degrees) if $\Delta G < 0$ or $\gamma_{LS} + \gamma_{LG} < \gamma_{SG}$.

When a metallic surface has a high value of γ_{SG} (e.g., Fe), it can be wet and consequently coated with other metals such as Zn, Al, and Sn. Given that wetting property does not only depend on γ_{SG}, but on the equilibrium mentioned, other metals such as Pb do not wet Fe: when a coating of Pb is required on a surface of Fe it is necessary to coat it first with Sn and once this lining has solidified, Pb can then be applied as it does wet Sn. Another possibility is coating (plating) Fe with a liquid Sn-Pb alloy.

The wetting ability of a metal or liquid alloy is useful when brazing metals: two metallic surfaces can be welded using another metal or liquid alloy (*filler metal*) with a melting temperature lower than the ones of the surfaces (forming solid solution with them during solidification).

Low temperature welding processes include *soldering* when a filler metal that solidifies at temperatures lower than 450°C (i.e., Sn-Pb alloy), and *brazing* when the filler solidifies at temperatures higher than 450°C such as copper fillers (with T_M between 710°C and 1100°C, for example 60/40 brass), silver fillers (with at least 20% Ag, such as the Zn-27% Ag-38% Cu-9.5% Mn-5.5% Ni alloy, for working temperatures lower than 850°C), etc.

A necessary condition for brazing is that the filler metal or alloy wets the metallic surfaces of the parts, and it depends, as mentioned, on the relationship between γ_{LS}, γ_{LG}, and γ_{SG}. Not all values of γ are known given the amount of possible combinations between fillers (metals and alloys) and base alloys. However, it can be pointed out that a wetting action is common when the base alloy is susceptible to dissolution in a filler liquid alloy (analog to the behavior of a NaCl bar in contact with water, which will be wet by water easier than one made of glass), when a solid solution can be formed with the filler alloy in the base alloy or when there is affinity to form intermetallic compounds with both alloys. In general, combinations of alloys (filler and base) that can form intermetallic

TABLE 2.2 Interfacial Free Energy Between Solid and Liquid Metals

Solid	Liquid	γ_{LS} (mJ/m^2)	Solid	Liquid	γ_{LS} (mJ/m^2)	Solid	Liquid	γ_{LS} (mJ/m^2)
Zn	Sn	119	Al	Sn	158	Mo	Sn	1000
Zn	In	122	Cu	Pb	390	W	Sn	1000
Zn	Bi	148	Fe	Cu	430	Fe	Pb	1415
Zn	Pb	185	Nb	Cu	428	Fe	Ag	1370
Zn	Sn	150	Cr	Ag	540	Fe	Na	2000
Ag	Pb	160	W	Cu	980			

compounds are avoided due to their fragile behavior, while combinations that form solid solutions are preferred. Table 2.2 shows the values of γ_{LS} for some common combinations of solid and liquid; and it is important to point out that wetting is easier for low values of γ_{LS}.

EXERCISE 2.3

Calculate for graphite, the solid-gas superficial tension, and explain why the brazing of castings with Cu as a filler metal is not possible.

Data: $\Delta H_{SG} = 6270\text{J/mol},\ a = 2.3\text{Å},\ c = 6.7\text{Å}$.

Solution

Just as the interfacial solid-liquid energy is proportional to $\Delta H_m/2$, the solid-gas interfacial energy is also proportional to $\Delta H_{SG}/2$ (sublimation energy). From Fig. 1.5, it is evident that the hexagon (base) has three atoms associated to it, thus:

$$A_{hexagon} = \frac{6\,(2.3)\,2.3\sqrt{3}}{4} = 13.74\text{Å}^2$$

$$\gamma_{SG} = \frac{\Delta H_{SG}}{2} = \frac{(6270\,\text{J/mol})(3\,\text{atoms})}{2\,(6.023 \times 10^{23}\,\text{atoms/mol})\,(13.74 \times 10^{-20}\,\text{m}^2)}$$

$$= 0.11\,\text{J/m}^2 = 110\,\text{mJ/m}^2$$

From Table 2.2, for Fe (solid) and Cu (liquid), $\gamma_{LS} = 430\text{mJ/m}^2$ and from Eq. (2.2):

$$\cos\theta = \frac{\gamma_{SG} - \gamma_{LS}}{\gamma_{LG}} = \frac{110 - 430}{\gamma_{LG}}$$

In other words, the value of γ_{LG} between liquid Cu and gaseous C does not matter, the cosine of θ is negative (meaning $\theta > 90$degrees), which is translated to Cu not wetting graphite (main constituent of gray castings).

Comment: A liquid metal cannot wet a solid with a low solid/gas interfacial energy value, which is why when brazing gray cast irons, their surface must be electrochemically cleaned to remove graphite (phase constituent).

Graphite has hexagonal crystals and the bond energy of atoms located in the same basal plane is 418–500 kJ/mol, while the bond energy between atoms of adjacent planes is considerably lower: 4.18–8.36 kJ/mol. Its *c/a* ratio is higher than 1.633, resulting in a noncompact lamellar structure, causing its thermal conductivity in the directions parallel to the basal planes 100 times higher than in a direction perpendicular to them. Furthermore, the crystals exfoliate very easily, making graphite a good lubricant and a material easy to form.

The wetting and filling ability of a liquid metal or alloy is also of interest in powder metallurgy: during the liquid phase sintering stage or when a liquid metal infiltrates a porous compact in order to fill the pores of the metallic skeleton (previously obtained by sintering of powders in solid phase). Table 2.3 shows some values of interfacial energy (γ_{LS}) for some liquid metal and solid oxides or carbides of interest for the liquid phase sintering for powder metallurgy industries.

When a liquid wets a solid metal, a total disaggregation of the grains of the solid can occur if the grain boundary energy γ_{GB} has a value of twice γ_{LS}. As shown in Fig. 2.3, for the liquid to penetrate between grains (intergranular), the surface free energy ΔG of the assembly must be smaller or equal to zero. The diedric equilibrium angle in the partial penetration is:

$$\cos\frac{\phi}{2} = \frac{\gamma_{GB}}{2\gamma_{LS}} \tag{2.3}$$

Total disaggregation (e.g., intergranular corrosion, hot shortness, etc.) will happen when $\phi = 0$; and thus $\gamma_{GB} = 2\gamma_{LS}$.

TABLE 2.3 Interfacial Free Energies Between Solid (Oxides or Carbides) and Liquid (Metal)

Solid-Liquid	γ_{LS} (mJ/m^2)
Al_2O_3-Ni	1850
ZrO_2-Ni	970
UO_2-Ni	1280
TiC-Co	505
HfC-Co	385
VC-Co	465
NbC-Co	480
UC-U	141

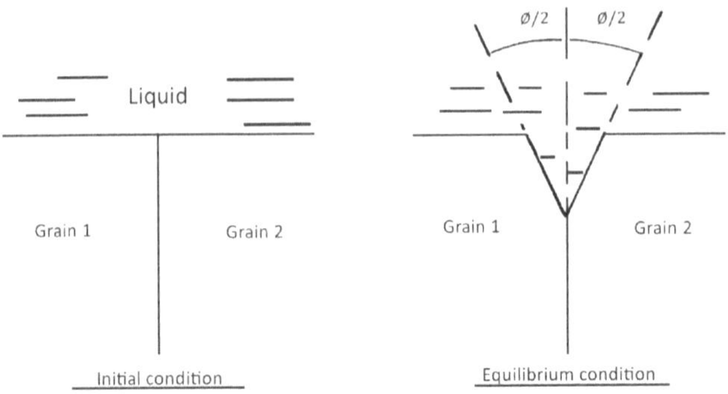

FIG. 2.3 Initial stages of intergranular wetting. Balance of liquid surface and grain boundary tensions.

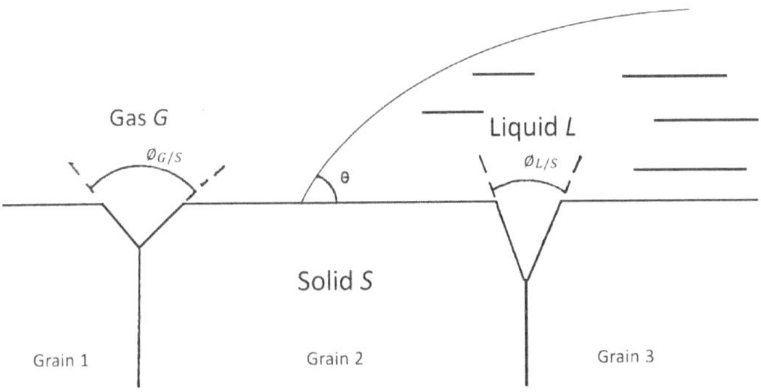

FIG. 2.4 Equilibrium between a polycrystalline solid, liquid, and gas.

The micrographic determination of penetration ϕ and wetting θ angles is a technique used to empirically obtain the values of γ_{LS} between a liquid and a solid, as the equilibrium conditions in the triple points (Fig. 2.4) can be determined by:

$$\gamma_{SG} = \gamma_{LS} + \gamma_{LG} \cdot \cos\theta \tag{2.4}$$

$$\gamma_{GB} = 2\gamma_{LS} \cdot \cos\left(\frac{\phi_{L/S}}{2}\right) \tag{2.5}$$

$$\gamma_{GB} = 2\gamma_{sv} \cdot \cos\left(\frac{\phi_{G/S}}{2}\right) \tag{2.6}$$

which results in:

$$\gamma_{LS} = \gamma_{LG} \cdot \cos\theta \frac{\cos\left(\frac{\phi_{G/S}}{2}\right)}{\cos\left(\frac{\phi_{L/S}}{2}\right) - \cos\left(\frac{\phi_{G/S}}{2}\right)} \tag{2.7}$$

EXERCISE 2.4

The interfacial energy in the grain boundaries of copper is 500 mJ/m^2, while the copper (solid) and bismuth (liquid) interfacial energy is 50 mJ/m^2. Would liquid bismuth wet copper?

Solution

Considering Eq. (2.3):

$$\cos\frac{\phi}{2} = 2\frac{\gamma_{GB}}{\gamma_{LS}}$$

where $\gamma_{GB} = \gamma_{Cu} = 500\,\text{mJ/m}^2$ and $\gamma_{LS} = \gamma_{Bi-Cu} = 50\,\text{mJ/m}^2$, then

$$\cos\frac{\phi}{2} = \frac{500\,\text{mJ/m}^2}{2(50\,\text{mJ/m}^2)} = 5$$

The maximum value that the cosine of an angle can have is 1, therefore $\phi = 0$ degrees, which means that all the bismuth would penetrate the grain boundary causing intergranular corrosion.

2.2 HOMOGENEOUS NUCLEATION AND CRITICAL NUCLEUS SIZE

During solidification, the variation in free energy is negative (*driving force*) when a crystalline solid forms at a certain temperature $T_1 < T_M$, and its value (e.g., Joules per volume unit) equals $G_S - G_L$ as shown in Fig. 1.7. Therefore:

$$\begin{aligned}\Delta G_{SL} &= G_S - G_L = (H_S - H_L) - T_1(S_S - S_L) = -\Delta H_f - T_1\Delta S_{SL} \\ &= -\Delta H_f + T_1\Delta S_{SL} = -\Delta H_f + T_1\frac{\Delta H}{T_M} = -\Delta H_f\frac{T_M - T_1}{T_M} = \Delta H_f\frac{\Delta T_1}{T_M}\end{aligned} \tag{2.8}$$

where ΔH_f is the melting latent heat and ΔT_1 the undercooling.

EXERCISE 2.5

Calculate the driving force, or $G_L - G_S$, for the solidification of Ni, considering an undercooling of 10 K(10°C) and data from Tables 1.2–1.4.

Solution

Using Eq. (2.8) and knowing that $\Delta H_f = 17{,}556\,\text{J/mol}$, $T_M = 1726\,\text{K}$, the atomic weight is 58.71 g/mol and $\rho = 8902\,\text{kg/m}^3$:

$$\begin{aligned}G_L - G_S &= \Delta H_f\frac{\Delta T_1}{T_M} = (17{,}556\,\text{J/mol})(8902\,\text{kg/m}^3)\left(\frac{1\,\text{mol}}{58.71\times10^{-3}\,\text{kg}}\right)\left(\frac{10\,\text{K}}{1726\,\text{K}}\right) \\ &= 15.42\times10^6\,\text{J/m}^3 = 15.42\,\text{MJ/m}^3\end{aligned}$$

As indicated in the previous section, and given that atoms at the interface do not have the same coordination level as the interior ones, the variation in free energy to solidify a nucleus (cluster) with radius r must consider the addition of two terms: ΔG_V which corresponds to the formation of volume (supposing that all atoms have the maximum coordination index) and the term ΔG_S for the surface free energy. This is:

$$\Delta G = \Delta G_V + \Delta G_S \tag{2.9}$$

where:

$$\Delta G_V = -\Delta H_f \cdot \frac{\Delta T_1}{T_M} \cdot \frac{4}{3}\pi r^3 \tag{2.10}$$

$$\Delta G_S = 4\pi r^2 \cdot \gamma_{LS} \tag{2.11}$$

Fig. 2.5 shows the variation of ΔG_V and ΔG_S, at T_1, as a function of the nucleus radius. The sum of both terms, ΔG, presents a maximum for the values of r^* and ΔG^* as follows:

$$r^* = \frac{2\gamma_{LS} \cdot T_M}{\Delta H_f \Delta T_1} \tag{2.12}$$

$$\Delta G^* = \frac{16\pi}{3} \frac{\gamma_{LS}^3}{\Delta H_f^2} \left(\frac{T_M^2}{\Delta T_1^2}\right) \tag{2.13}$$

Furthermore, it can be deduced that at T_1, a cluster of atoms with radius $r_m < r^*$ formed by metallic atoms will not grow, given that free energy variation in order to attract a new atom and increase its radius would mean a positive value of ΔG. On the contrary, a cluster with radius $r_M > r^*$ will have a tendency to increase its radius by attracting more atoms given that an infinitesimal increase in radius r_M will result in a negative variation of free energy.

At T_1, the spontaneous clustering of atoms that do not reach the critical size (higher than r^*) will disappear by collision with other atoms or small clusters.

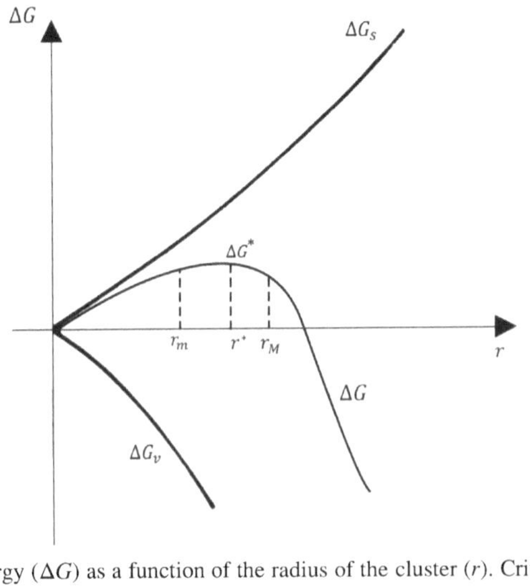

FIG. 2.5 Free energy (ΔG) as a function of the radius of the cluster (r). Critical radius is r^* and critical free energy is ΔG^*.

Only those clusters with radius higher than r^* can spontaneously grow. The critical radius of the nucleus at a certain temperature, and consequently the number of atoms necessary to form that cluster, is inversely proportional to the ΔT_1 undercooling, as indicated in Eq. (2.12).

Thermodynamic considerations aside, it is known that in solid state an atom is surrounded by less neighboring atoms if the curvature radius of the cluster is small; therefore, the probability for a surface atom to become detached from the solid in order to join the liquid is inversely proportional to the radius of the nucleus. On the other hand, if T_1 is low (large ΔT_1 undercooling), the critical radius of the nucleus will be smaller, since the bond between atoms allows peripheral atoms to be fixed (bond energy increases), in a more stable manner, to the solid nucleus.

EXERCISE 2.6

Calculate the number of atoms in a spherical nucleus of Ni, with critical radius, as a function of the undercooling.

Data: $a = 3.5\text{Å}$, $\Delta H_f = 17{,}556\,\text{J/mol}$, $T_M = 1726\,\text{K}$, the atomic weight is 58.71 g/mol, $\rho = 8902\,\text{kg/m}^3$ and $\gamma_{LS} = 255\,\text{mJ/m}^2$ (Tables 1.2–1.4 and 2.1).

Solution

Considering Eq. (2.12):

$$r^* = \frac{2\gamma_{LS}\cdot T_M}{\Delta H_f \Delta T_1} = \frac{2(0.255\,\text{J/m}^2)(1726\,\text{K})}{(17{,}556\,\text{J/mol})\left(\dfrac{1\,\text{mol}}{58.71\times 10^{-3}\,\text{kg}}\right)(8902\,\text{kg/m}^3)\Delta T_1} = \frac{3.31\times 10^{-7}\,\text{m}}{\Delta T_1}$$

$$= \frac{3310\,\text{Å}}{\Delta T_1}$$

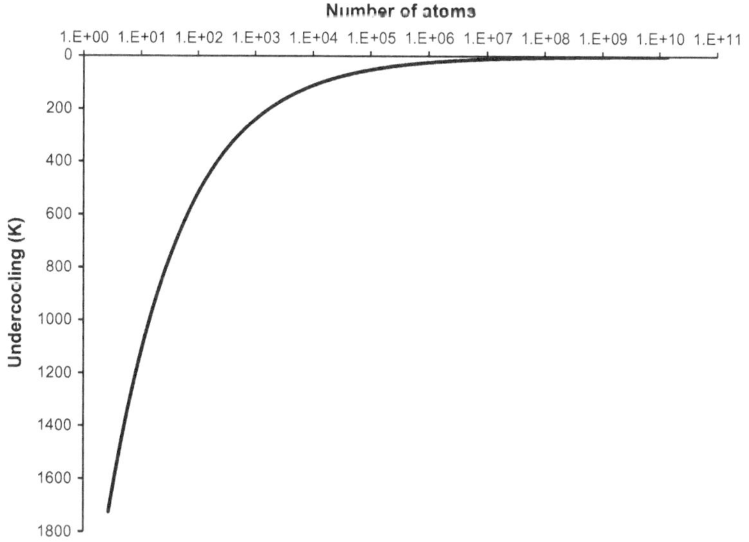

FIG. 2.6 Number of atoms as a function of undercooling in Ni.

Thus the number of atoms is:

$$n_{atoms} = \frac{16}{3}\pi\left(\frac{3310}{3.52}\right)^3 \frac{1}{\Delta T_1^3} = \frac{1.39 \times 10^{10}}{\Delta T_1^3}$$

which are plotted in Fig. 2.6, where the horizontal axis is in logarithmic scale.

For $\Delta T = 0.2\,T_M = 345.2\,\mathrm{K}$, the number of atoms will be $337.69 \approx 338$ atoms.

2.3 DISCONTINUOUS NATURE OF SOLIDIFICATION

At temperatures close to the melting point, a metallic liquid can be considered as a statistical ensemble of atomic pregroupings or clusters that have incipient crystalline structures (Fig. 2.7).

The solidification process implies that there are nanocrystals with a radius higher than r^* (supercritic nuclei) with the ability to grow, which were formed due to thermal fluctuations that help to overcome the activation energy barrier ΔG^*. The formation of the nucleus is possible as the total free energy variation (caused by the formation of nuclei) is indeed negative as mixing with other nuclei results in an entropy increase.

From a thermodynamic point of view, the number of possible nuclei n_r with radius r^*, at T_1 will be the one that accomplishes the minimum free energy value; in other words, the most negative value of the following expression:

$$\Delta G = n_r\,\Delta G^* - T_1\,\Delta S_r \tag{2.14}$$

where ΔS_r is the variation in configuration or mixing entropy of clusters with radius r^* in a liquid of N atoms. Thus, supposing that $n_r \ll N$, the resulting free energy value is:

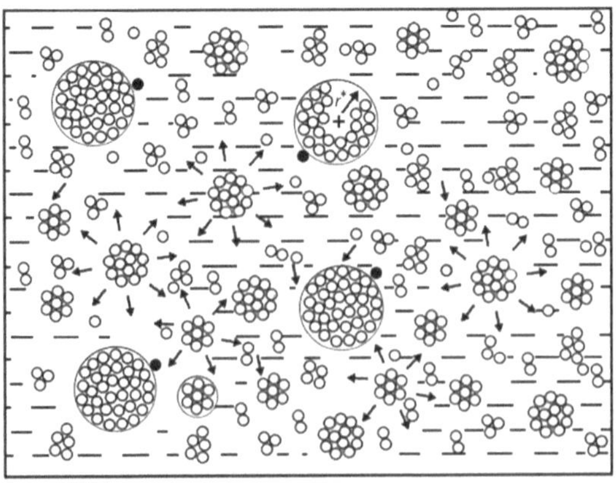

FIG. 2.7 Clusters or groups of atoms in a pure metal liquid at a temperature below T_M.

$$\Delta G = n_r \, \Delta G^* - T_1 \Delta S_r = n_r \, \Delta G^* - T_1 \cdot K \cdot n_r \left[1 - \ln \frac{n_r}{N}\right] \tag{2.15}$$

Differentiating Eq. (2.15) with respect to n_r and making it equal to zero, the following is obtained:

$$n_{r^*} = N \cdot \exp\left(\frac{-\Delta G^*}{kT_1}\right) \tag{2.16}$$

If the value of ΔG^* indicated in Eq. (2.13) is considered, the number of nuclei with radius r^* (critical radius) during homogeneous nucleation is inversely proportional to the $T_1 \Delta T_1^2$ product.

When the metal fully solidifies, the polycrystalline structure formed is the result of previous nuclei that grew rapidly: the material will have a higher number of grains if solidification takes place at lower temperatures (the lower the temperature, the higher the number of nuclei will be). Lowering the solidification temperature will always result in smaller final grain size.

A way to promote solidification of the liquid is by adding *nucleating agents*, which act as substrate for clusters of atoms. This process is known as *heterogeneous nucleation* (the lack of an external nucleating agent defines *homogeneous nucleation*). The value of n_r when nucleation is heterogeneous will always be higher than the one previously calculated for the homogeneous process. As shown in Fig. 2.8, the presence of a substrate results in a lower number of metallic atoms necessary to form a nucleus with curvature R_1 (compared to homogeneous nucleation). The qualitative relation between the number of atoms necessary to form, in each case, a nucleus, and the undercooling ΔT, is also presented in Fig. 2.8.

A grain size as small as possible is frequently sought as properties such as strength, good elongation and, above all, toughness, are increased: precisely because grain boundaries are regions of the metal where a local disorder exists because of the presence of large amounts of linear crystalline defects or dislocations (*dislocation tangles*), which act as barriers to plastic deformation and increase both, mechanical resistance and toughness.

In the interior of each grain there are also structural defects (vacancies and dislocations) produced during the solidification process; but the behavior of the grain boundary is considerably different from the behavior of the crystalline grain interior (the interior of the grain induces straining while the boundary hinders it).

Thus, to obtain a smaller grain size during solidification, an industrial practice is to promote heterogeneous nucleation by adding nucleating agents in the shape of fine solid particles (elements or chemical compounds) with good wetting ability when in contact with the liquid, and which also catalyze the nucleation due to their affinity and their crystallographic similarity with the metal. As an example, aluminum alloys use Ti and B; Mg alloys, Zr; gray cast irons, Si and Mg and ferrous alloys, Co and Zn.

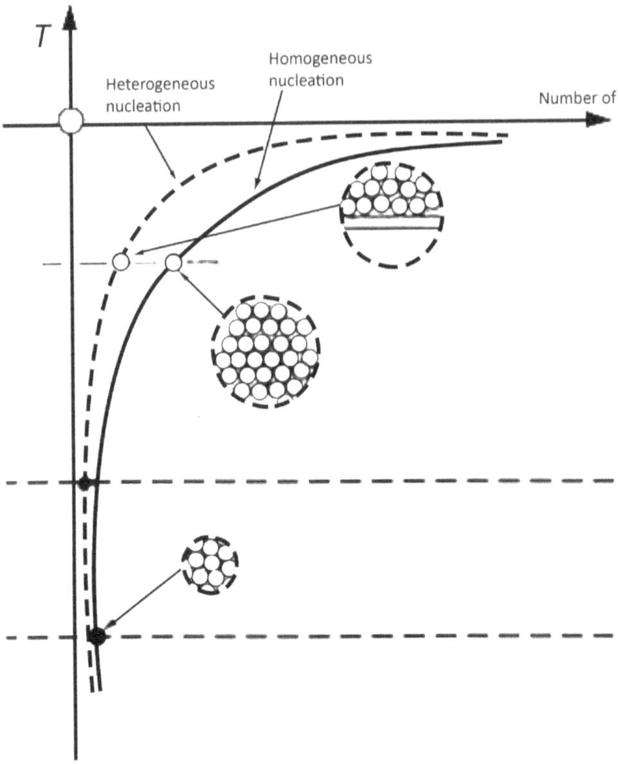

FIG. 2.8 Number of atoms in a cluster as a function of undercooling.

The effect of the nucleating agent depends, apart from its nature, on multiple factors: the exact moment it is added, the time it remains in the melted metal, etc. These parameters are important as the nucleating agents in the form of dispersed fine-size particles can suffer coalescence in the melted metal, and also undergo changes in their chemical characteristics and/or in their surface energy.

To obtain an even finer grain size, certain techniques such as vibration (mechanical, electromagnetic, or ultrasonic) of the solid-liquid mass are used. If stirring is applied to the liquid, finer grain structures are obtained, as the increase in nucleating agents produced by fragmentation of crystals in the growing stage results in broken solid nuclei of the same metal which also act as heterogeneous nucleating agents on the liquid.

EXERCISE 2.7

Calculate the minimum undercooling required for the homogeneous nucleation of Ni and obtain the critical radius of the nucleus and the number of atoms per critical nucleus (Tables 1.2–1.4 and 2.1).

Solution

(a) Undercooling

Supposing that there is 1 critical nucleus per cubic centimeter (or 1×10^6 nuclei per cubic meter) and that the number of atoms per cubic meter is:

$$N = \frac{N_{Av}}{V_s} = \frac{6.023 \times 10^{23}\,\text{atoms/mol}}{6.6 \times 10^{-6}\,\text{m}^3/\text{mol}} = 9.13 \times 10^{28}\,\text{atoms/m}^3$$

The latent heat per cubic meter is:

$$\Delta H_f = \frac{17{,}556\,\text{J/mol}}{6.6 \times 10^{-6}\,\text{m}^3/\text{mol}} = 2.66 \times 10^9\,\text{J/m}^3$$

Eqs. (2.13), (2.16) can be used:

$$\frac{n_{r^*}}{N} = \exp\left(\frac{-\Delta G^*}{kT_1}\right)$$

$$\ln\frac{n_{r^*}}{N} = \frac{-\Delta G^*}{kT_1}$$

$$\ln\frac{n_{r^*}}{N} = \frac{-\pi}{\Delta T_1^2 \cdot 3}\left(\frac{16 \cdot \gamma_{LS}^3 \cdot T_M^2}{\Delta H_f^2 kT_1}\right)$$

and then solving for $T_1 \Delta T_1^2$:

$$T_1 \Delta T_1^2 = -\frac{16\pi \cdot \gamma_{LS}^3 \cdot T_M^2}{3 \cdot k \cdot \Delta H_f^2 \cdot \ln\left(\frac{n_{r^*}}{N}\right)} = -\frac{16\pi\,(0.255)^3 (1726)^2}{3(13.8 \times 10^{-24})(2.66 \times 10^9)^2\ \ln\left(\frac{1 \times 10^6}{9.13 \times 10^{28}}\right)}$$

$$= 1.60 \times 10^8\,\text{K}^3$$

Then, the undercooling is 340.13 K ($\cong 0.2\,T_M$).

(b) Critical radius

To obtain the critical radius, Eq. (2.12) is used:

$$r^* = \frac{2\gamma_{LS} \cdot T_M}{\Delta H_f \Delta T_1} = \frac{2(0.255)(1726)}{(2.66 \times 10^9)(340.13)} = 9.73 \times 10^{-10}\,\text{m} = 9.73\,\text{Å}$$

(c) Number of atoms per nucleus

Considering the lattice parameter of Ni as 3.52 Å:

$$n_{atoms} = \frac{16}{3}\pi\left(\frac{9.73}{3.52}\right)^3 = 353.77\ \text{atoms} \cong 354\ \text{atoms}$$

Comment: For any metal, with an undercooling of $\sim 0.2\,T_M$, critical radius nuclei have sizes of ~ 10Å (1 nm) and the number of atoms per nucleus lies between 100 and 1000.

2.3.1 Nucleation Rate

Statistical thermodynamics explain the conditions that promote the formation of nuclei and regulate their size; but do not explain their kinetic phenomenon (time necessary to form nuclei and their growth). When analyzing the time necessary to build a supercritic nucleus (size higher than r^*), it is not enough to have some clusters with radius r^*. From the initial instant $t = 0$, there are clusters with radius $r \leq r^*$, in continuous formation and disaggregation. For one of these groups of radius r^* to become supercritic, it must also be oriented in such a way that acquiring an additional atom is possible. Until that happens, solidification will not start at T_1, because the nuclei formation rate per volume unit and time unit (or nucleation rate) depends on various factors:

1. The number of nuclei n_r (that can be calculated thermodynamically): the higher the number of critical nuclei, the higher the probability that one of them starts to grow and initiates solidification.
2. Though not as important, the number of atoms ω^* in the nucleus-liquid interface: for a spherical nucleus with radius r^*, the number ω^* is proportional to the surface with a value $4\pi r^{*2}$. As a simplification ω^* is considered to have a value of 1.
3. The frequency ν_{LS} at which the atoms jump through the liquid-solid interface: the nucleation rate depends on ν_{LS}, which is proportional to the atomic diffusion coefficient D_L of the liquid. There are viscous materials, such as glasses and polymers, whose diffusivity decreases considerably with temperature, yet, in general, the value of D_L for metals decreases very little, with the only exception being fast cooling processes to generate "*frozen*" liquids (metallic glasses) at very low temperatures.

Furthermore, the three factors that influence the nucleation rate

$$I = n_{r^*}\omega^* \nu_{LS} \tag{2.17}$$

just mentioned, are usually divided into two groups according to their function: the force that induces solidification ($n_{r^*}\omega^*$) which is proportional to the undercooling ΔT_1; and the diffusivity ν_{LS} which generally decreases with temperature. The product of both functions results in a bell-shaped curve (Fig. 2.9), where the nucleation rate is slow at temperatures close to the melting point (small undercooling), as well as when undercooling is considerable. Fig. 2.9 also shows the curves corresponding to the homogeneous and heterogeneous nucleation (since n_{r^*} has higher values at any given temperature, heterogeneous nucleation is faster than homogeneous one).

The following formula is used to calculate the homogeneous nucleation rate:

$$I_{homogeneous} = n \cdot \exp\left[-\frac{\Delta G^*}{kT}\right] \cdot \frac{D_L}{\mathbf{b}^2} \cdot \omega^* \tag{2.18}$$

where n is the number of atoms per volume unit, D_L is the diffusion coefficient at a temperature T and $\mathbf{b}$ is the Burgers vector (atomic diameter). In the case of heterogeneous nucleation, Eq. (2.17) is almost identical

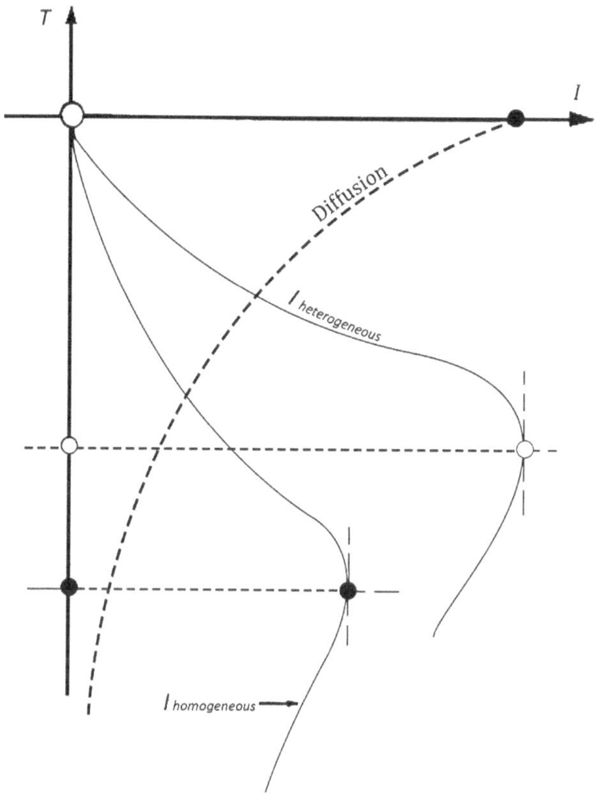

FIG. 2.9 Homogeneous and heterogeneous nucleation rate curves as a function of undercooling.

though it includes the shape factor of the nuclei as a function of the wetting angle (θ):

$$f(\theta) = \frac{(2 + \cos\theta)(1 - \cos\theta)^2}{4} \tag{2.19}$$

$$\Delta G^*_{heterogeneous} = \Delta G^*_{homogeneous} \cdot f(\theta) \tag{2.20}$$

EXERCISE 2.8

Considering that two nucleating agents are being evaluated for their addition to the metal of Exercise 2.7, one with $\theta = 30$degrees and the other with $\theta = 10$degrees, determine which one will require the smallest undercooling for its heterogeneous nucleation.

Solution

(a) $\theta = 30$degrees

The first step is to calculate the wetting equilibrium factor (Eq. 2.19):

$$f(\theta = 30°) = \frac{(2 + \cos 30°)(1 - \cos 30)^2}{4} = 1.28 \times 10^{-2}$$

On the other hand $\Delta G^*_{heterogeneous} = \Delta G^*_{heterogeneous} \cdot f(\theta)$ and considering the results of Exercise 2.7 ($T_1 \Delta T_1^2 = 1.6033 \times 10^8 \text{K}^3$), this heterogeneous undercooling must satisfy:

$$T_1 \Delta T_1^2 = 1.60 \times 10^8 \left(1.28 \times 10^{-2}\right) = 2{,}052{,}224 \text{K}^3$$

And the undercooling will then be: $\Delta T_1 = 34.84 \text{K}$

(b) $\theta = 10$ degrees

Following the same procedure for this new wetting angle:

$$f(\theta = 10°) = \frac{(2 + \cos 10°)(1 - \cos 10)^2}{4} = 1.72 \times 10^{-4}$$

$$T_1 \Delta T_1^2 = 1.60 \times 10^8 \left(1.72 \times 10^{-4}\right) = 27{,}576.76 \text{K}^3$$

And the undercooling will then be: $\Delta T_1 = 3.85 \text{K}$

Therefore, the lower the wetting angle, the smaller the undercooling required for nucleation to take place.

As previously mentioned, for viscous nonmetallic materials (including oxides and organic polymers), the coefficient D_L diminishes exponentially with temperature and can reach zero for low values of T_1. On the other hand, for metallic liquids D_L is almost constant ($\sim 10^{-9} \text{m}^2/\text{s}$) at common industrial operation temperatures. However, the bell-shaped curve shown in Fig. 2.9 will be valid for both metallic and nonmetallic liquids (those with lower diffusivity).

Supposing that $\omega^* = 1$, $N = 10^{29} \text{atoms}/\text{m}^3$, $D_L = 10^{-9} \text{m}^2/\text{s}$, and $\mathbf{b}^2 \approx 10^{-19} \text{m}^2$, the nucleation rate for metallic liquids can be expressed in a simpler manner:

$$I = 10^{39} \cdot \exp\left(-\frac{\text{constant}}{\Delta T_1^2}\right) \left[\frac{\text{nuclei}}{\text{m}^3 \text{s}}\right] \qquad (2.21)$$

where the exponential function equals approximately zero for small undercooling and for low values of T_1. Furthermore, in order for the nucleation to take place, a critical undercooling ΔT_c is required to promote a very fast nucleation process.

According to Turnbull and others, the critical undercooling ΔT_c for homogeneous nucleation in pure metals, is approximately $0.2 T_M$. In the case of heterogeneous nucleation, this critical value is considerably smaller ($0.02 T_M$).

EXERCISE 2.9

Calculate the homogeneous and heterogeneous ($\theta = 30$degrees) nucleation rates as a function of undercooling

Solution

(a) Homogeneous nucleation rate

Following Eq. (2.21)

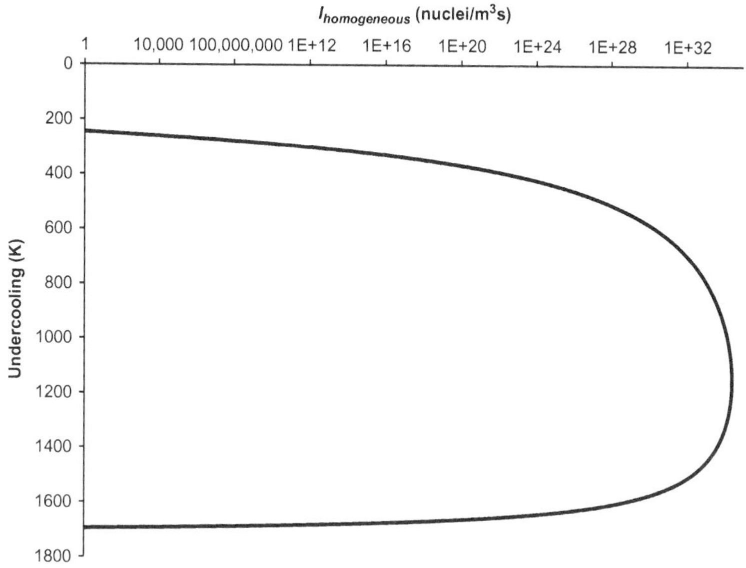

FIG. 2.10 Homogeneous nucleation rate as a function of undercooling.

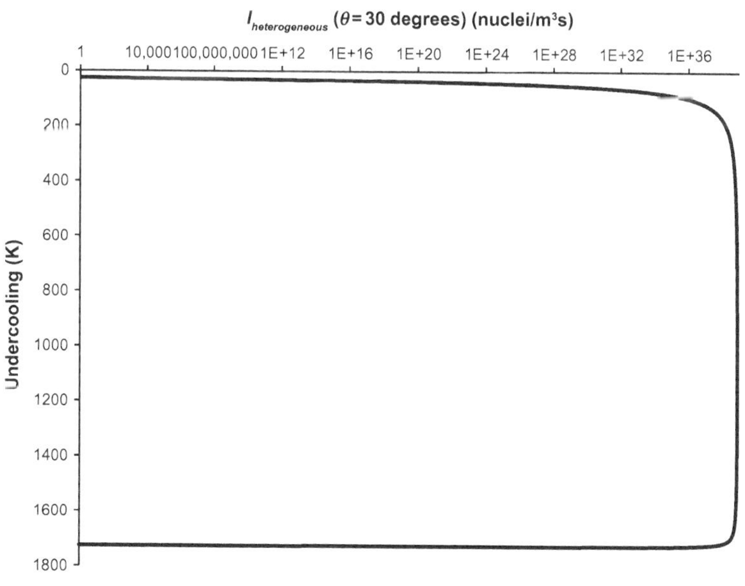

FIG. 2.11 Heterogeneous nucleation rate as a function of undercooling, considering a wetting angle of 30 degrees.

$$I_{homogeneous} \cong 1 \times 10^{39} \cdot \exp\left(\frac{-8 \times 10^9}{T\,\Delta T^2}\right) \text{nuclei/m}^3\text{s}$$

The graphical representation of this expression can be seen in Fig. 2.10.

(b) Heterogeneous nucleation rate (wetting angle of 30 degrees)

$$I_{heterogeneous} \cong 1 \times 10^{39} \cdot \exp\left(\frac{-8 \times 10^9 \cdot f(\theta)}{T\,\Delta T^2}\right) \text{nuclei/m}^3\text{s}$$

The corresponding plot of this type of nucleation rate is observed in Fig. 2.11.

Comment: When comparing both figures, it is evident that the heterogeneous nucleation rate is considerably faster (four orders of magnitude), and the shape of the curve is less rounded.

EXERCISE 2.10

Demonstrate that nucleation rate has a maximum value when undercooling equals 2/3 T_M

Solution

$$I = A \cdot \exp\left(-\frac{B}{T\,\Delta T^2}\right)$$

In order to maximize I, the product $T\Delta T^2$ has to be maximized or the term $-B/T\Delta T^2$ minimized.

Differentiating $T\Delta T^2 = T(T_M - T)^2$ and equal to zero:

$$-2(T_M - T)T + (T_M - T)^2 = 0$$

$$(T_M - T)(-2T + T_M - T) = 0$$

with two roots:

$$T_1 = T_M, \quad T_2 = \frac{1}{3} T_M$$

The first solution gives the minimum value and the second one the maximum, thus $\Delta T = T_M - T_2 = 2/3\,T_M$ except when manufacturing metallic glasses which require a large undercooling.

2.3.2 Growth Rate

Once the nucleus has formed, it keeps growing by adding atoms from the liquid. Furthermore, this growth of the nuclei is not isotropic, as could be expected, but of the directional type. Metals grow, preferentially, according to certain orientations: metals with cubic crystalline system (both face-centered and body-centered) have a $\langle 1\,0\,0 \rangle$ preferential growth, while hexagonal closed-packed metals grow in the $\langle 0\,0\,0\,1 \rangle$ direction.

When nuclei grow, ramifications known as *dendrites* (arborescent shape, the term comes from the Greek word, meaning tree) are formed. The dendritic axis appears in the shape of branches with certain spacing between them; this

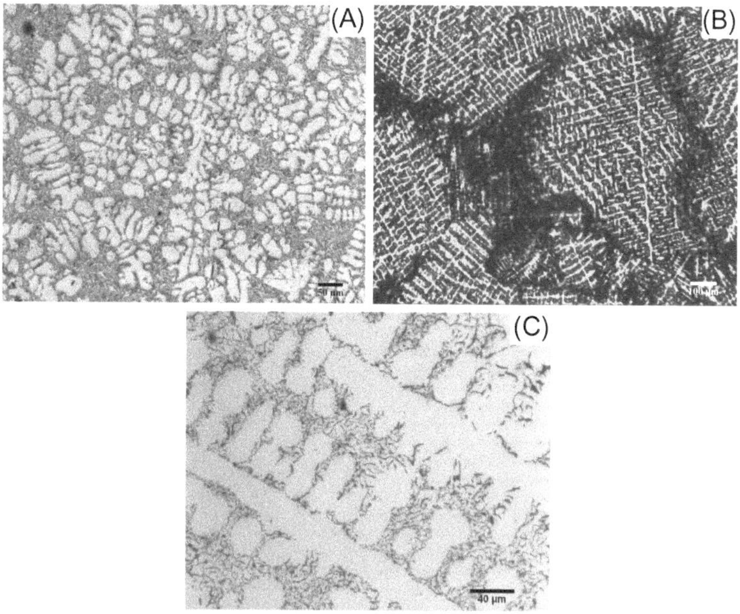

FIG. 2.12 Dendritic microstructure in: (A) pressure die casting Al-Si alloy, (B) Ag-Cu alloy, and (C) gray casting.

spacing is a product of local heat release during each solidification stage, resulting also in local decrease in the undercooling (Fig. 2.12).

Due to the directional nature of solidification, it is difficult to establish an accurate model for the three-dimensional growth rate of the nuclei and similar considerations to the ones for nucleation must be made. The driving force to add an atom from the liquid to the solid nucleus is proportional to the undercooling ΔT_1; on the other hand, the ability of the atoms to diffuse and reach an existing nucleus considerably diminishes with temperature in nonmetallic liquids.

The growth rate shown in Fig. 2.13 can be calculated with the following formula:

$$v = \Omega \cdot \rho_{planar} \cdot \frac{D_L}{\mathbf{b}^2} \cdot \frac{\Delta H_f}{RT_M} \cdot \frac{\Delta T_1}{T_1} \tag{2.22}$$

where Ω is the atomic volume, ρ_{planar} the number of atoms per surface unit of a dense plane, D_L the diffusion as previously mentioned, $\mathbf{b}$ the Burgers vector, ΔH_f the melting latent heat, ΔT_1 the $T_M - T_1$ undercooling, and R the ideal constant for gasses ($R = kN_A$).

In the case of metals in liquid phase, two cases can be considered:

- Constant diffusion in liquid state:

$$D_L \approx 1 \times 10^{-9}\,\text{m}^2/\text{s} \tag{2.23}$$

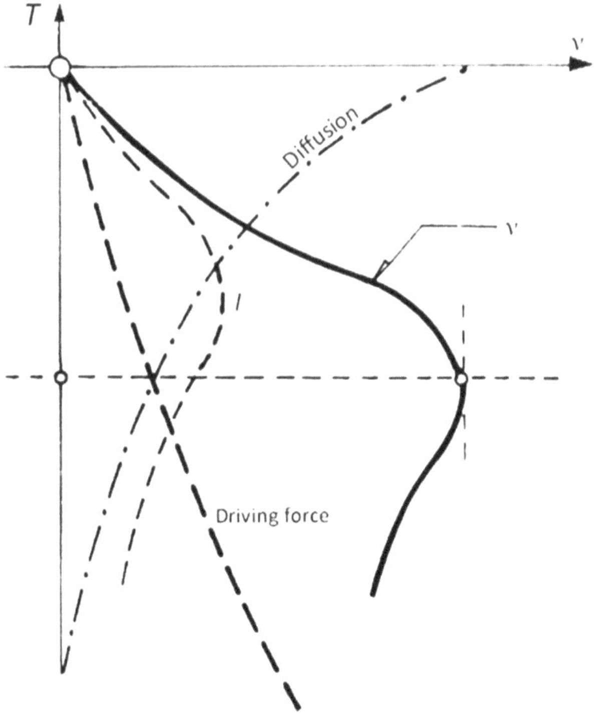

FIG. 2.13 Diffusion-controlled nucleation and nuclei growth rates as a function of undercooling.

- Variable diffusion in liquid state, thermally activated, of the Arrhenius type:

$$D_L \approx 1 \times 10^{-4}\left(-11.5\,\frac{T_M}{T}\right)\mathrm{m^2/s} \tag{2.24}$$

However, for industrial conditions, the growth rate, just as the nucleation rate, is barely affected by diffusion, as seen in Fig. 2.13. In other words, the only barrier for the solidification of metals is nucleation of crystals with critical radius.

EXERCISE 2.11

Calculate the growth rate of Ni considering the following data: lattice parameter $a_{\mathrm{Ni}} = 3.51\,\text{Å}$, diffusion coefficient $D_L = 1 \times 10^{-9}\,\mathrm{m^2/s}$, Burgers vector $\mathbf{b} = 2.49\,\text{Å}$, and an undercooling of 5°C (Tables 1.3 and 1.4).

Solution

The atomic volume is defined by the volume that an atom occupies in the crystalline lattice. Nickel is an fcc metal, thus:

$$\Omega = \frac{a^3}{\text{number of atoms per unit cell}} = \frac{\left(3.51 \times 10^{-10}\right)^3}{4} = 1.08 \times 10^{-29}\,\mathrm{m^3/atom}$$

On the other hand, the planar density must be calculated on the most dense plane of the cell, in this case the (1 1 1), which is an equilateral triangle with sides $\sqrt{2}a$, thus its area is:

$$A=\frac{\sqrt{3}}{2}a^2=\frac{\sqrt{3}}{2}\left(3.51\times10^{-10}\,\text{m}\right)^2=1.07\times10^{-19}\,\text{m}^2$$

Consequently, the planar density is calculated as the number of atoms in the (1 1 1) plane per unit area: there are 1/6 of atom at each corner (total of 1/2 atom) and 1/2 atom at each side of the triangle (3/2 atoms), making it a total of 2 atoms in this plane:

$$\rho_{planar}=\frac{\text{number of atoms in the plane}}{\text{area}}=\frac{2\text{ atoms}}{1.07\times10^{-19}\,\text{m}^2}=1.87\times10^{19}\,\text{atoms/m}^2$$

And finally Eq. (2.22) can be used, knowing that an undercooling of 5°C = 5 K which makes the temperature $T_1=1726-5=1721\,\text{K}$:

$$v=\Omega\cdot\rho_{planar}\cdot\frac{D_L}{b^2}\cdot\frac{\Delta H_f}{RT_M}\cdot\frac{\Delta T_1}{T_1}$$

$$v=\frac{\left(1.08\times10^{-29}\,\text{m}^3/\text{atom}\right)\left(1.87\times10^{19}\,\text{atoms/m}^2\right)\left(1\times10^{-9}\,\text{m}^2/\text{s}\right)(17{,}556\,\text{J/mol})(5\,\text{K})}{\left(2.49\times10^{-10}\,\text{m}\right)^2(8.314\,\text{J/mol K})(1726\,\text{K})(1721\,\text{K})}$$

$$v=0.0116\,\text{m/s}=1.16\,\text{cm/s}$$

If this model for continuous growth rate (Eq. 2.22) is applied to industrial conditions, different behaviors will be found for metallic liquids (D_L has approximately constant values) compared to oxides or organic polymers: for metallic liquids, continuous growth rate can be expressed as an equation of the $v=\text{constant}\cdot\Delta T_1/T_1$ type and only viscous liquids would be ruled out by the continuous growth law with a bell shape.

Finally, it may be considered, as a good approximation, that the percentage of solidified liquid with time (function of the convolution of nucleation and growth rates) follows a TTT curve (time, temperature, and transformation) of the "C type," as the one shown in Fig. 2.14 (time required to solidify 1% and 99% of a metal in isothermal conditions). The expressions to calculate solidification times follow the Johnson-Mehl-Avrami law:

$$dx'=I\,d\tau(1-x')[v(t-\tau)]^3 \tag{2.25}$$

where $Id\tau$ is the number of nuclei that appear in the time interval $(\tau,\ \tau+d\tau)$, x' is the solidified fraction, $v(t-\tau)$ implies the growth from the instant considered until time τ, and the exponent 3 accounts for growth in 3 dimensions.

Reordering terms:

$$\frac{dx'}{1-x'}=Iv^3(t-\tau)^3d\tau \tag{2.26}$$

and then integrating both sides of the equation:

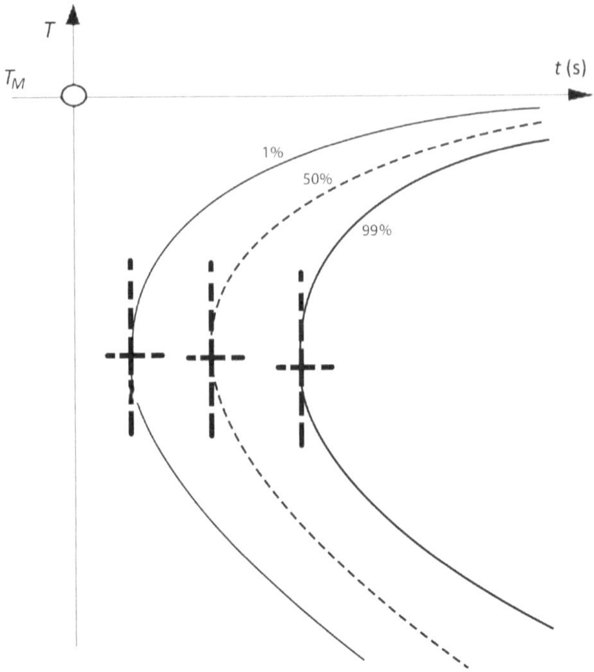

FIG. 2.14 Time-temperature-transformation (TTT) curves (controlled by nucleation and diffusion).

$$\int_0^x \frac{dx'}{1-x'} = Iv^3 \int_0^{t_x} (t-\tau)^3 d\tau \tag{2.27}$$

$$-\ln(1-x) = Iv^3 \frac{t_x{}^4}{4} \tag{2.28}$$

finally solving for t_x:

$$t_x = \sqrt[4]{\frac{-4\ln(1-x)}{Iv^3}} \tag{2.29}$$

EXERCISE 2.12

Obtain the TTT curves for Ni in the following cases: (a) homogeneous nucleation and constant diffusion, (b) heterogeneous nucleation ($\theta = 30$degrees) and constant diffusion, (c) homogeneous nucleation and variable diffusion, and (d) heterogeneous nucleation ($\theta = 30$degrees) and variable diffusion

Solution

(a) Homogeneous nucleation and constant diffusion

$$I \cong 1\times 10^{39} \exp\left(-\frac{8\times 10^9}{T\Delta T^2}\right) \text{nuclei/m}^3\,\text{s}$$

$$v \cong 4.01\left(\frac{\Delta T}{T}\right)\text{m/s}$$

The graphical representation can be seen in Fig. 2.15.

(b) Heterogeneous nucleation ($\theta = 30$ degrees) and constant diffusion

$$I \cong 1\times10^{39}\exp\left(-\frac{8\times10^{9}f(\theta)}{T\Delta T^{2}}\right)\text{nuclei/m}^3\text{s}$$

$$v \cong 4.01\left(\frac{\Delta T}{T}\right)\text{m/s}$$

This plot can also be observed in Fig. 2.15.

(c) Homogeneous nucleation and variable diffusion

$$I \cong 1\times10^{44}\exp\left(-\frac{8\times10^{9}}{T\Delta T^{2}}-\frac{11.5T_M}{T}\right)\text{nuclei/m}^3\text{s}$$

$$v \cong 4.01\times10^{5}\left(\frac{\Delta T}{T}\right)\exp\left(-\frac{11.5\,T_M}{T}\right)\text{m/s}$$

The schematic representation can be seen in Fig. 2.16.

(d) Heterogeneous nucleation ($\theta = 30$ degrees) and variable diffusion

$$I \cong 1\times10^{44}\exp\left(-\frac{8\times10^{9}f(\theta)}{T\Delta T^{2}}-\frac{11.5\,T_M}{T}\right)\text{nuclei/m}^3\text{s}$$

$$v \cong 4.01\times10^{5}\left(\frac{\Delta T}{T}\right)\exp\left(-\frac{11.5\,T_M}{T}\right)\text{m/s}$$

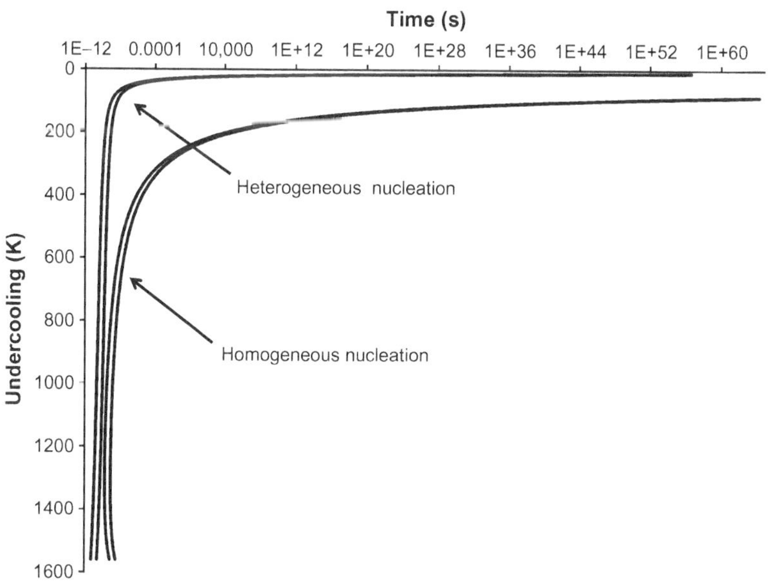

FIG. 2.15 Ni homogeneous and heterogeneous ($\theta = 30$ degrees) nucleation with constant diffusion as a function of undercooling, 1 and 99% TTT curves.

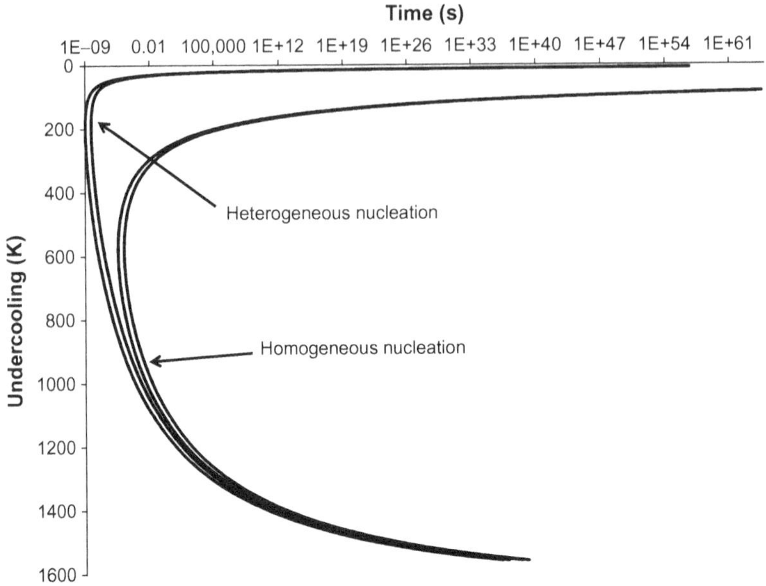

FIG. 2.16 Ni homogeneous and heterogeneous ($\theta = 30$ degrees) nucleation with variable diffusion as a function of undercooling, 1 and 99% TTT curves.

This plot can also be observed in Fig. 2.16.

Comment: Following the same procedure to the solidification kinetics for Al, it can be seen that this last one is slower than Ni.

Phase transformation models with "C" curves can only be applied to liquids which steeply change their diffusivity with temperature. Therefore, in the case of metals and alloys, only the upper zone of the "C" curve is valid (Fig. 2.17). Furthermore, "C" curves of diffusion-controlled transformations are not only valid for viscous solid solidification, but also for the kinetic analysis of some solid state reactions which occur in many alloys where transformation is caused by solubility variations or by allotropic changes in the phases forming the alloy.

2.4 SOLIDIFICATION OF A METAL BY CONTINUOUS COOLING

2.4.1 Solidification Curves

Considering the isothermal solidification curve of Fig. 2.17, that same curve is also valid for continuous cooling. Fig. 2.18 shows the starting points (M, N, and P) of solidification, for continuous cooling rates (CCR) with different

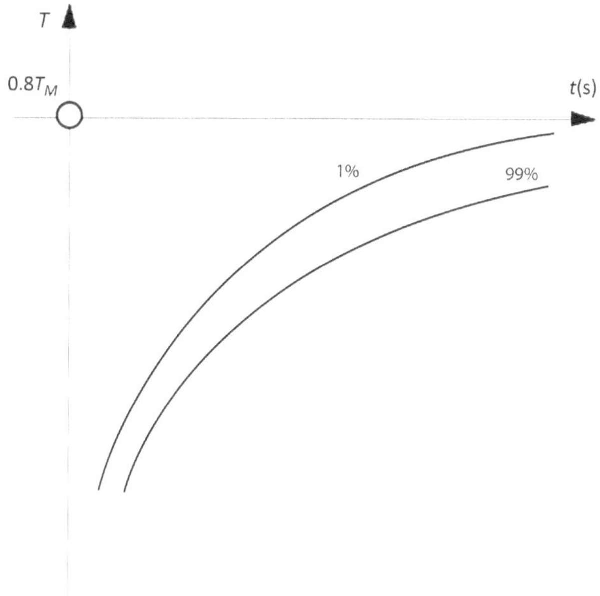

FIG. 2.17 TTT curve of a metal (controlled by nucleation): 1% and 99% of solidification.

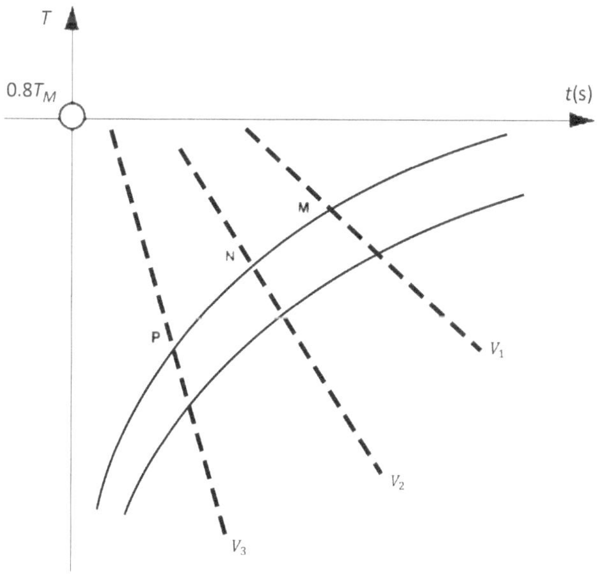

FIG. 2.18 Start and end of freezing in continuous cooling rates (controlled by nucleation).

speeds $V_1 < V_2 < V_3$. The values for time at the M, N, and P points (1% curve) correspond to the inverse values of the nucleation rate for continuous cooling.

Given that for high nucleation rates the resulting solid structure will present finer grains caused by the high number of nuclei formed, high cooling rates will also produce very fine grains for the same reason.

Besides, it must be mentioned that in equal cooling conditions (e.g., the solidification in the chill cast or metallic mold, with the same shape and size, and identical external cooling fluid) the solidification will be slower as the *thermal capacity* of the metal is high (amount of Joules to be removed to achieve a decrease in temperature of 1°C). Thermal capacity (Table 2.4) is a property of each metal, which approximately follows the Dulong-Petit law:

$$P_A(\mathrm{kg/mol}) \cdot c_m(\mathrm{J/kg\,°C}) \approx 3R \approx 25\ \mathrm{J/°C\ mol}\ (\text{at } 298\mathrm{K}) \tag{2.30}$$

where P_A is the atomic weight, c_m is the specific heat.

If a metallic solid is cooled at a rate higher than a critical value (necessary to avoid the nose of the TTT curve or making it as steep as possible) and avoids the

TABLE 2.4 Thermal Capacity of Some Metals

Metal	At 2000 K (J/kg °C)	At 298 K (J/kg °C)	Metal	At 2000 K (J/kg °C)	At 298 K (J/kg °C)
Beryllium	3260.40	1822.48	Niobium	346.94	267.52
Sodium		1224.74	Molybdenum	372.02	250.80
Magnesium	1337.60	1015.74	Silver	284.24	238.26
Aluminum	1086.80	898.70	Cadmium	263.34	229.90
Potassium		752.40	Tin	242.44	225.72
Silicon	907.06	710.60	Antimony	259.16	209.00
Titanium	785.84	522.50	Tantalum	167.20	142.12
Vanadium	865.26	484.88	Mercury		137.94
Manganese	836.00	476.52	Platinum	179.74	133.76
Chromium	936.32	459.80	Tungsten	167.20	133.76
Iron	823.46	451.44	Plutonium	171.38	133.76
Nickel	731.50	443.08	Lead	137.94	129.58
Cobalt	685.52	418.00	Gold	146.30	129.58
Zinc		388.74	Bismuth	150.48	125.40
Copper	493.24	384.56	Uranium	200.64	117.04

nucleation and growth of clusters, the result is a frozen liquid, long-range disordered, known as *metallic glass*. This vitreous solid belongs thermodynamically and structurally to the liquid phase and to obtain it, very high cooling rates are required: most of the metallic glasses are produced at cooling rates higher than 10^5 K/s.

Research on metallic glasses has considerably increased since the last century. This is related to their interesting mechanical properties (high tenacity along with high mechanical strength) and chemical ones (resistance to corrosion), as well as their acoustic and thermal behavior at very low temperatures. They are excellent materials for electric and magnetic applications.

2.4.2 Considerations on the Nuclei Growth With Unidirectional Heat Flow (Chvorinov Law)

Real solidifications do not happen isothermically and always show a temperature gradient in the liquid. Supposing that the liquid metal is in a mold, there is also a real temperature gradient through its wall, as shown in Fig. 2.19.

If the solid-liquid interface moves at a steady rate, the heat removed from the solid must equal the heat released by the liquid plus the melting latent heat (since solidification is taking place) as follows (known as Chvorinov formula):

$$k_S \cdot T'_S = k_L \cdot T'_L + \Delta H_{f_V} \cdot v \cdot A \tag{2.31}$$

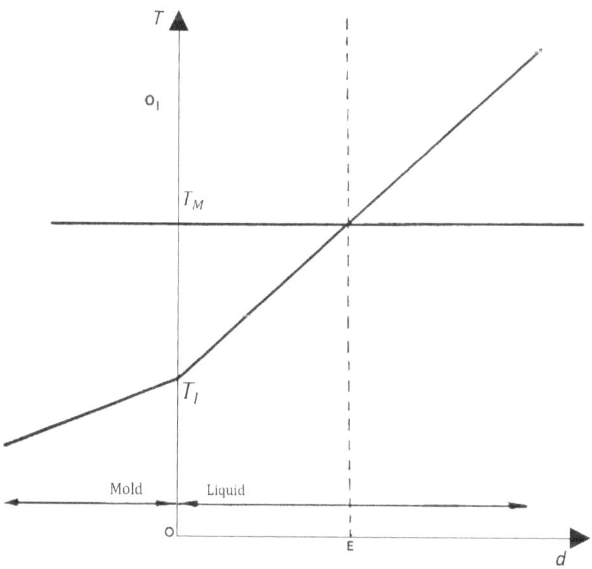

FIG. 2.19 Initial temperature profile vs. distance to the mold wall.

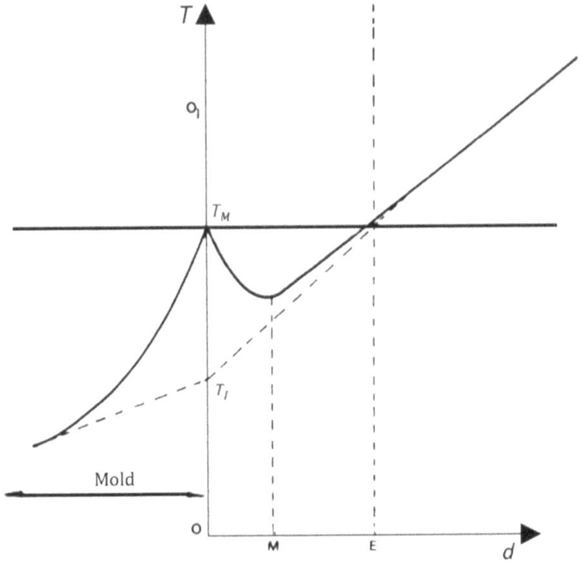

FIG. 2.20 Temperature profile vs. distance to the mold wall as solidification begins.

where k_S and k_L are the thermal conductivities of solid and liquid respectively, T_S' and T_L' are the thermal gradients of solid (mold) and liquid respectively, ΔH_{fV} is the latent heat per volume unit, v is the rate of solidification, and A is the cross-section area.

Because this continuity condition must be maintained, the rate of heat removal from the solid is the variable, controlling the speed of advancement of the interface.

The liquid between O and E (Fig. 2.20) can solidify given that it has a temperature lower than T_M. Solidification will begin at the $O-O_I$ wall, since it is the liquid region with the higher undercooling as it has a temperature T_1. In that $O-O_I$ wall, the highest number of nuclei will appear, higher than the one expected from the undercooling alone, because the heterogeneous nucleation effect has to be considered as the wall of the mold influences the process. As a consequence of the formation of these nuclei, the temperature at the $O-O_I$ wall increases from T_1 to T_M. At the same time, the latent heat ΔH_f generated by the solidification is absorbed by both the metallic mold and the liquid (reheating it). Thus, the real temperature gradient in the liquid will have a profile like the one shown in Fig. 2.20; therefore, between O and M, the temperature gradient in the liquid is negative.

The smaller the undercooling in the liquid interface, the lower the fraction of the latent heat removed through the ingot mold will be, as presented in Figs. 2.21 and 2.22:

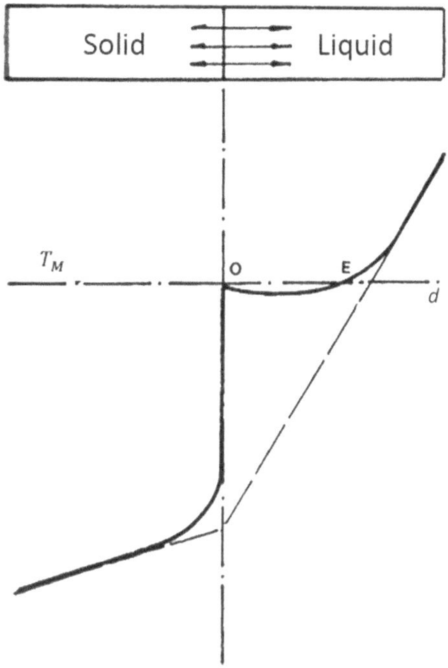

FIG. 2.21 Sand cast and liquid temperature vs. distance profile (model 1).

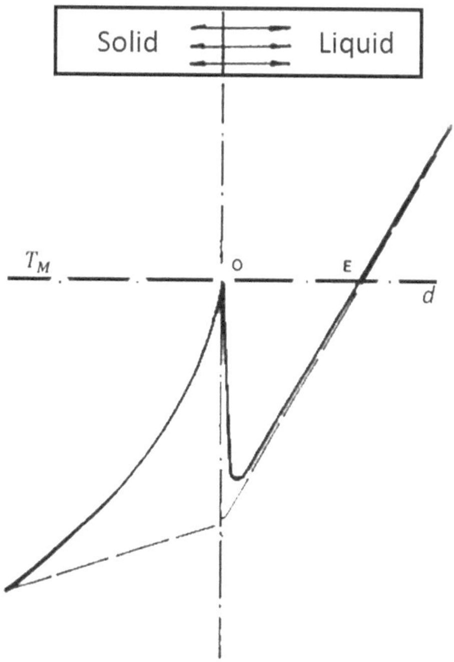

FIG. 2.22 Chill cast and liquid temperature vs. distance profile (model 2).

- Model 1 (Fig. 2.21) corresponds to the solidification of a metal inside a very refractory mold (e.g., ceramic or sand mold).The growth of peripheral nuclei towards E will be slow due to the small undercooling. Meanwhile, new nuclei will be formed inside the liquid between O and E, that, by simple mechanical obstacles, will stop the growth of peripheral nuclei. The solidified structure would be more or less equiaxed.
- If the real conditions of solidification match those of Model 2 (Fig. 2.22), for example in die-casting or metallic molds, with heat transfer preferentially occurring through the solid, and the nuclei formed at the wall growing fast towards E, with no time available to form other nuclei at the interior of the liquid between O and E.

The nuclei formed, with random orientations, at the mold wall (*chill* grains) have a tendency to grow faster at certain preferential directions (Section 2.3.2). Therefore, if the preferential growth orientation is very different from the one of the directional heat flow, the elongated grain will be overgrown by neighbors oriented with the preferred direction.

The selective and anisotropic grain growth is schematized in Fig. 2.23 and corresponds to a metal of the face-center cubic crystalline system. It shows that every grain that has an $\langle 1\,0\,0\rangle$ axis not orthogonal to the isotherms, shows very restricted growth.

The grain growth with latent heat transfer only to the solid, creates not only elongated grains orthogonal to the isotherms (forming a columnar or basaltic structure) but also with a common crystallographic orientation, known as *solidification texture*.

In practice, the grain structure of a casting or ingot follows model 1 instead of model 2: outer *chill crystals*, intermediate *columnar grains*, and central *equiaxed grains* as shown in Fig. 2.24.

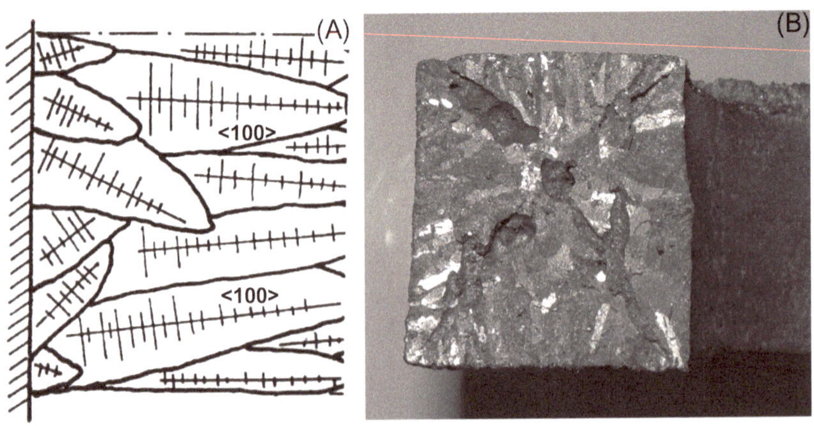

FIG. 2.23 (A) Selective dendritic grain growth and (B) diagonal fissures present in a refractory casting.

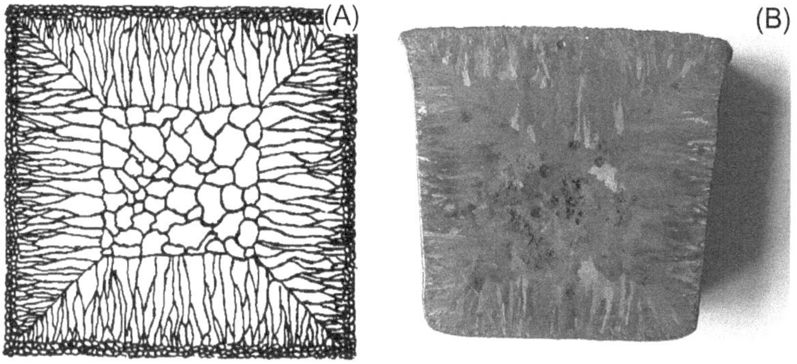

FIG. 2.24 Grains in an ingot: (A) schematic cross-section representation and (B) macrograph of a 1% C-13% Mn Hadfield steel.

EXERCISE 2.13

Calculate the minimum speed of solidification to form dendritic structures when freezing a steel ingot.

Data: The temperature gradient in the solid-liquid interface is 5000°C/m, $k_S = 29.26\,\text{J/m°Cs}$, $\Delta H_f = 2.72 \times 10^5\,\text{J/kg}$, $\rho_L = 7800\,\text{kg/m}^3$.

Solution

Eq. (2.31) requires that $T_L' \leq 0$ (negative temperature gradient in the liquid) in order to form columnar or equiaxed dendritic structures (*see* solidification models 1 and 2). If $T_L' = 0$, the minimum speed per area unit is:

$$v_{min} = \frac{k_S \cdot T_S'}{\Delta H_f v} = \frac{(29.26\,\text{J/m°Cs})(5000\,°\text{C/m})}{(2.72 \times 10^5\,\text{J/kg})(7800\,\text{kg/m}^3)} = 6.90 \times 10^{-5}\,\text{m/s}$$

The Chvorinov equation is used to compare solidification times of simple-shaped castings of a pure metal poured in molds with flat and smooth walls at T_M:

$$t_f = C\left(\frac{V}{A}\right)^2 \tag{2.32}$$

where t_f is the total time of solidification of such castings, V the volume, A the cross-section area, and C a constant for a given metal-mold material and mold temperature T_0, calculated by:

$$C = \frac{\pi}{4}\left(\frac{\rho_s \Delta H_f}{T_M - T_0}\right)^2 \frac{1}{k_m \rho_m c_m} \tag{2.33}$$

where ρ_s is the density of the metal, ΔH_f the latent heat of melting, T_M the melting point of the metal, k_m the thermal conductivity of the mold, ρ_m the density of the mold, and c_m the specific heat of the mold. Furthermore the $k_m\,\rho_m\,c_m$ product, known as heat diffusivity, is a measure of the rate at which the mold can absorb heat.

EXERCISE 2.14

Compare the solidification times of a thin slab of steel with a thickness of 10 mm in a metallic mold casting (or chill casting) and in a ceramic mold (sand mold casting).

Data:

- Sand mold: $k_m = 0.61$ J/m°Cs, $\rho_m = 1500\,\mathrm{kg/m^3}$, $c_m = 1128.6$ J/kg°C
- Metallic mold (Cu): $k_m = 392.92$ J/m°Cs, $\rho_m = 9000\,\mathrm{kg/m^3}$, $c_m = 376.2$ J/kg°C
- Steel: $T_M = 1538$°C, $\Delta H_f = 15{,}290.44$ J/mol, $\rho = 8000\,\mathrm{kg/m^3}$, atomic weight = 5.58×10^{-2} kg/mol

Solution

The first step in order to calculate the solidification times is to obtain the values of C for both molds, supposing the mold is at room temperature:

$$C_{sand} = \frac{\pi}{4}\left(\frac{\rho_s \Delta H_f}{T_M - T_0}\right)^2 \frac{1}{k_m \rho_m c_m}$$

$$= \frac{\pi}{4}\left[\frac{(8000\,\mathrm{kg/m^3})(15{,}290.44\,\mathrm{J/mol})\left(\dfrac{1}{5.58\times 10^{-2}}\,\mathrm{mol/kg}\right)}{1538 - 23°\mathrm{C}}\right]^2$$

$$\frac{1}{(0.61\,\mathrm{J/m°Cs})(1500\,\mathrm{kg/m^3})(1128.6\,\mathrm{J/kg°C})}$$

$$C_{sand} = 1.60 \times 10^6\,\mathrm{s/m^2}$$

$$C_{copper} = \frac{\pi}{4}\left(\frac{\rho_s \Delta H_f}{T_M - T_0}\right)^2 \frac{1}{k_m \rho_m c_m}$$

$$= \frac{\pi}{4}\left[\frac{(8000\,\mathrm{kg/m^3})(15{,}290.44\,\mathrm{J/mol})\left(\dfrac{1}{5.58\times 10^{-2}}\,\mathrm{mol/kg}\right)}{1538 - 23°\mathrm{C}}\right]^2$$

$$\frac{1}{(392.92\,\mathrm{J/m°Cs})(9000\,\mathrm{kg/m^3})(376.2\,\mathrm{J/kg°C})}$$

$$C_{copper} = 1236.09\,\mathrm{s/m^2}$$

Since it is not specified, two types of solidification can occur, either through one face (V/A = thickness = 10mm) or through two faces (V/A = thickness/2 = 5mm). The solidification times can now be obtained, for both cases:

- Through one face:

$$t_{f_{sand}} = C\left(\frac{V}{A}\right)^2 = (1.60\times 10^6\,\mathrm{s/m^2})(0.01\,\mathrm{m})^2 = 160\,\mathrm{s}$$

$$t_{f_{copper}} = C\left(\frac{V}{A}\right)^2 = (1236.09\,\mathrm{s/m^2})(0.01\,\mathrm{m})^2 = 0.12\,\mathrm{s}$$

- Through two faces:

$$t_{f_{sand}} = C\left(\frac{V}{A}\right)^2 = (1.60\times 10^6\,\mathrm{s/m^2})(0.005\,\mathrm{m})^2 = 40.07\,\mathrm{s}$$

$$t_{f_{copper}} = C\left(\frac{V}{A}\right)^2 = (1236.09\,\mathrm{s/m^2})(0.005\,\mathrm{m})^2 = 0.03\,\mathrm{s}$$

Comment: The solidification is very slow in the sand mold while very fast in copper. Furthermore, solidification through two faces shortens the times by a factor of 4. In all cases, the solidification times are related to the massivity of the part (*V*/*A*).

On the other hand, the equivalence between the diffusion coefficient D(m^2/s) for matter transport and 1/C(m^2/s) for heat transport is noticeable, considering in both cases $t = x^2/D$, where x is the distance of transport.

The solidification times for Al are, approximately, 25% higher than those for steel, which means that Al is a metal that solidifies relatively "bad."

BIBLIOGRAPHY

Burke, J., 1965. The Kinetics of Phase Transformations in Metals. Pergamon Press, Oxford.

Chadwick, G., 1972. Metallography of Phase Transformations. Butterworth, London.

Chalmers, B., 1977. Principles of Solidification. Krieger Publishing, New York, NY.

Chen, S., 1980. Glassy Metals. Rep. Prog. Phys. 43 (4), 353–432.

Clark, J., Flemings, M., 1986. Nuevos Materiales y Economía. Investigación y Ciencia 123, 15–22.

Davies, G., 1973. Solidification and Casting. Applied Science Publishers, London.

Eustathopoulos, N., 1983. Energetics of solid/liquid interfaces of metals and alloys. Int. Met. Rev. 28 (1), 189–210.

Flemings, M., 1974. Solidification Processing, fifth ed. McGraw-Hill, New York, NY.

Flinn, R., 1963. Fundamentals of Metal Casting. Addison-Wesley Publishing Company, Reading, MA.

Hollomon, J., Turnbull, D., 1953. Nucleation. Prog. Met. Phys. 4, 333–388.

Kurz, W., Fisher, D., 1989. Fundamentals of Solidification, third ed. Trans Tech Publications, Aedermannsdorf.

Petty, E., 1973. Physical Metallurgy of Engineering Materials. George Allen & Unwin, London.

Porter, D., Easterling, K., 2009. Phase Transformations in Metals and Alloys, third ed. CRC Press, Boca Raton, FL.

Turnbull, D., 1949. Thermodynamics in Metallurgy. American Society for Metals, Metals Park, OH.

Verdeja, L., Sancho, J., Ballester, A., Gonzalez, R., 2014. Refractory and Ceramic Materials. Editorial Síntesis, Madrid.

Winegard, W., 1964. An Introduction to the Solidification of Metals. The Institute of Metals, London.

Chapter 3

Total Insolubility and Solubility in Alloys

Alloys are materials with metallic bond, formed by more than one metal. Binary alloys are formed by two elements; ternary ones formed by three; quaternary by four, and so on.

Between the two metals that form a binary alloy, there can be some affinity or none. When affinity exists, it can be of various types: intermetallic, eutectic, solid solution, etc., and more than one of these types can occur simultaneously.

In the case of a solid solution affinity, two metals in their solid state are soluble between them when the crystalline system of one among them (solvent) can randomly substitute some of its atoms with those of the other metal (*substitutional solid solutions*); or can acquire some foreign atoms in its interatomic spaces (*interstitial solid solutions*).

For both types of solid solutions, there can be two opposite end scenarios:

1. When two metals in liquid state form a homogeneous mix, but do not present solid solution affinity at all are, totally insoluble between them in solid state (very unusual).
2. When two metals form a solid solution for any amount of metallic solvent and solute.

3.1 TOTAL INSOLUBILITY BETWEEN TWO METALS (A AND B)

When A and B (two metals, or one metal and one metalloid) are totally insoluble between them in solid-state, it does not mean that any other type of affinity is also excluded. There are cases where elements do not present any type of affinity. In other cases, it is possible for metal A and metal B to present affinity of the chemical type which produces an intermetallic with composition A_xB_y; or another type of affinity known as eutectic.

In the following sections, the three cases are analyzed: first when A and B do not present any type of solid affinity; then two cases where metals are insoluble between them but with chemical affinity and finally with eutectic affinity.

Solidification and Solid-State Transformations of Metals and Alloys. http://dx.doi.org/10.1016/B978-0-12-812607-3.00003-6

3.1.1 A and B Do Not Have Affinity Between Them

Phase diagrams can be constructed from a series of cooling curves for alloys formed by the same elements and only varying their respective composition. As such, T_A and T_B, being the solidification temperatures of A and B respectively, Fig. 3.1 shows the cooling curves obtained by thermal analysis techniques for each of the pure metals. The starting point is an homogeneous melt of A and B, at a temperature higher than T_A. If this melt (liquid alloy) cools, it is expected that metal A begins to solidify on reaching temperature T_A, but this does not happen. The atoms of B, which possess high kinetic energy due to thermal vibrations, act as obstacles for the solidification of A and it is necessary to cool the material to T_A' for the metal A to begin the solidification process.

This modifying behavior of B in the solidification of A is not only caused by the higher kinetic energy of the atoms of B but also by the barrier effect suffered by the atoms of A of the liquid (surrounded by atoms of B) in the process of adhering themselves to the previously formed solid A. At this point, the phase transformation kinetics must be considered: solid A will grow at the expense of

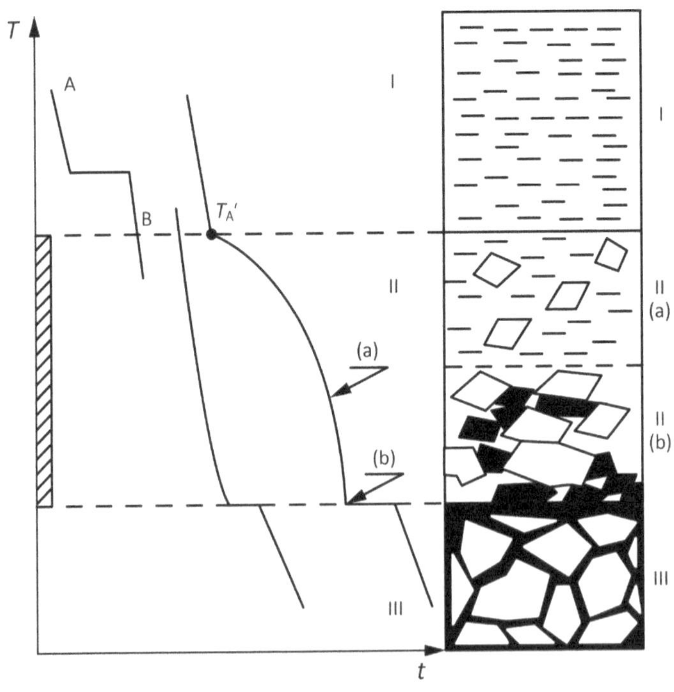

FIG. 3.1 Cooling curves and microstructures for an alloy of a totally insoluble system.

the liquid whenever more atoms of A pass from the liquid to the solid than on the contrary, throughout the solid/liquid interface per time unit.

The higher the proportion of atoms of B in the liquid, the lower the temperature T_A' will be. Furthermore, solidification of A will not happen until the energy of atomic bonds between atoms of A counteracts not only the existing repulsion among the atoms themselves (due to their own kinetic energy), but also the disturbances produced by the atoms of B.

Solidification (just as the one presented in Fig. 3.1, in the II (a) zone) will start by forming some crystals of the pure metal A (given that no atom of B is retained in the lattice of A because there is no affinity between A and B in solid phase). The remaining liquid, rich in B with respect to the initial composition, will require a new temperature drop for solidification to continue (by attraction of new atoms of A in the liquid towards the already formed solid). The lower the temperature, the higher the number of attracted atoms will be. If the temperature drops continuously and slowly (in equilibrium), the amount of solidified metal A will increase until the remaining liquid only has atoms of B.

The temperature drop must be slow (before the cooling continues) for the liquid to be perfectly homogeneous at all temperatures. The nonhomogeneity is caused by the richness of atoms of B at the interface compared to the rest of the liquid. Therefore, at each temperature, a period of time is required to increase the possibility of uniformity (by diffusion) of atoms of B at all points of the liquid (the curve considers a group of isothermal steps that correspond to an infinitesimal drop of temperature).

When temperature reaches T_B, the solidification of B starts, forming crystals that solidify at that constant temperature in a kind of *matrix pattern* where the grains of A are embedded (*disperse constituent*), as shown in Fig. 3.1 (III).

If another alloy with higher content of B is considered, the solidification of A would start at a temperature lower than T_A', but solidification, at the adequate conditions, would end in the same manner at T_B. Once all the equilibrium solidification curves are obtained for all possible alloys with compositions between 100% A and 100% B, they can be condensed in one diagram that would consider all the information of the start and end of the solidification for all possible alloys of A and B. Moreover, any alloy that solidifies over a range of temperatures, and at any temperature within this range, is a mix of solid and liquid phases.

Fig. 3.2 indicates the solidification curves for different alloys of germanium (A) and indium (B), with the horizontal axis showing the concentrations (either in weight or atomic percentages) of A and B, and the vertical axis showing the temperatures of start and end of the solidification of each of these alloys. In addition, it must be mentioned that total insolubility in a binary metallic system is not frequent, contrary to binary ceramic systems.

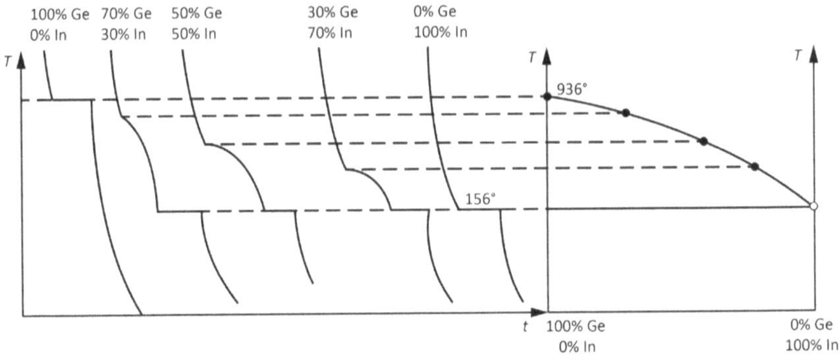

FIG. 3.2 Cooling curves and phase diagram for the Ge-In system.

EXERCISE 3.1

The Bi-Cu phase diagram of Fig. 3.3 indicates that both metals are totally insoluble and do not present any type of affinity in solid state. (a) Define the matrix and disperse constituents present in the microstructure at room temperature of a 10% Bi-Cu alloy and (b) determine the temperature at which the alloy would begin to melt during heating.

Solution

(a) Amount of disperse and matrix constituents

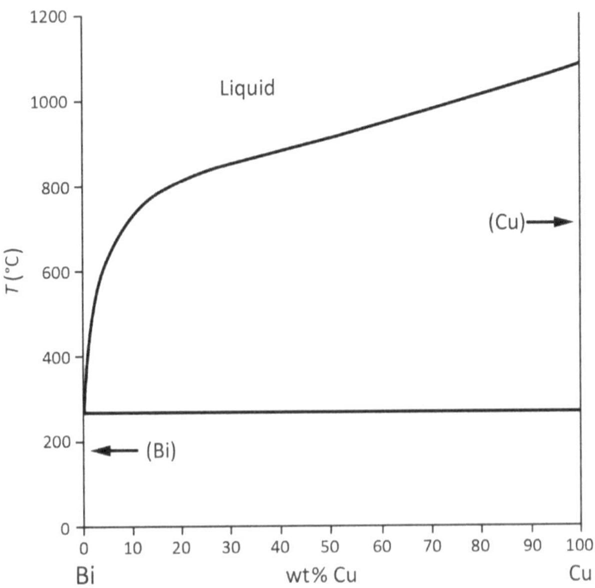

FIG. 3.3 Bi-Cu phase diagram (Chung, 2007).

Given the total solid-state insolubility of Cu and Bi, the alloy would be formed by dispersed crystals (90%) of pure Cu in a matrix (10%) of crystals of pure Bi.

(b) Temperature to begin melting

During heating, the alloy would begin to melt at 270.8°C (melting temperature of Bi), meaning that forming operations cannot occur at temperatures higher than 271°C to avoid the formation of either liquid droplets of the alloy or oxides.

3.1.2 A and B Form an Intermetallic Compound With Stoichiometric Composition A_xB_y

An intermetallic (intermediate phase) is a compound with both metallic and another type of bond, differing from chemical compounds which only have nonmetallic bonds.

When there is chemical affinity between two elements that form a solid compound $C = A_xB_y$, the crystalline lattice of the compound C is usually different and more stable than the lattices of A and B. This happens because the grouping of atoms of A and B has attraction between them (ionic, covalent, or hybrid bonds) and is higher than the bonding energy between atoms of A and between atoms of B separately.

Two cases can be analyzed: the solidification of a liquid alloy corresponding to the stoichiometric composition A_xB_y and the solidification of an alloy with an excess of one of the components:

1. The liquid alloy has exactly the A_xB_y composition

   In liquid phase, the atoms of A and B are freely mixed, as their kinetic energy (function of temperature) is higher than the affinity (chemical) that tends to group (coordinate) them in the specific solid lattice of the compound.

   Solidification starts when temperature has decreased enough that the kinetic energy of the atoms (repulsion) has reached an equilibrium with the attraction forces (chemical affinity) that tries to group them in a solid lattice. Given that the attraction between the atoms of A and B is higher than the bonding energy of each element, the melting temperature T_C of the compound is higher than T_A and T_B. When reaching T_C, the first fraction that solidifies leaves the liquid with the same proportion of A and B atoms (X : Y) compared to that of the initial liquid, and therefore, solidification will continue at constant temperature, resulting in a structure formed by grains of the intermetallic C (this occurs by a process of nucleation and growth similar to the solidification of a pure metal).

   The left side of Fig. 3.4 shows the solidification curves of A and B (pure metals or metal and metalloid). In the middle section is the solidification

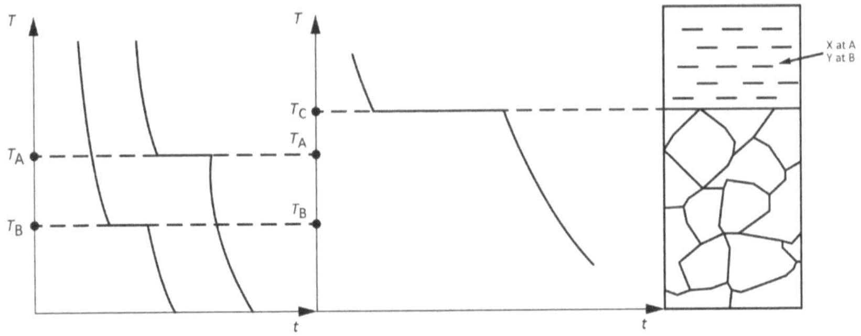

FIG. 3.4 Cooling curves and microstructures for an intermetallic compound A_xB_y (total insolubility system).

curve of the compound C (A_xB_y) and at the right side are the corresponding microstructures.

2. The liquid alloy has an excess of atoms of A with respect to the A_xB_y stoichiometric composition ($T_A > T_B$)

   This type of solidification is similar to the one for two insoluble metals A and C (C being in this case, the intermetallic A_xB_y compound). As presented in Fig. 3.5, the solidification of C starts at T_C', lower than T_C (the higher the excess of atoms of A, the lower T_C' will be). The remaining liquid, enriched with A (compared to its initial composition), will require a new temperature drop to reduce the kinetic energy of the atoms of A in the liquid (and, as a consequence, a disturbance) so that solidification of C can continue.

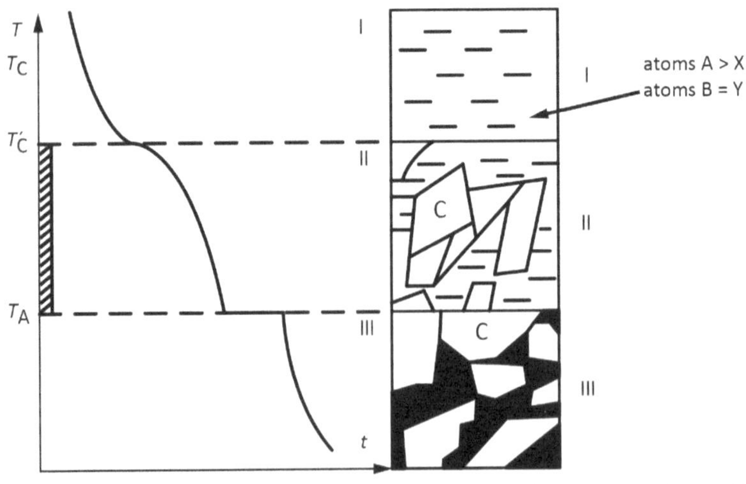

FIG. 3.5 Cooling curve and microstructure of an intermetallic compound A_xB_y with an excess of A.

The solidification of C takes place until T_A is reached; at that moment, the residual liquid, formed only by A, solidifies freely at a constant temperature like a pure metal. The solidified alloy will be formed by grains of C which solidified first (primary constituent), surrounded by a matrix of grains of A which solidified at T_A.

On the other hand, if the liquid alloy had an excess of atoms of B compared to the A_xB_y stoichiometric composition, the results would be analogous: changing A for B in the previous paragraphs. The alloy would start its solidification with the formation of grains of the compound C at a temperature lower than T_C; the solidification of C will happen in a temperature range ending at T_B, resulting in a microstructure formed by grains of C (primary or disperse constituent), embedded in a matrix constituent formed by grains of B.

Fig. 3.6 shows the Ga-As binary system. In this system, elements are insoluble between them and form an intermetallic compound with a 50% Ga–50% As (at%) composition.

It is important to point out that by slow heating, the solid alloys with Ga composition higher than 50% start to melt at 29.8°C, while any other alloy with composition lower than 50% Ga remains solid until reaching at least 810°C.

The nature of the matrix constituent always determines both the thermal behavior of the alloy and its mechanical properties (e.g., in tension, the straining will begin in the matrix constituent which will be the one that first

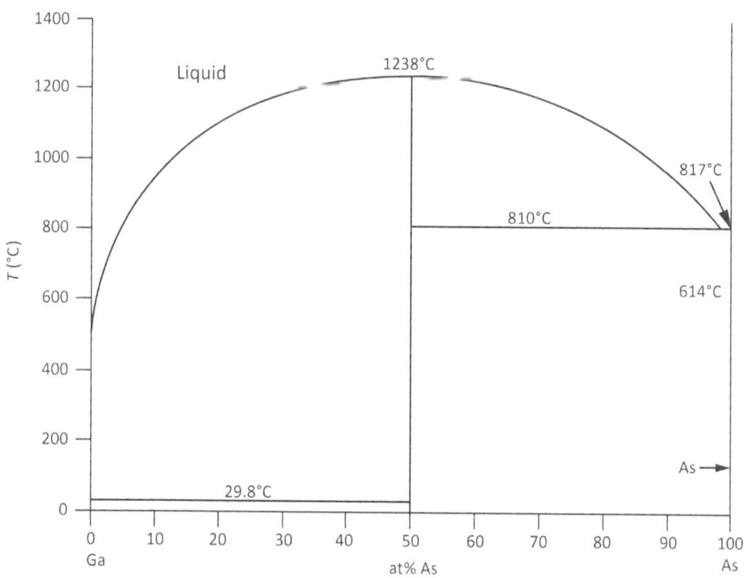

FIG. 3.6 Ga-As phase diagram with the GaAs intermetallic compound (Chung, 2007).

absorbs the mechanical stresses). It is evident that if the amount of primary intermetallic compound phase is high enough so that the grains of C are in contact with each other and the excess of B (solidified at T_B) had to solidify at isolated interdendritic spaces; from the mechanical point of view the matrix constituent properties will now be those of compound C (even if, during heating, the alloy will always begin to partially melt upon reaching T_B).

3.1.3 A and B, are Insoluble in Solid State, but Present Eutectic Affinity

Two metals A and B with no affinity in solid phase (and in consequence, will solidify separately), are known to have *eutectic affinity* if they tend to form a "liquid compound" with composition A_xB_y, stable in liquid state at temperatures lower than T_A and T_B (thermodynamically, the eutectic is a failed chemical compound).

As an example of eutectic affinity, during winter it is a common practice to deposit NaCl (salt) over frozen roads in order to liquefy the layer of ice. These two solids in contact, ice and NaCl, are liquefied at the temperature of the ice, and can remain at liquid phase, even below 0°C. In fact, if the weight proportion is 23.5% NaCl and 76.5% ice, the liquid remains in that state without any of the components solidifying until −22° C (at this temperature the mix would solidify in the shape of fine particles of salt and ice). This phenomenon is explained through the eutectic affinity of ice and NaCl that tends to form a liquid with stoichiometric composition between them.

The term "*eutectic*" (from the Greek "*easy melting*") indicates that the melting temperature of the compound with eutectic composition is lower than those of its components.

The solidification of the eutectic liquids may be analyzed similar to intermetallic compounds, first considering solidification of the liquid alloy formed by A and B with exactly the composition A_xB_y. Later the alloys with an excess of one of the components can be considered:

1. The alloy has the A_xB_y composition

   With T_A and T_B being the solidification temperatures of the pure metal A and the pure metal B, respectively, and having the liquid exact composition A_xB_y, supposing that $T_A > T_B$, if the temperature drops to T_A, the atoms of A will not start to solidify, given that the eutectic affinity induces that the proportion in the liquid remains as X atoms of A and Y atoms of B. Therefore, none of the atoms of A will solidify until B can solidify as well. However, the solidification will not start even on reaching T_B.

   The eutectic affinity constrains the atoms of A and B to stay in liquid state; in this sense, it must be classified among the repulsion forces that oppose the attraction of the metallic bond. The solidification will only start when the attraction forces (caused by the metallic bond of A and B atoms, separately) equal those of repulsion forces (thermal vibration caused by the kinetic energy of the atoms, from both A and B, in liquid phase, plus the eutectic affinity energy of the X/Y proportion of atoms of A and B in liquid state).

The solidification starts at T_E (eutectic temperature) lower than both T_A and T_B, which balances the attraction and repulsion forces, in other words, a temperature where the metallic bond forces also neutralize the eutectic affinity of the liquid.

Given that at T_E, metal A is found at liquid phase, but considerably undercooled compared to the theoretical solidification temperature T_A, when reaching T_E, some atoms of A start to solidify, with a high nucleation rate and with a tendency to form a very fine structure. But immediately, and since the liquid needs to maintain the X/Y proportion of A and B, atoms of B, that are in proportional amount to those already solidified atoms of A, also solidify (Y atoms of B per X atoms of A).

Provided that the freezing conditions for the remaining liquid at the temperature T_E are identical to the initial ones, the solidification will continue at a constant temperature as shown at the center part of Fig. 3.7, in the end, resulting in eutectic colonies (zone II on the right side of Fig. 3.7).

2. The alloy has an excess of one component

   In the case of a liquid alloy with excess of A, its solidification is similar to that of two insoluble metals A and E (where E would be the eutectic solid). According to Fig. 3.8, solidification starts at T_A', lower than T_A. The higher the excess of A in the alloy, the closer T_A' to T_A will be, with the formation of grains of A (disperse constituent).

   The process continues with the formation of a larger amount of solid A when temperature drops. When A solidifies, the residual liquid is enriched with B, becoming closer to the eutectic composition. Once the composition A_xB_y is reached, the liquid will solidify (at the constant temperature T_E

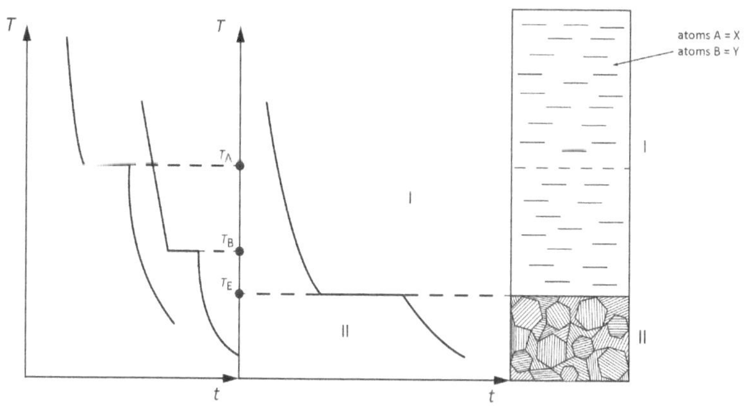

FIG. 3.7 Cooling curves and microstructures for an eutectic (total insolubility system).

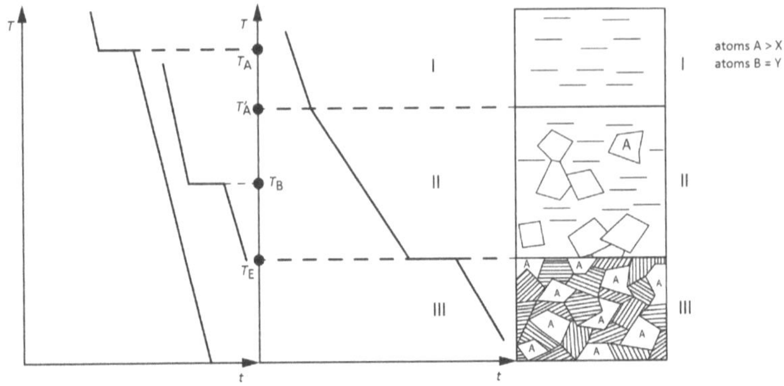

FIG. 3.8 Cooling curves and microstructures of an eutectic with an excess of A.

when cooling reaches this value) as a matrix with the previously solidified grains of A (in the temperature range between T_A and T_E) embedded in it, as shown in part III of Fig. 3.8.

If the composition of the liquid alloy had an excess of B compared to the A_xB_y composition, the results would be analogous: B would start solidifying at a temperature $T_B' < T_B$, the solidification of the excess of B would take place in the temperature range of T_B' and T_E, and finally, the result would be a solid with a microstructure formed by grains of B (in this case, the primary or disperse constituent) embedded in an matrix of eutectic constituent formed by fine juxtaposed crystals of A and B.

EXERCISE 3.2

Determine the temperature at which a brine, having water with 10% salt and equal weight proportions of solid and liquid, would form. Consider the liquidus lines in Fig. 3.9 as straight lines that converge in the eutectic point.

Solution

The first step is obtaining the fractions of both liquid (f_L) and solid (f_S) for water with 10 wt% salt, through a balance of matter:

$$1 \cdot C_0 = f_S \cdot 0 + f_L \cdot 10$$

resulting in:

$$f_L = \frac{C_0}{10}$$

and

$$f_S = 1 - f_L = \frac{10 - C_0}{10}$$

Since both phases have the same weight proportions:

$$f_L = f_S \Longrightarrow \frac{10 - C_0}{10} = \frac{C_0}{10}$$

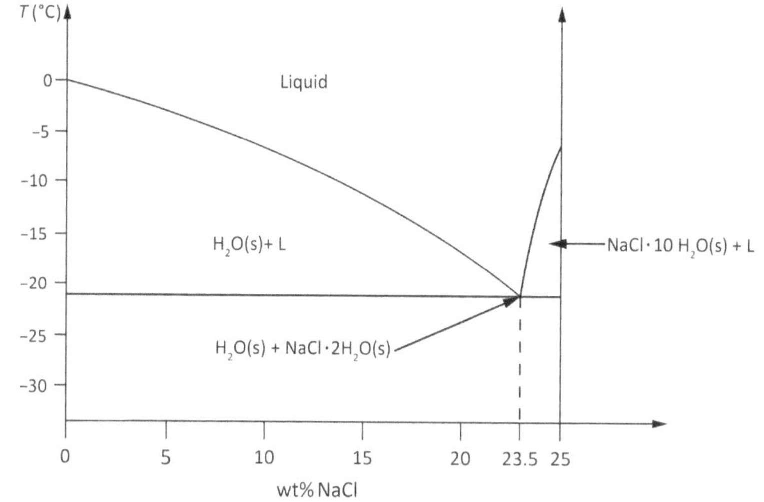

FIG. 3.9 Ice-salt phase diagram (Mesplede, 2000).

then

$$10 - C_0 = C_0 \Rightarrow C_0 = 5\%$$

meaning a 5 wt% of $NaCl \cdot 10H_2O$. Using similar triangles, this brine would begin to form at a temperature of:

$$\frac{0-(-22)}{23.5} = \frac{0 - T_L}{10} \Rightarrow T_L = -9.4°C \approx -9°C$$

resulting, at this temperature, in ice surrounded by liquid.

While eutectic alloys are always a fine mixture of two solids, each with their own crystallographic structure, the distribution of these solids in the eutectic grain (or colony) may vary considerably from one binary system to another. Eutectics can be classified into *lamellar* (sandwich), *globular*, rod-like, or *acicular* (needles), depending on the appearance and distribution of the two solid phases that compose the eutectic alloy. Some examples are mentioned in Chapter 4.

Fig. 3.10 shows the solidification diagram of the system formed by Au and Si that fulfills the condition of being two metals insoluble between themselves in solid phase, but with affinity to form an eutectic with the atomic composition of approximately 80% gold and 20% silicon. Even though the melting temperatures of both gold and silicon are 1064° C and 1414° C respectively, any solid alloy of this system, when heated, would start to melt at 363° C and, as a consequence, would be useless as a structural solid material above that temperature (shortness effect).

Furthermore, it should be pointed out that alloys with close to 100% A (or 100% B) result in such primary constituent A (or B) amounts that their grains

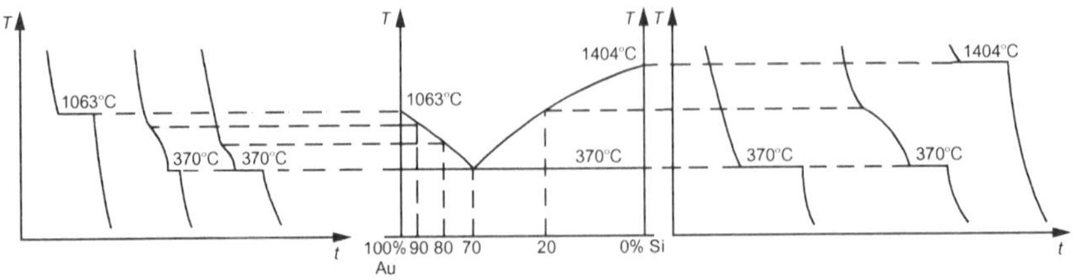

FIG. 3.10 Au-Si phase diagram and cooling curves.

will be in contact with each other before the liquid eutectic solidifies, resulting in lose of its role of continuous matrix constituent by solidifying at isolated regions (at the interdendritic spaces left by the primary constituent previously solidified). From a thermal point of view, these alloys would also be useless as structural materials if the working temperature is higher than the eutectic one.

3.2 SOLUBILITY

3.2.1 Substitutional Solid Solutions

This type of solutions are formed when atoms of a solvent metal are randomly replaced from their lattice positions by atoms of a solute (foreign element or metal) while maintaining the cohesion between them caused by the metallic bond. For two metals, A and B, to be limitlessly or totally (any proportion of A and B) soluble between themselves in solid state, some conditions known as Hume-Rothery rules must be met (Hume-Rothery, 1969):

(a) Firstly, it is necessary that both metals crystallize in the same system. Otherwise, the solubility would be, in the best case, partial. For example, if metal A crystallizes in the face-centered cubic system and B in the tetragonal system, the solution of certain atoms of B in the crystalline lattice of A (alpha solid solution, with the same lattice as A) is possible, though this would only occur with a certain amount of B. Inversely, if the solvent metal is B, they could form solid solution (beta solid solution) with some atoms of A, but just until a certain proportion is reached. In general, the solubility limits of B in A and of A in B are different and increase exponentially with temperature (Arrhenius' law).

(b) Secondly, in order for A and B to form partial or total solid solutions it is necessary for both to have the same valence. Given that the bond in solid solutions is of the metallic type, the number of electrons given to the electronic cloud by the atoms of both A and B must be the same. If substitution is made with a metal with different valence, the electronic cloud loses stability, which increases to a point where thermodynamic instability makes solid solution impossible; from this proportion onwards, other compounds with lesser energy are formed (e.g., electronic compounds and extreme solid solutions).

(c) Thirdly, both elements must have the same electrochemical characteristics. Two metals with pronounced electrochemical differences tend to form intermetallic compounds, but not two solid solutions (Table 1.1 shows the relative electronegativity of metals).

(d) Fourthly, a condition to the relative size difference must be met. In order to form total solid solutions, atomic diameters must not differ by $\pm 15\%$ (Table 1.3).

3.2.2 Interstitial Solid Solutions

When the size difference between atoms is considerable, interstitial solid solutions can be formed: solute atoms with small diameter are inserted randomly in the interatomic spaces or interstices of the solvent.

The magnitude of the interatomic space where the solute atom can be inserted depends, mostly, on the crystalline system of the solvent metal. Moreover, for the face-centered cubic system, if the insertion is made in the center of the edge or in the center of the cube (equivalent positions), the ratio of diameters between solute and solvent is 0.414 (octahedral insertion in the face-centered cubic system, Fig. 3.11). If the insertion takes place at the center of the tetrahedron as shown in Fig. 3.12, the diameter ratio is 0.225.

Given the analogy between the face-centered cubic and the hexagonal close-packed systems, the ratios for octahedral and tetrahedral insertions are also 0.414 and 0.225. In the octahedral insertion, the atom of solute is located at the center of a regular octahedron, whose side is the diameter of the solvent atom; in the tetrahedral insertion, the solute is located at the center of a regular tetrahedron, whose side is also the diameter of the solvent atom (as indicated in Fig. 3.13).

The body-centered cubic system has 0.154 and 0.291 as ratios for octahedral and tetrahedral insertions, respectively (Figs. 3.14 and 3.15).

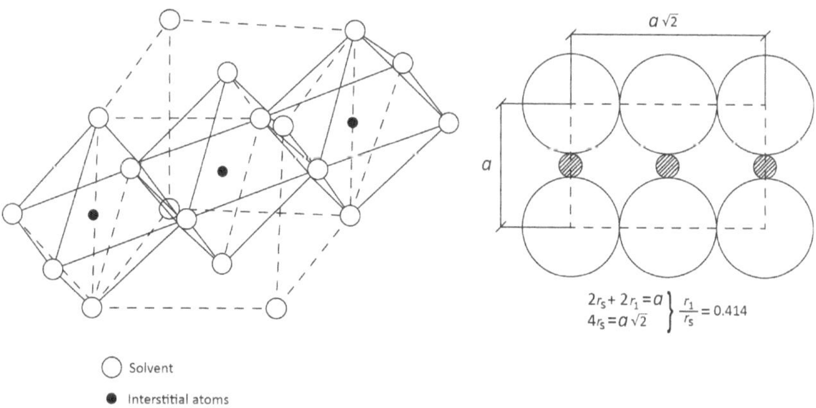

FIG. 3.11 Octahedral interstices in a face-centered cubic system.

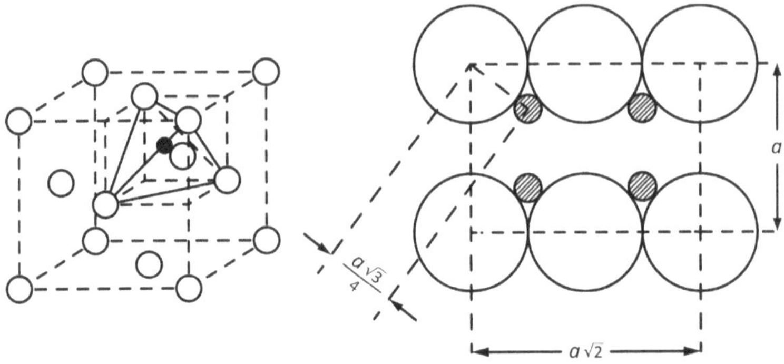

$$\left.\begin{aligned}2(r_s+r_1)&=\frac{a\sqrt{3}}{2}\\4r_s&=a\sqrt{2}\end{aligned}\right\}r_1/r_s=0.225$$

FIG. 3.12 Tetrahedral interstices in a face-centered cubic system.

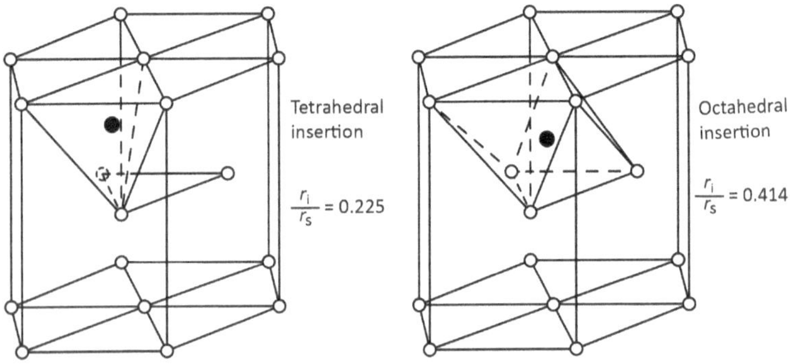

FIG. 3.13 Tetrahedral and octahedral interstices in a hexagonal system.

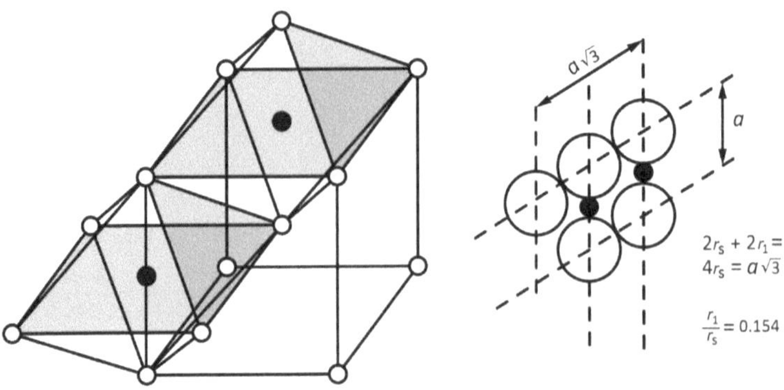

FIG. 3.14 Octahedral interstices in the body-centered cubic system.

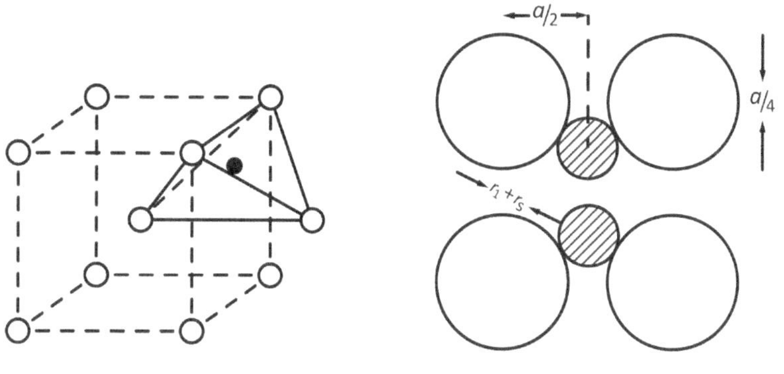

$$\left.\begin{array}{l}(r_1+r_s)^2=\frac{a^2}{4}+\frac{a^2}{16}\rightarrow r_1+r_s=\frac{a\sqrt{5}}{4}\\ 4r_s=a\sqrt{3}\end{array}\right\} r_1/r_s=0.291$$

FIG. 3.15 Tetrahedral interstices in the body-centered cubic systems.

Therefore, interstitial solid solutions can only be formed between a metal and a solute element, usually a nonmetal with small diameter, which, in general, is carbon (atomic diameter 1.544 Å), nitrogen (atomic diameter 1.44 Å), oxygen, or hydrogen. However, this insertion creates a distortion in the lattice of the solvent, because the space available per interstice is smaller than the atomic diameter of the solute. Thus, and due to the valence differences, interstitial solid solutions are always of the partial type and with very limited amounts of solvent-solute.

EXERCISE 3.3

Obtain the octahedral insertion in α-Fe (ferrite) and γ-Fe (austenite) of carbon, comparing them to the theoretical ones.

Data: $r_C = 0.77$ Å, $a_{\alpha\text{-Fe}} = 2.86$ Å, and $a_{\gamma\text{-Fe}} = 3.64$ Å.

Solution

(a) For γ-Fe:

In γ-Fe, carbon can be located at the center of the cell and at the center of the edges, and knowing that it is an fcc cell:

$$4r_{Fe} = \sqrt{2}a \Rightarrow r_{Fe} = \frac{\sqrt{2}}{4}a$$

Considering the lattice parameter is 3.64 Å, then the radius is:

$$r_{Fe} = \frac{\sqrt{2}}{4}a = \frac{\sqrt{2}}{4}(3.64) = 1.29\,\text{Å}$$

$$\left(\frac{r_C}{r_{Fe}}\right)_{real} = \frac{0.77}{1.29} = 0.60$$

while the ideal ratio is:

$$\left(\frac{r_C}{r_{Fe}}\right)_{ideal} = 0.414$$

which means that available space is smaller than the carbon atom, therefore C would deform the crystalline lattice.

Furthermore, in a fcc crystal, there are 4 atoms of Fe (6 faces with 1/2 atom plus 8 corners with 1/8 atom, equals 4 atoms in total) and theoretically 4 atoms of C (12 edges with 1/4 atom plus 1 atom at the center, 4 in total), meaning 50% of Fe atoms and 50% of C atoms (all interstices occupied):

$$\frac{4\mathrm{C}}{4\mathrm{Fe}+4\mathrm{C}}=50\%\ \text{at C}$$

However, it is known that γ-Fe does not have more than 2% weight of C, so:

$$\frac{\%\mathrm{C}/\text{atomic weight of C}}{\%\mathrm{C}/\text{atomic weight of C}+\%\mathrm{Fe}/\text{atomic weight of Fe}}=\frac{2/12}{2/12+98/56}=8.70\%\ \text{at C}$$

$$\frac{8.70}{50}=0.17$$

which means that less than one fifth of the available interstices are occupied by C.

(b) For α-Fe

In α-Fe, carbon can be located at the center of the faces and at the center of the edges, and knowing that it is a bcc cell:

$$4r_{\mathrm{Fe}}=\sqrt{3}a\Rightarrow r_{\mathrm{Fe}}=\frac{\sqrt{3}}{4}a$$

Considering the lattice parameter is 2.86 Å, then the radius is:

$$r_{\mathrm{Fe}}=\frac{\sqrt{3}}{4}a=\frac{\sqrt{3}}{4}(2.86)=1.24\,\text{Å}$$

$$\left(\frac{r_{\mathrm{C}}}{r_{\mathrm{Fe}}}\right)_{real}=\frac{0.77}{1.24}=0.62$$

while the ideal ratio is:

$$\left(\frac{r_{\mathrm{C}}}{r_{\mathrm{Fe}}}\right)_{ideal}=0.15$$

which means that available space is smaller than the carbon atom, therefore C would also greatly deform the alpha crystalline lattice.

Furthermore, in a bcc crystal, there are 2 atoms of Fe (8 corners with 1/8 atom plus 1 atom at the center, 2 in total) and theoretically 6 atoms of C (12 edges with 1/4 atom plus 6 faces with 1/2 atom, 6 in total), meaning:

$$\frac{6\mathrm{C}}{2\mathrm{Fe}+6\mathrm{C}}=75\%\ \text{at C}$$

25% of Fe atoms and 75% of C atoms (all interstices occupied). However, it is known that α-Fe does not have a solubility higher than 0.02 wt% C:

$$\frac{\%\mathrm{C}/\text{atomic weight of C}}{\%\mathrm{C}/\text{atomic weight of C}+\%\mathrm{Fe}/\text{atomic weight of Fe}}$$

$$=\frac{0.02/12}{0.02/12+99.98/56}=0.10\%\ \text{at C}$$

$$\frac{0.10}{75}=1.3\times10^{-3}$$

which means that less than one in 750 of the available interstices are occupied by C.

3.2.3 Total and Partial Solubility

As mentioned in Section 3.2.1, in order for two metals, A and B, to be able to form a *solid-state total solution* it is necessary that both crystallize in the same system, have similar atomic diameter, same valence, and analogous electrochemical character. It is evident that if a metal can crystallize in more than one system, as a function of temperature, a change in the crystalline system will be accompanied with its specific solubility (Exercise 3.3).

It is not common for two metals to have all the necessary conditions to form total solid solutions. For example, aluminum (after iron and its alloys, Al is the next metal in importance due to its global consumption) does not form total solid solution with any other metal while commercial copper-nickel alloys, copper-gold alloys, and silver-gold alloys can all form total solid solutions. However, most metals form partial solid solutions between them.

Table 3.1 indicates the solubility limits, for binary alloys of some industrial metals, as well as the maximum solubility temperatures. Below these, as mentioned, the solubility on the solvent metal decreases exponentially.

TABLE 3.1 Some Limited Binary Solid Solutions of Industrial Metals

Solvent	Solute	Solubility Limits (% Weight of Solute in the Solvent Lattice)
Al	Si	1.65% at 577°C (1.01% at 200°C) by solubility loss, Si precipitates
	Mg	14.9% at 451°C (0.8% at 100°C)
	Cu	5.65% at 548°C (0.2% at 200°C)
	Zn	82.00% at 382°C (31.6% at 275°C; 4.4% at 100°C)
	Mn	1.82% at 659°C (0.4% at 500°C)
	Sn	0.10% at 600°C (0.06% at 500°C)
Cu	O	< 0.004% at 1066°C (0.001% at 700°C)
	P	1.75% at 714°C (0.6% at 350°C)
	Cd	3.72% at 549°C (0.1% at 300°C)
	Cr	0.65% at 1076°C (0.05% at 400°C)
	Be	2.7% at 866°C (2.00% at 700°C; 1.55% at 605°C)
	Si	5.3% at 863°C (4.65% at 555°C)
	Zn	32.5% at 903°C (39.00% at 456°C; 35% at 200°C)
	Sn	13.5% at 799°C (15.8% from 586°C to 520°C; 11.00% at 350°C; 1.3% at 350°C)
	Al	7.4% at 1036°C (9.4% at 565°C)
	Pb	0.00% at any temperature

Continued

TABLE 3.1 Some Limited Binary Solid Solutions of Industrial Metals—cont'd

Solvent	Solute	Solubility Limits (% Weight of Solute in the Solvent Lattice)
Zn	Sn	0.1% at 198°C
	Al	1.00% at 382°C (0.6% at 275°C)
	Cu	2.7% at 424°C (0.3% at 100°C)
	Mg	0.16% at 364°C (0.008% at 200°C)
Pb	Sn	19.00% at 183°C (5.00% at 100°C)
	Sb	3.5% at 251.2°C (0.44% at 100°C)
Ni	S	0.00%
	C	0.55% at 1316°C (0.1% at 900°C)
	Cr	47.00% at 1345°C (37.00% at 500°C)
Sn	Pb	2.5% at 183°C (0.00% at 13°C)
Fe-γ	H	0.0009% at 1394°C (0.0004% at 912°C)
	C	2.11% at 1148°C (0.77% at 727°C)
	N	2.8% at 650°C (2.35% at 590°C)
	S	0.065% at 1370°C (0.01% at 914°C)
	Mn	100% at 1200°C
	Ni	100% at 1300°C
	Co	100% at 1300°C
	Cr	12.00% at 1130°C (7.00% at 830°C)
	Mo	2.5% at 1150°C (decreasing until 910°C)
	W	3.5% at 1100°C (decreasing until 910°C)
	V	1.4% at 1154°C (decreasing until 910°C)
	Ti	0.7% at 1175°C (decreasing until 910°C)
	Al	0.625% at 1150°C (decreasing until 910°C)
Fe-α	H	0.0003% at 910°C (0.0001% at 100°C)
	C	0.218% at 727°C (0.005% at 575°C)
	N	0.095% at 585°C (0.05% at 500°C)
	S	0.02% at 914°C

Continued

TABLE 3.1 Some Limited Binary Solid Solutions of Industrial Metals—cont'd

Solvent	Solute	Solubility Limits (% Weight of Solute in the Solvent Lattice)
	V	100% at 1400°C
	Cr	100% at 1400°C
	Mo	30.00% at 1400°C (10.00% at 910°C)
	W	35.5% at 1554°C (6.00% at 888°C)
	Ti	9.00% at 1291°C (2.2% at 800°C)
	Co	75.00% at 770°C
	Al	36.00% at 1094°C (33.3% at 100°C)
	Si	10.9% at 1275°C (3.5% at 500°C)
	Mn	3.00% at 700°C
	Ni	7.00% at 500°C

Temperature is actually an important factor on the partial solubility of two metals: a temperature increment corresponds to an increment on the lattice parameter (Exercise 1.2). Therefore, two metals partially soluble between them, with a solvus at a specific temperature, will decrease their relative solubility as temperature drops.

3.2.4 Total Solubility Binary Phase Diagram

Fig. 3.16 indicates the cooling curve of a liquid alloy formed by 60% Ni (metal A) and 40% Cu (metal B) that when solidified, formed a solid solution. In this case, the analysis is made for the solidification of binary alloys (A and B metals) that form total solid solution and with melting temperatures such that $T_A > T_B$. During solidification the atoms of B insert themselves into the crystalline lattice of A at temperatures higher than T_B, because there is *solution affinity* between A and B that tends to capture the atoms of B in the lattice of A.

The solidification of a solid solution starts at T_L, and the larger the ratio of atoms of B compared to the one the liquid originally had, the lower the T_L temperature will be. The first solid formed is richer in A than the mean composition of the liquid. In fact, at T_L a number of atoms of B less than the average (40% Cu) are attached to the solid because at that temperature, the atoms of B in the liquid are at a temperature higher than T_B; consequently, the solid-liquid

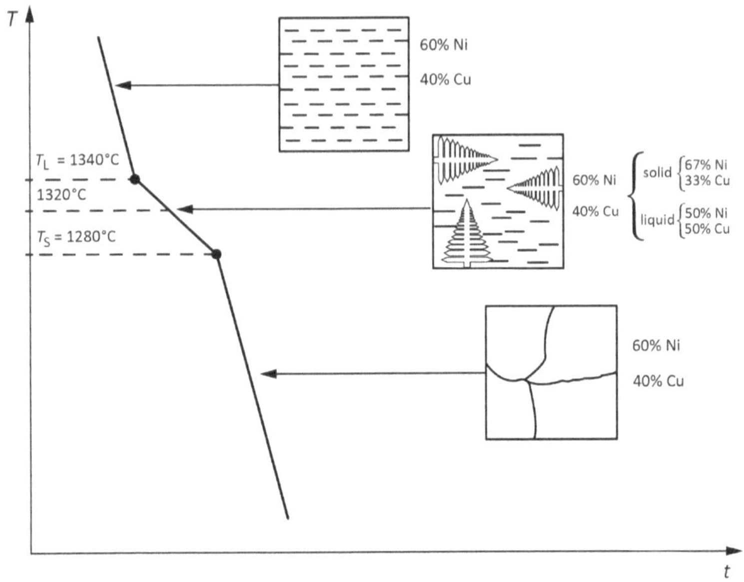

FIG. 3.16 Equilibrium cooling curve and microstructures of the 60% Ni-40% Cu (wt%) alloy.

interface is enriched in B; therefore, and considering the disturbance produced by B in the solidification of A, it is necessary to decrease the temperature in order for the alloy to continue its solidification.

The alloy will finish freezing when all the atoms of B are incorporated into the solid; this occurs (because of the solution affinity existing between B and A) above T_B. This process is valid for two metals that form total or partial solid solutions (with B compositions lower than the solubility limit).

The solidification is of the *equilibrium* type when, for each temperature, the solid has (at all points) the same C_S chemical composition, and the liquid also has a uniform C_L chemical composition. At T_1, the uniform C_{S1} composition of the solid is different from the uniform C_{S2} composition at another T_2 temperature. The same occurs for the liquid.

In order for the solidification to take place at equilibrium conditions, cooling rates must be extremely low, allowing diffusion of B atoms towards the interior of the solid, so uniformity of both solid and liquid can be reached before decreasing to a lower temperature. However, solid state diffusion is a slow process, orders of magnitude slower than the liquid one, and thus, it is the controlling factor in maintaining equilibrium. In practice, total homogeneity of the solid is not reached at each temperature, and a "*core*" structure will appear (*coring*).

Fig. 3.17 presents different cooling curves, obtained for Cu-Ni alloys with different compositions from 100% Ni to 100% Cu. The resulting solidification diagram is also shown, with the composition of the alloy in the horizontal axis

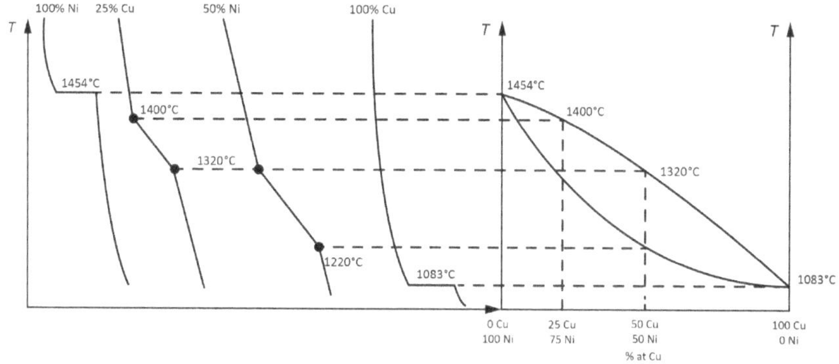

FIG. 3.17 Cooling curves and phase diagram in the Ni-Cu system (simple phase).

and the beginning (*liquidus* line) and ending (*solidus* line) temperatures of the solidification in the vertical axis.

For the 60% Ni-40% Cu alloy (Fig. 3.16), at temperatures between 1340°C and 1280°C, a combination of liquid and solid phases is observed. For a temperature between these two (e.g., 1320°C) the composition of solid and liquid phases would be: 67% Ni-33% Cu for the solid and 50% Ni-50% Cu for the liquid (*equilibrium-phase rule* or *horizontal rule*). Solid cannot have Ni content higher than 67%: all the alloys with a composition higher than 67% are in solid phase above 1320°C; but in contact with the liquid, atoms of Cu of the liquid continue to move towards the solid until the 67% Ni uniform composition of the solid is reached. The solid cannot have a Ni content below 67% because at 1320°C it would partially melt.

On the other hand, the liquid in equilibrium with the solid cannot have at 1320°C a Ni composition below 502%, because part of the atoms of the solid Ni would pass to the liquid enriching it; and at this temperature, the liquid can have up to 50% of atoms of Ni. The liquid phase cannot have more than 50% Ni because it would partially solidify for analogous reasons.

Consequently, the compositions of the solid and liquid phases in equilibrium at 1320°C would be, respectively, the ones corresponding to the alloy whose solidification ends at this temperature and the alloy whose solidification starts at the same temperature.

Generalizing for an alloy with $m\%$ composition, the solid and liquid compositions (in equilibrium) at a certain T_1 temperature, correspond to the intersection of the horizontal line at the T_1 temperature with the *solidus* and *liquidus* lines (*horizontal rule* or *tie-line rule*).

From Fig. 3.18, the proportions of the solid phase of $s\%$ composition and liquid phase of $l\%$ composition can also be deduced at a certain temperature. With W_S being the total weight of the solid at the defined temperature, W_L the total weight of the liquid, and W_T the total weight of the alloy, then:

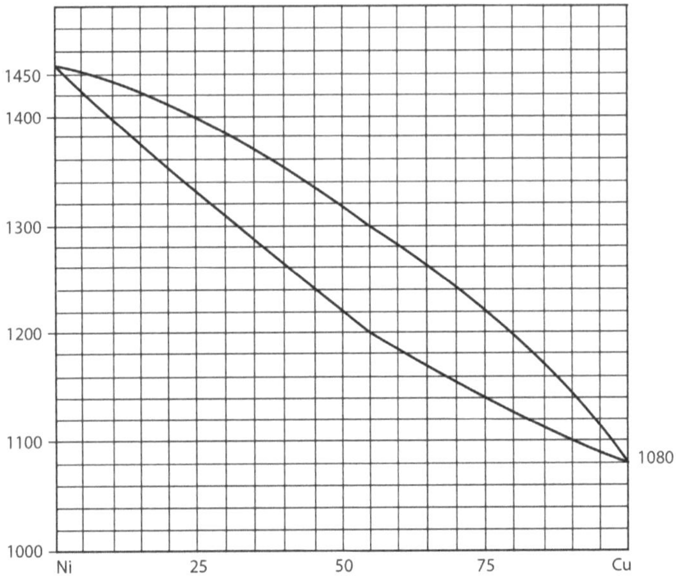

FIG. 3.18 Ni-Cu phase diagram (total solubility).

$$W_S + W_L = W_T \tag{3.1}$$

$$W_S \cdot s + W_L \cdot l = W_T \cdot m \tag{3.2}$$

this last expression being the result of making a matter balance on B.

From Eqs. (3.1), (3.2), the *lever rule* can be obtained:

$$\frac{W_S}{W_L} = \frac{l-m}{m-s} \tag{3.3}$$

Likewise, the solid to liquid, liquid to total, and solid to total ratios may be deduced:

$$\frac{W_S}{W_L} = \frac{\overline{lm}}{\overline{ms}}; \frac{W_L}{W_T} = \frac{\overline{ms}}{\overline{ls}}; \frac{W_S}{W_T} = \frac{\overline{lm}}{\overline{ls}} \tag{3.4}$$

In a binary phase diagram, the *partition coefficient* K is defined as the ratio of the concentration of solute in the solid to the concentration of solute in the liquid as follows:

$$K = \frac{C_S}{C_L} \tag{3.5}$$

and this coefficient will be higher or lower than 1 according to the shape of the diagram (Fig. 3.19). If the liquidus and solidus are straight lines in a temperature interval (this can be true, without considerable error, when the compositions are close to 100% A or 100% B), the partition coefficient is constant along that interval as deduced by geometric analysis.

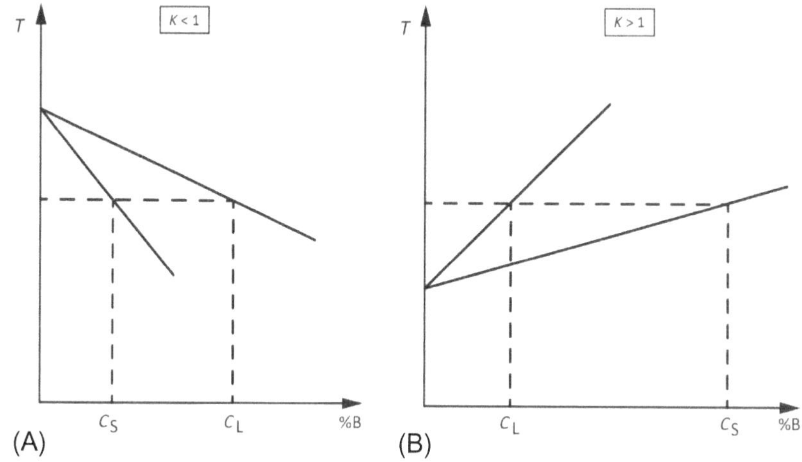

FIG. 3.19 Partition coefficient on phase diagrams where (A) $K < 1$ and (B) $K > 1$.

The partition coefficient is the result of the difference in composition between the solid and liquid phases during solidification. It is also responsible for segregation (or *coring*) of solute that exists in the final solidified alloy (Chapter 5).

Even though the characteristic diagram is Fig. 3.18, another total solubility diagram is shown in Fig. 3.20. This diagram can be considered as two total solid

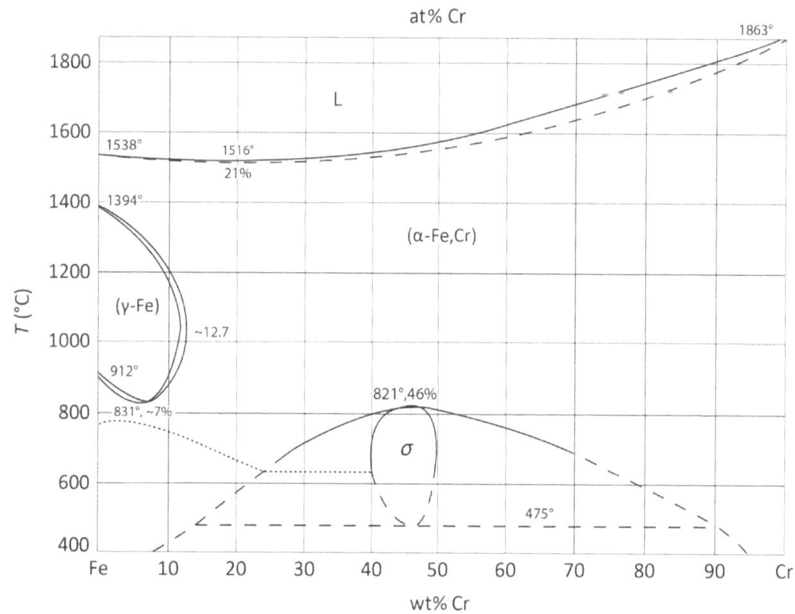

FIG. 3.20 Fe-Cr phase diagram (Chung, 2007).

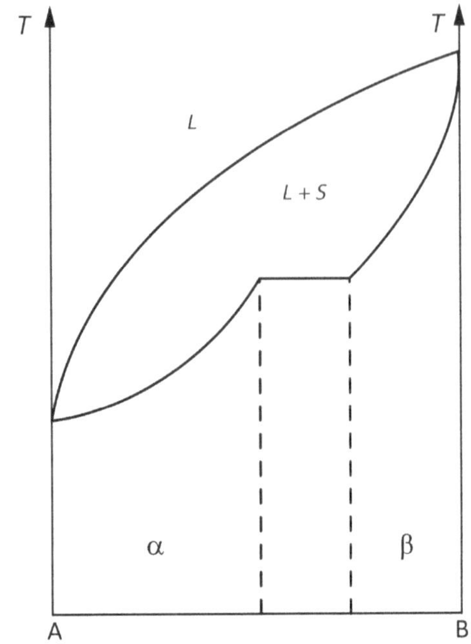

FIG. 3.21 Horizontal portion of solidus line separating α and β phases.

solution diagrams joined together at a certain composition (46% Cr). An alloy with a composition below 46% Cr can be analyzed in the same way as an alloy of the Ni-Cu system, and the same applies to an alloy with composition above 46% Cr.

As a general rule, it can be asserted that when a diagram has, either from the 100% A side or 100% B side, a solidus line which is continuous and derivable (in other words smooth) through all its length, corresponds to a solid solution alloy. For example, Fig. 3.21 shows two segments of the solidus line separated by an isothermal segment. The α-solution would have A as the solvent metal and the β-solution would have B as the solvent metal (Section 4.4).

EXERCISE 3.4

Calculate the solid and liquid fractions for the following alloys using data from Fig. 3.18: (a) Ni-25% Cu (at 1360°C), (b) Ni-50% Cu (at 1300°C), and (c) Ni-75% Cu (at 1200°C).

Solution

(a) Ni-25% Cu at 1360°C

At this temperature $c_L = 38\%$Cu and $c_S = 18\%$Cu and fractions can be calculated as:

$$\%L = \frac{x - c_S}{c_L - c_S} \cdot 100 = \frac{25 - 18}{38 - 18} \cdot 100 = 35\%$$

$$\%S = \frac{c_L - x}{c_L - c_S} \cdot 100 = \frac{38-25}{38-18} \cdot 100 = 65\%$$

(b) Ni-50% Cu at 1300°C

In this case $c_L = 55\%$Cu and $c_S = 332.5\%$Cu and the fractions are:

$$\%L = \frac{x - c_S}{c_L - c_S} \cdot 100 = \frac{50-32.5}{55-32.5} \cdot 100 = 78\%$$

$$\%S = 100 - 78 = 22\%$$

(c) Ni-75% Cu at 1200°C

And finally, for this alloy $c_L = 80\%$Cu and $c_S = 55\%$Cu, with fractions calculated as:

$$\%L = \frac{x - c_S}{c_L - c_S} \cdot 100 = \frac{75-55}{80-55} \cdot 100 = 80\%$$

$$\%S = 100 - 80 = 20\%$$

REFERENCES

Chung, Y., 2007. Introduction to Materials Science and Engineering. CRC Press, Boca Raton, FL.

Hume-Rothery, W., 1969. The Structure of Metals and Alloys. The Institute of Metals, London.

Mesplede, J., 2000. Chimie. Thermodynamique, Matériaux Inorganiques. Bréal éditions, France.

BIBLIOGRAPHY

A.S.M., 1973. Metallography Structures and Phase Diagrams, eighth ed. American Society for Metals, Metals Park, OH.

Bhadeshia, H., Honeycombe, R., 2006. Steels. Microstructure and Properties, third ed. Butterworth-Heinemann, London.

Cahn, R., Haasen, P., 1996. Physical Metallurgy, fourth ed. North Holland Publishing Co., Amsterdam, Netherlands.

Calvo Rodes, R., 1948. Metales y Aleaciones. I.N.T.A., Madrid, Spain.

Chipman, 1951. Basic Open Hearth Steelmaking. Iron & Steel Division AIME, New York, NY.

Hansen, M., Elliot, R., 1965. Constitution of Binary Alloys, second ed. McGraw-Hill, New York, NY.

Hatch, J. (Ed.), 1984. Aluminum: Properties and Physical Metallurgy. American Society for Metals, Metals Park, OH.

Levin, F., Robbins, C., McMurdie, H., 1969. Phase Diagrams for Ceramists. The American Ceramic Society, Columbus, OH.

Massalski, T., 2001. Binary Alloy Phase Diagrams, second ed. American Society for Metals, Metals Park, OH.

Pero-Sanz, J., 2006. Ciencia e ingeniería de materiales, fifth ed. CIE Dossat, Spain.

Porter, D., Easterling, K., 2009. Phase Transformations in Metals and Alloys, third ed. CRC Press, Boca Raton, FL.

Rhines, F., 1956. Phase Diagrams in Metallurgy Their Development and Application. McGraw-Hill, New York, NY.

TAPP Database, 1994. Thermochemical and Physical Properties. Cincinnati, OH.

Verdeja, L., Sancho, J., Ballester, A., Gonzalez, R., 2014. Refractory and Ceramic Materials. Editorial Síntesis, Madrid, Spain.

Winegard, W., 1964. An Introduction to the Solidification of Metals. The Institute of Metals, London.

Chapter 4

Invariant Solidification

4.1 INTRODUCTION

Considering all possible alloys formed with Ni and Cu (from 100% Ni to 100% Cu) and as shown in Chapter 3 (Figs. 3.17 and 3.18), the state of each alloy depends, in some way, on the constitution of the rest. The set of all possible Cu-Ni alloys is known as Cu-Ni *system*.

A group of bodies forms a system when the state of each body depends on the constitution of the others,[1] and the basic chemical compounds that can establish all the equilibrium relations of the system are referred to as *system components*. For example, in the case of the Cu-Ni system, the components are Cu and Ni.

The components of a system are found in the form of *phases*, which are the group of *physical and chemical homogeneous zones* at all points, and physically separated from the other phases by means of *interfaces*; in other words, zones limited by surfaces, with properties that differ from one side to the other.

A zone or substance is physically and chemically homogeneous when, at all points, it has the same physical state (e.g., amorphous, face-centered cubic, body-centered cubic, liquid, gas, alpha solid solution, etc.) and the same composition.

If an alloy of the Cu-Ni system, at 1320° C, is formed by dendrites of solid solution with a composition of 67% Ni–33% Cu and liquid with uniform composition of 50% Ni–50% Cu (Fig. 3.16), it can be stated that there are two phases: solid solution with 67% Ni and liquid with 50% Ni. Furthermore:

- A phase can be constituted by either one fragment or by many fragments. For example if inside the Cu-Ni liquid there are a lot of solid solution Cu-Ni dendrites with uniform composition of 67% Ni, then all these dendrites are considered as one solid phase.
- A phase can be formed by various components, either because there is solubility between them or because they form a chemical compound.

1. It is important to distinguish between a system and a simple grouping of bodies. A mixture of sand, shavings, and water would be a group of bodies, but not a system because the constitution of each one is independent of the state of the others.

Solidification and Solid-State Transformations of Metals and Alloys. http://dx.doi.org/10.1016/B978-0-12-812607-3.00004-8

- A compound can form more than one phase, for example, it can be in liquid, solid, or gaseous states, or present different allotropic structures in solid state, etc.
- The eutectic cannot be considered as one phase because it is formed by crystals that are physically and chemically differentiated between them. Therefore, a binary eutectic (formed by two metals, A and B) has two phases: metal A and metal B.

If a system of three components (A, B, and C metals) has an eutectic formed by C and alpha solid solution of metal B in metal A, that ternary eutectic has two phases: alpha solid solution and metal C.

A system is at *equilibrium* when over time, irrespective of the time interval, neither a new phase appears nor does an existing one disappear. The equilibrium of a system can be influenced by physical factors outside the components, such as temperature or pressure. These physical factors are known as *equilibrium factors*.

When none of the factors can be changed without them promoting the creation of a new phase or the disappearance of an existing phase, the equilibrium is known to be *unstable*. If a factor does not break the equilibrium though it should (e.g., the creation of a new phase), the equilibrium is known to be *metastable*.

4.2 SYSTEM AT EQUILIBRIUM AND THE PHASE RULE

In order for a number of phases to form a system in equilibrium (i.e., the state of each phase depends on the state of others and no phases are appearing or disappearing), two conditions must be met:

1. For each phase, a state equation can indicate the relation of concentrations of the components with the physical (equilibrium) factors that influence the system.

   In industry, for metallic alloys, the only factor that influences the equilibrium besides concentrations is temperature. Pressure is constant (1 atm) and the effects of an increase or decrease of pressure compared to the atmospheric one are considered insignificant (except in the case of vacuum casting or pressure die casting).
2. The concentration of each component in any phase must depend on the concentration of the same component at another phase.

Furthermore, supposing that a system has n components and φ phases, the *degrees of freedom* of a system can be determined through the Gibbs' law (or *phase rule*):

$$V = n - \varphi + 1 \tag{4.1}$$

for constant pressure, otherwise the number 1 must be changed to 2. V represents the number of variables that can be set in a system without breaking the equilibrium (without the appearance of new phases or the disappearance of existing ones). Moreover, the equilibrium is of the *invariant* type if $V=0$; *monovariant* type if $V=1$; *bivariant* type if $V=2$; and so on.

On the other hand, when various phases form a system in equilibrium (e.g., all the equilibrium diagrams), Gibbs' law can be used. For example, during solidification of an alloy of the binary Cu-Ni system, with mean composition of 60% Ni (Fig. 3.16) and temperatures higher than 1340°C, there is only one phase, liquid, and therefore, it is a bivariant equilibrium:

$$V=2-1+1=2 \tag{4.2}$$

where two of the unknown variables can be changed in an arbitrary fashion; for example, composition and temperature, without breaking the equilibrium.

When reaching 1340°C, equilibrium is broken because a new phase appears: the formation of solid crystals begins. In the temperature/time curve (that represents cooling), the start of the formation of the solid is followed by a decrease of the *dT/dt* slope according to the Le Chatelier rule: when cooling a system, if equilibrium is broken, there is a release of heat at that temperature.

Between 1340°C and 1280°C, another equilibrium is reached that corresponds to two phases (solid and liquid), and is of the monovariant type:

$$V=2-2+1=1 \tag{4.3}$$

because neither temperature nor composition can be modified without breaking equilibrium (solidification of the solid solution does not happen at constant temperature).

On reaching 1280°C, the liquid phase disappears and total solidification is accomplished, breaking the previous equilibrium and establishing a new one (bivariant) which will also follow the expression (4.2).

Invariant reactions are phase changes taking place at constant temperature. In order for them to occur in a binary system it is necessary to have three phases in equilibrium, according to the Gibbs' law.

An eutectic formed by two insoluble metals needs constant temperature during solidification (Section 3.1.3). The same is required when a liquid phase reacts with a solid phase to produce a different solid phase (peritectic reaction), or when a liquid phase creates another liquid phase and a solid phase (monotectic reaction), or when two liquids react to form one solid (sintectic reaction).

In binary systems, other types of invariant reactions can occur. For example, transformations where one of the phases is a gas, or reactions where all phases in equilibrium are solid (eutectoid and peritectoid reactions).

Sections 4.3–4.7 describe different types of invariant reactions in binary systems and alloys of interest to the industry.

4.3 BINARY EUTECTIC REACTION

In general, a binary system presents an eutectic reaction when a liquid with constant composition is transformed into two solids with defined compositions (two solid solutions, one solid solution and an intermetallic compound, other solids, etc.). The eutectic reaction can be expressed as:

$$\text{liquid E} \rightleftharpoons \text{solid 1} + \text{solid 2} \tag{4.4}$$

This reaction occurs at constant temperature until all liquid disappears. This statement can also be corroborated with the Gibbs' law.

As an example, the Pb-Sn system can be analyzed. When a liquid with 38.1% Pb–61.9% Sn composition reaches 183°C, two solid phases are in equilibrium: alpha solid solution with 19% Sn and beta solid solution with 97.5% Sn (Fig. 4.1). The first of these solutions corresponds to the substitutional solid solution of Sn atoms in the crystalline lattice of Pb; meanwhile the second solid solution, also substitutional, corresponds to the solution of Pb atoms in the lattice of Sn (i.e., there are two solid solutions in the diagram, α and β).

Fig. 4.1 also shows the solidification and cooling curves (temperature vs. time) of different alloys of the Pb-Sn system. The alloys with compositions between 0% and 19% Sn present, after solidifying, a structure corresponding to alpha solid solution. The alloys having compositions between 19% and 61.9% Sn have a room-temperature microstructure of grains of solid solution α_m (proeutectic) embedded in a complex matrix constituent: eutectic formed by α_m and β_n. The proportions of disperse (proeutectic) and matrix constituent

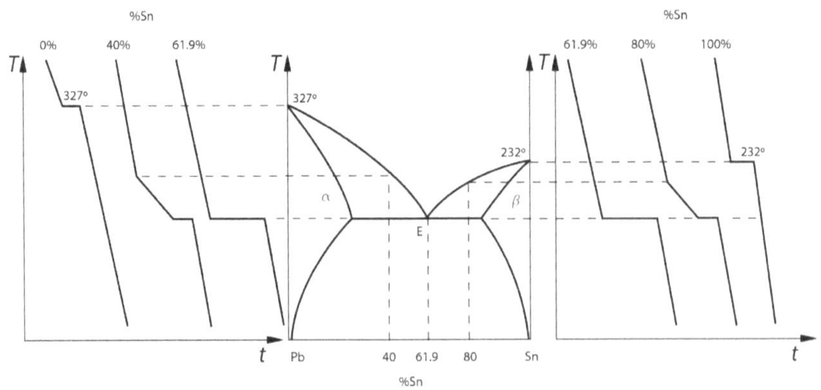

FIG. 4.1 Pb-Sn phase diagram and cooling curves.

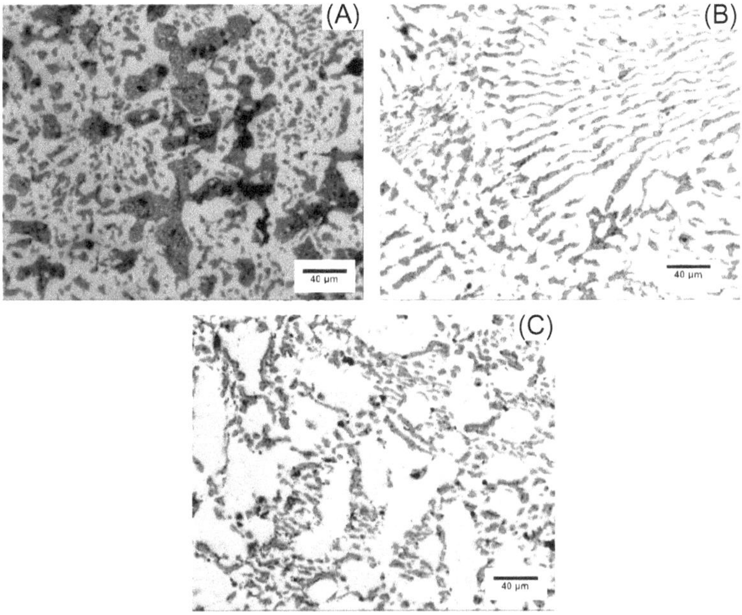

FIG. 4.2 Pb-Sn alloys: (A) hypoeutectic (19%–61.9% Sn) with disperse constituent α_m and matrix constituent $\alpha_m + \beta_n$ eutectic, (B) $\alpha_m + \beta_n$ eutectic (62% Sn), and (C) hypereutectic (62%–97.5% Sn) with disperse constituent β_n and matrix constituent $\alpha_m + \beta_n$ eutectic.

vary with the amount of Pb, as calculated by the lever rule. The same can be said about alloys with compositions between 61.9% and 97.5% Sn: its microstructure is formed by solid solution grains of β_n embedded in the $(\alpha_m + \beta_n)$ eutectic (Fig. 4.2). Alloys with compositions lower than the eutectic (61.9% Sn) are known as *hypoeutectics*, and those with compositions higher than the eutectic are known as *hypereutectics*.

An interesting property of the binary eutectics is that they do not present any type of chemical heterogeneity, or segregations (Chapter 5).

If solidification curves are analyzed, the horizontal segment length, for all compositions between 19% and 97.5% Sn, depends on the proximity to the eutectic composition: the closer to the eutectic, the larger the horizontal segment will be as the remaining eutectic liquid is higher.

This previous assessment is the basis of the Tammann method for the determination of the eutectic composition in a binary system. This consists of sketching, for equal amounts of four alloys (two hypoeutectic and two hypereutectic) whose compositions are known, the time required for the eutectic solidification; and therefore making a graphic representation (Fig. 4.3) of time versus the composition of the alloys, to obtain the eutectic as well as the α_m and β_n compositions.

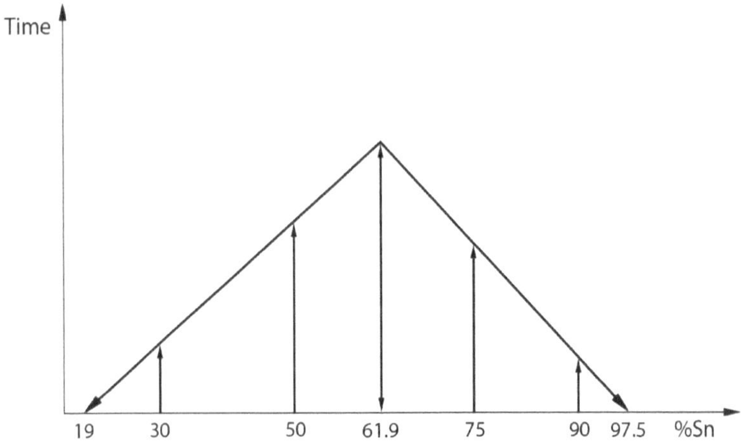

FIG. 4.3 Tammann method to determine the eutectic composition and the solubility limits.

EXERCISE 4.1

After experimental tests (thermal analysis) in the Pb-Sn system, the temperature arrests of the eutectic reaction of four alloys (two hypoeutectic and two hypereutectic) of equal weight obtained from the cooling curves (proportional to the solidification times) have the following lengths:

Alloy	Temperature arrest (mm)
Pb-30% Sn	25.6
Pb-40% Sn	48.8
Pb-70% Sn	77.5
Pb-80% Sn	49.3

Using the Tammann method, determine the approximate C_E, C_α, and C_β compositions.

Solution

In the vertical axis of the sketch (Fig. 4.4) the temperature arrests are represented, which are proportional to the eutectic fractions versus the respective compositions of Sn. Extending these lines until they intersect with the horizontal axis, the maximum solubility limits at the eutectic temperature of Sn in the crystalline lattice of Pb (19%) and of Pb in the crystalline lattice of Sn (2.5%), are found, while the intersection of both straight lines determines the composition of the eutectic (62% Sn).

Finally, to verify that the eutectic fractions in each alloy are proportional to the temperature arrest obtained experimentally:

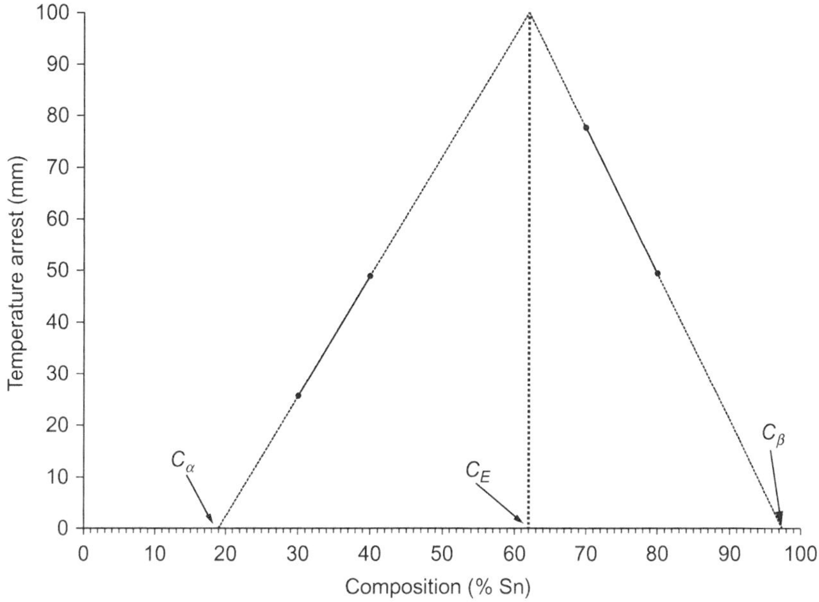

FIG. 4.4 Tammann method for the Pb-Sn system.

$$f_{E(30\%\ \text{Sn})} = \frac{30-19}{61.9-19} = 0.26\ (26\%)$$

$$f_{E(40\%\ \text{Sn})} = \frac{40-19}{61.9-19} = 0.49\ (49\%)$$

$$f_{E(70\%\ \text{Sn})} = \frac{97.5-70}{97.5-61.9} = 0.77\ (77\%)$$

$$f_{E(80\%\ \text{Sn})} = \frac{97.5-80}{97.5-61.9} = 0.49\ (49\%)$$

which are, in fact, similar to the experimental values and proportional to the length of the temperature arrests:

$$\frac{0.26}{0.49} = \frac{25.6}{48.8};\ \frac{0.77}{0.49} = \frac{77.5}{49.3};\ \text{etc.}$$

EXERCISE 4.2

In a binary eutectic system AB, the composition of the three phases in equilibrium at the eutectic temperature are: $\boldsymbol{\alpha} = 15\%$ **B**, $\boldsymbol{L} = 75\%$ **B**, and $\boldsymbol{\beta} = 95\%$ **B**. If the 50% A–50% B alloy has solidified under equilibrium conditions and does not present transformations in solid state, calculate for a temperature slightly lower than the eutectic:

(a) The primary phase and eutectic phase $(\alpha+\beta)$ proportions

(b) The total α and β proportions

Solution

(a) Primary phase and eutectic phase proportions

Since $C_\alpha = 15\%$, $C_L = 75\%$, and $x = 50\%$, applying the lever rule above the eutectic temperature:

$$\%\alpha_{\text{proeutectic}} = \frac{C_L - x}{C_L - C_\alpha} = \frac{75 - 50}{75 - 15} = 0.42\ (42\%)$$

and the remaining liquid (58%) will afterwards (at the eutectic temperature) transform into eutectic phase which will have the proportions calculated as:

$$\%\alpha_{\text{eutectic}} = \frac{C_\beta - x}{C_\beta - C_\alpha} = \frac{95 - 75}{95 - 15} = 0.25\ (25\%)$$

and

$$\%\beta_{\text{eutectic}} = 100 - \%\alpha_{\text{eutectic}} = 75\%$$

(b) Total proportions

Now the lever rule must be applied below the eutectic temperature:

$$\%\alpha_{\text{total}} = \frac{C_\beta - x}{C_\beta - C_\alpha} = \frac{95 - 50}{95 - 15} = 0.56\ (56\%)$$

which makes the remaining material beta phase: $\%\beta_{\text{total}} = 44\%$

To corroborate these values:

$$\alpha_{\text{total}} = \alpha_{\text{proeutectic}} + \alpha_{\text{matrix}}$$

$$56 \approx 42 + 58\ (0.25)$$

and

$$\beta_{\text{total}} = \beta_{\text{proeutectic}} + \beta_{\text{matrix}}$$

$$44 \approx 0 + 58\ (0.75)$$

Table 4.1 presents some binary eutectics of industrial interest. Some binary systems present more than one eutectic, like the Al-Cu system which has two, one of them is the base for the Al-Cu alloys and the other is the complex $\alpha + \beta$ constituent of copper-aluminum alloys.

Fig. 4.5 shows the full Al-Cu diagram. Besides the two eutectics, at solid state, alloys undergo changes in solubility that originate solubility limit transformation lines in the diagrams (solvus lines).

Generally, an equilibrium diagram is formed by: a solidification diagram (corresponding to the freezing range) and another one corresponding to the transformations that the alloy can experience in solid state.

The proportions in weight for each of the constituents of an eutectic can be easily calculated through the lever rule, when (as in the cases indicated in Table 4.1) the composition of the eutectic and those of its constituents are known. For example, the Zn-Al eutectic (industrially known as *Zamak*) is formed by two solid solutions: beta (82.8% in weight of Zn atoms substituted in the face-centered cubic crystalline lattice of Al) and alpha (1% in weight of Al substituted in the hcp lattice of Zn). Here W_α is the weight of the alpha solid

TABLE 4.1 Some Binary Eutectics of Industrial Value

Composition of the Eutectic (wt %)		Melting Temperature (°C)	Eutectic Constituents (wt %)		
87.4% Al	12.6% Si	577	Sol. Sol. in Al (1.65% Si)	+	Si
65.0% Al	35.0% Mg	451	Sol. Sol. in Al (14.9% Mg)	+	$\beta(Al_3Mg_2)$ 35.5% Mg
66.8% Al	33.2% Cu	548	Sol. Sol. in Al (5.65% Cu)	+	$\theta(Al_2Cu)$
5.0% Al	95.0% Zn	382	Sol. Sol. in Al (82.8% Zn)	+	Sol. Sol. in Zn (1.0% Al)
93.9% Al	6.1% Ni	640	Sol. Sol. in Al (0.05% Ni)	+	$\beta(Al_3Ni)$
99.61% Cu	0.39% O	1066	Sol. Sol. in Cu (<0.004% O)	+	$\beta(Cu_2O)$
91.6% Cu	8.4% P	714	Sol. Sol. in Cu (1.75% P)	+	$\beta(Cu_3P)$
91.7% Cu	8.3% Al	1083	Sol. Sol. in Cu (7.4% Al)	+	$\beta(AlCu_3)$
69.0% Cu	31.0% Sb	645	Sol. Sol. in Cu (11.0% Sb)	+	$\beta(Cu_3Sb)$
99.16% Cu	0.84% S	1068	Cu	+	$\delta(Cu_9S_5)$
8.9% Zn	91.9% Sn	198	Zn (<0.1%)	+	Sol. Sol. in Sn (1.7% Zn)
99.5% Zn	5.0% Al	382	Sol. Sol. in Zn (1.0% Al)	+	Sol. Sol. in Al (82.8% Zn)
78.0% Ni	22.0% S	635	Ni	+	$\beta(Ni_3S_2)$
98.03% Ni	1.97% C	1316	Sol. Sol. in Ni (0.55% C)	+	Graphite
88.5% Ni	11.5% Si	1154	Sol. Sol. in Ni (8.0% Si)	+	$\gamma(Ni_5Si_2)$
64.0% Ni	36.0% Sb	1104	Sol. Sol. in Ni (16.0% Sb)	+	$\gamma(Ni_5Sb_2)$
67.5% Ni	32.5% Sn	1132	Sol. Sol. in Ni (19.0% Sn)	+	$\beta(Ni_3Sn)$
95.7% Fe γ	4.3% C	1148	Sol. Sol. in Fe γ (2.11% C)	+	C_3Fe

Continued

TABLE 4.1 Some Binary Eutectics of Industrial Value—cont'd

Composition of the Eutectic (wt %)		Melting Temperature (°C)	Eutectic Constituents (wt %)		
95.74% Fe γ	4.26% C	1154	Sol. Sol. in Fe γ (2.08% C)	+	Graphite
77.09% Fe γ	22.91% O	1373	Fe	+	Wustite (FeO)
68.4% Fe γ	31.6% S	989	Fe	+	ϵ pyrrhotite ($Fe_{1-x}S$)
89.87% Fe δ	10.13% P	1049	Sol. Sol. in Fe δ (2.55% P)	+	δ(Fe_3P)
48.0% Fe δ	52.0% Sb	1003	Sol. Sol. in Fe δ (35.0% Sb)	+	ϵ(FeSb)
86.0% Fe δ	14.0% Ti	1291	Sol. Sol. in Fe δ (9.0% Ti)	+	Fe_2Ti
43.0% Fe δ	57.0% U	1057	Fe	+	Fe_2U
76.5% Mg	23.5% Ni	507	Mg	+	Mg_2Ni
98.62% Mg	1.38% Si	638.8	Mg	+	Mg_2Si
62.4% Mg	37.6% Sn	561.2	Sol. Sol. in Mg (14.85% Sn)	+	Mg_2Sn
32.6% Mg	67.4% Pb	466.2	Sol. Sol. in Mg (41.8% Pb)	+	Mg_2Pb

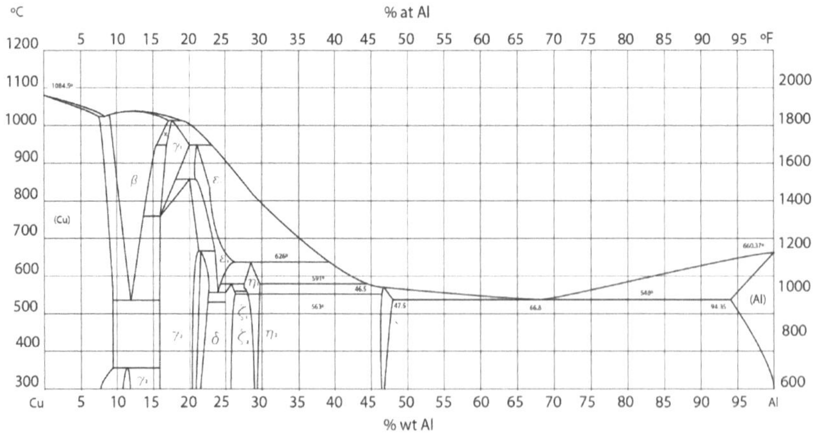

FIG. 4.5 Cu-Al phase diagram (ASM Handbook, Vol. 3, 1992).

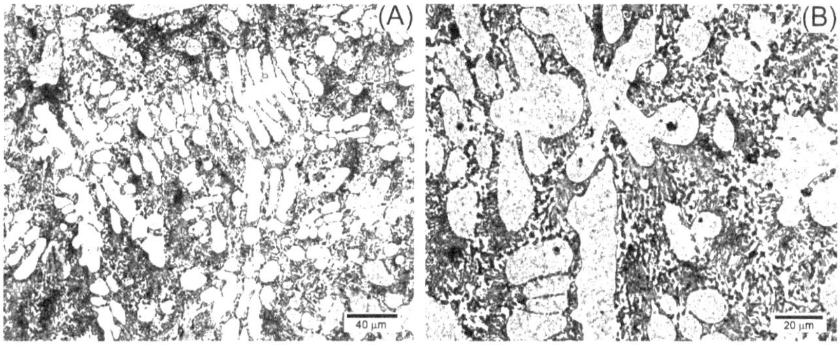

FIG. 4.6 (A) Microstructure of an hypoeutectic Zamak alloy (pressure die casting) and (B) seen at higher magnification.

solution, W_β the beta solid solution, and W_T (100) the weight of the eutectic (95% Zn) then:

$$W_T(100) = W_\alpha + W_\beta \tag{4.5}$$

and the balance of matter in Al:

$$5(100) = 1(W_\alpha) + 17.2(W_\beta) \tag{4.6}$$

From both expressions, it is deduced that $W_\beta/W_\alpha = 4/12.2$ and $W_\beta/W_T = 4/16.2$ (lever rule). Therefore, the eutectic being considered is formed by 24.69% in weight of β solid solution and 75.31% in weight of α solid solution. In the case of the alloy being an hypoeutectic of Zn, just as in the case of *Zamak* three to four formed by primary α in an eutectic matrix of α and β (Fig. 4.6), the weight proportions of the disperse constituent α and the α-β matrix can also be calculated with this procedure (lever rule).

If the area fraction occupied by a phase (as seen in micrographs) equals its volume fraction (determined by quantitative metallographic techniques), the relationship between the constituents' densities can also be calculated through the lever rule.

With a simple operation, analogous to the one for the Al-Zn system, the composition of the eutectic known as *ledeburite* (Table 4.1) can be deduced: a 4.3% C liquid in the Fe-C metastable system, at 1148°C, will be transformed into 51.9% austenite (interstitial solid solution of 2.11% in γ-Fe) and 48.1% cementite (Fe_3C, 6.67% C). Binary ferrous alloys with C content between 2.11% and 6.67% (matrix constituent will be ledeburite) are known as *white cast irons*, because of their typical fracture color.

EXERCISE 4.3

Calculate the equilibrium concentrations at the eutectic temperature of the following system: A melts at 1000°**C**, B melts at 700°**C**; the 25% B alloy solidifies fully at 500°**C** and, under equilibrium conditions, has 73.3% of primary phase and

26.7% of eutectic constituent ($\boldsymbol{\alpha}+\boldsymbol{\beta}$); the 50% B alloy, at the same temperature, has 40% of primary phase and 60% of eutectic mix, and has a total of 50% $\boldsymbol{\alpha}$.

Solution

The first assumption to be made, since it is not specified, is that both given alloys are hypoeutectic. Applying the lever rule, the following equations can be formulated:

$$25\%\ \mathrm{B} \Rightarrow \%\alpha_{\text{proeutectic}} = \frac{c_L - x}{c_L - c_\alpha} = \frac{c_L - 25}{c_L - c_\alpha} = 0.733\ (73.3\%)$$

$$50\%\ \mathrm{B} \Rightarrow \%\alpha_{\text{proeutectic}} = \frac{c_L - x}{c_L - c_\alpha} = \frac{c_L - 50}{c_L - c_\alpha} = 0.4\ (40\%)$$

$$50\%\ \mathrm{B} \Rightarrow \%\alpha_{\text{total}} = \frac{c_\beta - x}{c_\beta - c_\alpha} = \frac{c_\beta - 50}{c_\beta - c_\alpha} = 0.5\ (50\%)$$

Solving the system of the first two equations:

$$c_\alpha = 4.95\%\ \mathrm{B}$$

and

$$c_L = 80.03\%\ \mathrm{B}$$

Finally the α concentration can be substituted in the third equation to obtain the concentration of β:

$$c_\beta = 95.06\%\ \mathrm{B}$$

On the other hand, if one of the alloys is hypoeutectic and the other is hyper-eutectic, equations would be:

$$\%\alpha_{\text{proeutectic}} = \frac{c_L - x}{c_L - c_\alpha} = \frac{c_L - 25}{c_L - c_\alpha} = 0.733\ (73.3\%)$$

$$\%\beta_{\text{proeutectic}} = \frac{x - c_L}{c_\beta - c_L} = \frac{50 - c_L}{c_\beta - c_L} = 0.4\ (40\%)$$

$$\%\beta_{\text{total}} = \frac{x - c_\alpha}{c_\beta - c_\alpha} = \frac{50 - c_\alpha}{c_\beta - c_\alpha} = 0.5\ (50\%)$$

These three equations form a 3×3 system and the solution will be:

$$c_\alpha = 22.56\%\ \mathrm{B}$$
$$c_\beta = 77.44\%\ \mathrm{B}$$
$$c_L = 31.7\%\ \mathrm{B}$$

4.3.1 Abnormal Eutectics. Al-Si System

Eutectics are of the "abnormal" type[2] when the constituents of the eutectic have considerably different melting temperatures (660°C for Al and 1414°C for Si) and also different weight proportions (11.13% Si and 88.87% of α solid solution

2. For example the 12.6% Al-Si alloy, known as a siluminum and commonly used in molding of lightweight parts because of its advantages compared to pure Al.

of Si in Al); both these aspects (melting temperature of the components and weight proportions between them) influence significantly the structure and the mechanical properties of the eutectic.

It is a fact that in all eutectics, the constituent with the highest melting point solidifies first (Si in the Al-Si eutectic), then the other constituent solidifies (α solid solution) in the necessary amount in order for the liquid to maintain the adequate proportion (12.6% Si).

Kinetics for the solidification of Si are faster than for Al, as the driving force that induces solidification is higher: for Si a 1414–577°C interval and for Al a 660.37–577°C interval. Thus, its time-temperature-transformation curve is closer to the origin in the time axis than the one for Al. This results in fast nucleation of Si developing freely in the shape of cuboids or needles before being surrounded by the corresponding amount of solidified Al. Needle or irregular arrangements of second phases occur in systems such as Al-Si, where the composition of the liquid eutectic is closer to the composition of one solid phase compared to the other, and where the less abundant phase (Si) grows in a highly anisotropic fashion. The resulting microstructure (Fig. 4.7) does not present the common aspect of an eutectic given that neither mutual coherent nor simultaneous growth of the constituents in the original liquid has occurred. The *notch-effect* of the cuboids, and especially of the needles, on the matrix, explains the low toughness of the alloys with eutectic compositions.

In 1920, Aladar Pacz added Na before the casting process, producing an eutectic microstructure in the shape of small particles (globules) of Si dispersed in the Al matrix. This alloy is known as *modified siluminum*. Na produces undercooling and displaces the eutectic composition to 14% Si, and thus turning the 12.6% Si alloy into a slightly hypoeutectic one, and

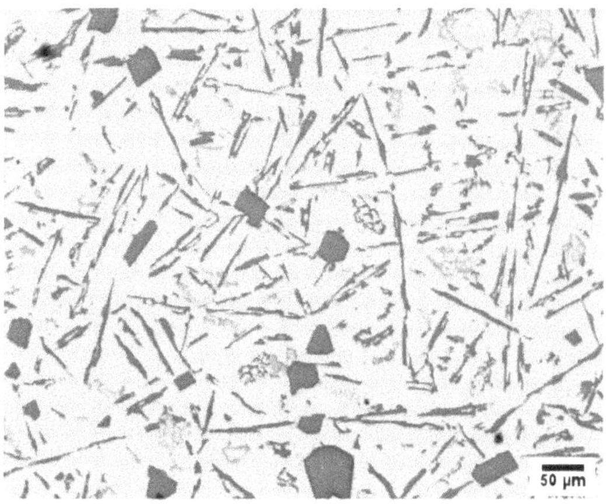

FIG. 4.7 Microstructure of an unmodified siluminum (12.6% Si-Al) alloy.

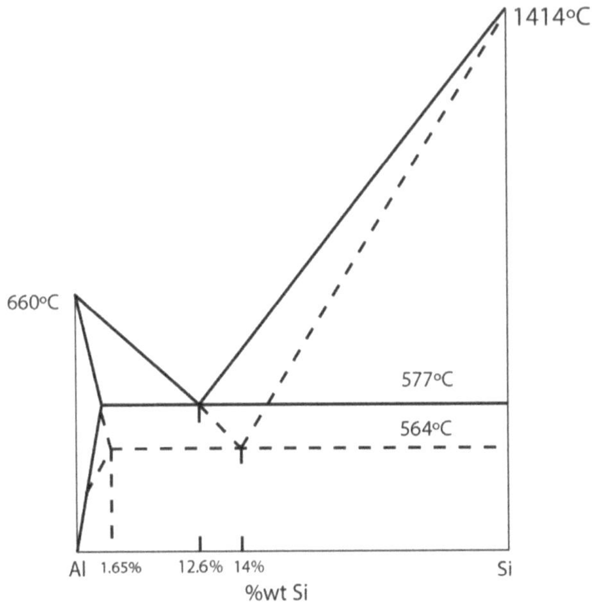

FIG. 4.8 Undercooling produced by Na in the Al-Si system.

tends to equal the solidification kinetics of the eutectic constituents (Fig. 4.8).

Furthermore, the early solidification of Al (since the 12.6% alloy is slightly hypoeutectic in Si) tends to equal the time necessary to solidify Al and Si. Thus, the Si needles and cuboids, quickly surrounded by the corresponding amount of Al, reach smaller sizes and produce a lesser notch-effect. The impact toughness is higher in the modified siluminum than in the unmodified one. Fig. 4.9 shows

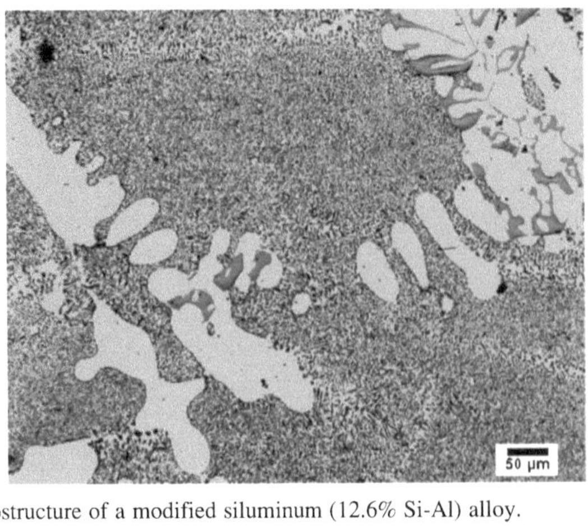

FIG. 4.9 Microstructure of a modified siluminum (12.6% Si-Al) alloy.

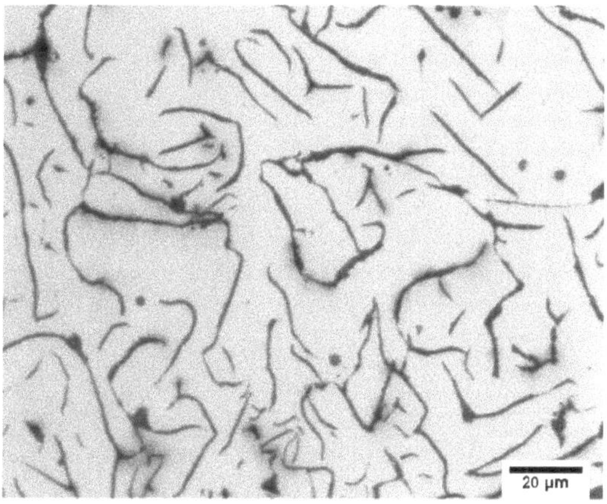

FIG. 4.10 Microstructure of an eutectic of graphite and austenite (gray casting).

the morphology of the 12.6% Si modified alloy that, in honor of Aladar Pacz, is commonly known as "*alpax*." Nowadays, the most common modifier element is not Na but Sr.

Another example of an abnormal eutectic is present in the "stable" Fe-C system. It is stable when the formation of Fe_3C is prevented by the addition of graphitizing elements such as Si. The eutectic (Table 4.1) is formed by austenite with 2.08% C and graphite (97.77% in weight of austenite and 2.33% of graphite, Fig. 4.10).

Both constituents are noticeably different at their melting temperatures (1154°C for austenite and above 2900°C for graphite) as well as in their weight proportions. As a consequence, it must be classified as an abnormal eutectic. Furthermore, these "stable" ferrous alloys with C percentage higher than 2.08% are known as *gray castings* because of the gray aspect that graphite gives to the fracture surface; its solidification has important analogies with the Al-Si alloys.

EXERCISE 4.4

The composition of the three phases in equilibrium of the Al-Si system at the eutectic temperature are: $\boldsymbol{L}=12.6\%$ **Si**, $\boldsymbol{\alpha}=1.65\%$ **Si**, and $\boldsymbol{\beta}=100\%$ **Si**. For an alloy with 8.5 % **Si** (the most widely used Al alloy for pressure die casting) calculate, at a temperature slightly lower than the eutectic one, the proportions of the matrix and disperse constituents, as well as the total proportions of $\boldsymbol{\alpha}$ and $\boldsymbol{\beta}$ phases.

Solution

Applying the lever rule above the eutectic temperature:

$$\%\alpha_{\text{proeutectic}}=\frac{C_L-x}{C_L-C_\alpha}=\frac{12.6-8.5}{12.6-1.65}=0.37\ (37\%)$$

$$\%\text{eutectic}=100-37=63\%$$

Thus the disperse constituent is proeutectic α phase (37%) and the matrix constituent is liquid (63%) which will transform into eutectic $(\alpha+\beta)$.

The eutectic will have following proportions:

$$\%\alpha_{\text{eutectic}}=\frac{C_\beta-L}{C_\beta-C_\alpha}=\frac{100-12.6}{100-1.65}=0.89\ (89\%)$$

$$\%\beta_{\text{eutectic}}=100-89=11\%$$

If the lever rule is also applied below the eutectic temperature:

$$\%\alpha_{\text{total}}=\frac{C_\beta-x}{C_\beta-C_\alpha}=\frac{100-8.5}{100-1.65}=0.93\ (93\%)$$

$$\%\beta_{\text{total}}=100-93=7\%$$

To corroborate the results:

$$\alpha_{\text{total}}=\alpha_{\text{proeutectic}}+\alpha_{\text{eutectic}}$$

$$93\cong 37+(0.63)(89)$$

4.3.2 Eutectics With S

It is common for industries to choose chemical compositions that result in alloys with better properties and a microstructure of the base metal (or one of its possible solid solutions), dispersed in an eutectic matrix. These alloys must also consider the enhanced casting properties of the eutectic, such as low melting point and invariance of freezing temperature (just as it happens when the chosen composition is an eutectic).

For the same reasons, in some cases, the formation of an eutectic results being detrimental for the alloy, just as it happens when Nickel or one of its alloys is exposed to hydrogen sulfide vapors.

As an example, the Nicrom alloys (60% Ni-16% Cr-24% Fe) are refractory metallic materials, resistant to heat and oxidation at high temperature. If exposed to $S^{=}$, Ni_3S_2 is formed by the reaction of S ions and Ni atoms; and if the service temperatures are higher than 635°C (Table 4.1), this Ni sulfide in contact with other Ni atoms produces local melting that make the alloy useless.

Another example is certain Fe-based alloys, such as steels. During its manufacture, the addition of S to the melted steel must be avoided in order to prevent, during solidification, the formation of the iron and iron sulfide eutectic (Table 4.1). That steel (since the eutectic would melt at that temperature) would be unviable at temperatures higher than 989°C (usually required in hot-working processes or heat treatments).

However, there are grades of special steels with S contents as high as 0.25%, known as *free-cutting steels* ($\%C<0.25\%$) or *improved-machinability steels* ($\%C>0.25\%$). Its S content serves the purpose of forming (by reaction with

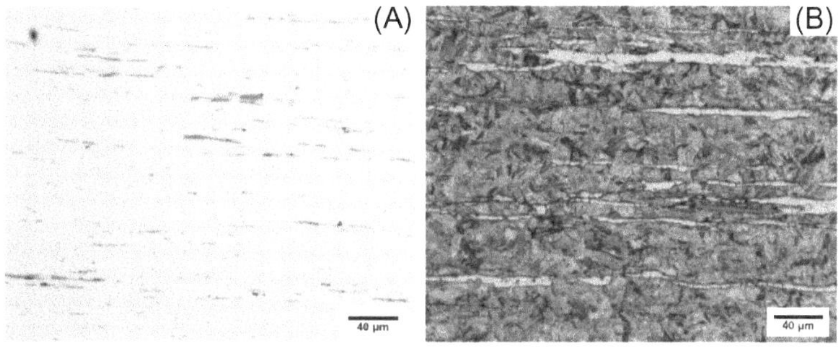

FIG. 4.11 Free-cutting (0.2% C–12% Cr-Fe) stainless steel: (A) as polished and (B) etched with an acidic solution of ferric chloride to distinguish its ferritic-martensitic structure.

the Mn of the melted steel) a high number of MnS inclusions (Fig. 4.11), which during cutting operations result in shavings being fragile and, therefore, small.

The manufacture of these steels is special, due to the risks of forming the mentioned eutectic. The proportion of S must be perfectly balanced with Mn in order for all S to form MnS; an excess of S would result in the iron and iron sulfide eutectic (Table 4.1). Even when the S/Mn ratio is adequate, S produces macro-segregations, and the improved-machinability steel ingots must be blunted to avoid problems during rolling (melting of the eutectic in the S-enriched zones because of macro-segregation).

4.3.3 Cu-O Eutectic. Types of Commercial Copper

Interpreting phase diagrams can, usually, give information on the manufacturing process of a metal or alloy and on the limits of the chemical composition to obtain adequate properties. It also indicates whether, by metallography, finding small amounts of a component is possible. The Cu-O system at atmospheric pressure is shown in Fig. 4.12.

Both black-copper (from oxides or native ores, or even scraps) and blister-copper (from sulfides, ore processing or copper matte) are fire-refined in order to eliminate impurities caused by oxidation. These impurities (like As, S, Sb, etc.) become volatile oxides (As_2O_3, SO_2, Sb_2O_3, etc.), while metals (Fe, Ni, etc.) become oxides (FeO, NiO, etc.) that pass to the silicate slag. After refining, the metallic copper has a high oxygen content (0.6%–0.9%) dissolved as an element or copper oxide which, when solidified, results in the copper being brittle and with a dark reddish color. According to the phase diagram, this copper is not useful as a metallic material since its structure is formed by grains of Cu_2O embedded in an eutectic matrix (of Cu_2O and α solid solution of O in Cu).

Therefore, liquid Cu with 0.6%–0.9% O_2 requires deoxidation, which can be made with hydrogen from hydrocarbons, green poles (poling), etc., until reaching less than 0.39%. In this way, the microstructure is formed by primary grains of α solid solution surrounded by eutectic. The best would be to achieve

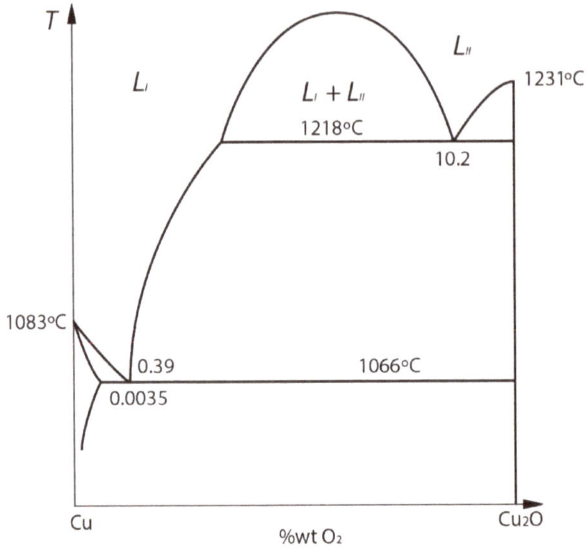

FIG. 4.12 Cu-O phase diagram (ASM Handbook, Vol. 3, 1992).

copper completely free of oxygen in order for the structure to be formed only by grains of Cu, or at least by grains of α solid solution (whose behavior will be metallic) without the eutectic. But it is not possible to obtain Cu without oxygen by deoxidation with hydrogen because of the physical-chemical equilibriums between hydrogen, oxygen, and liquid Cu.

It is noteworthy, that the limited solubility of oxygen in copper is less than 0.004% at 1066°C and less than 0.001% at 700°C (Table 4.1). It is enough for the cast Cu to have more than 0.004% of oxygen for the eutectic to be present, which is even detectable by optical microscopy: there is an eutectic matrix of Cu_2O and α surrounding the grains of primary α. In the micrograph of Fig. 4.13, the oxidules of Cu can be seen profiling the grain boundaries of the α

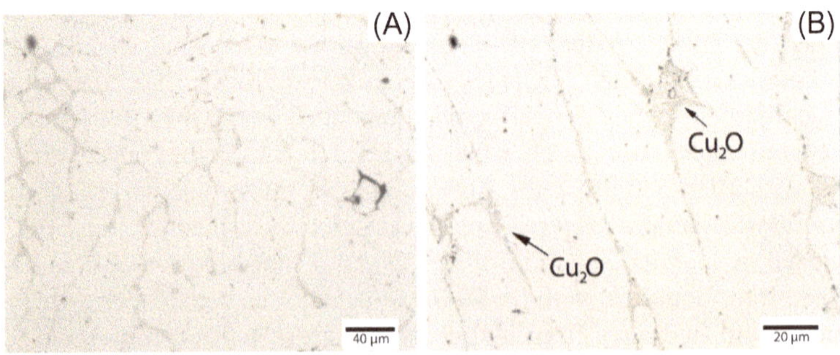

FIG. 4.13 (A) Microstructure of a molded copper, with 0.008% oxygen and (B) Cu_2O globules clearly surrounding primary grains.

constituent. It is important to point out that in the polished state, the eutectic α and the primary α constituents are indistinguishable.

Taking into account the brittle nature of the eutectic, copper alloys with more than $0.15\%O_2$ do not have the good formability properties of Cu and its solid solutions (consequence of the face-centered cubic crystalline system); on the contrary, they are unforgeable and only useful for applications that require remelting.

Copper alloys with oxygen content lower than 0.15% are known as *tough pitch copper* or TPC. If Cu is totally free of oxygen, it is known as *oxygen-free copper* or OFC. The total removal of oxygen can be achieved with a vacuum remelting of the electrolytic copper, known as *oxygen-free high conductivity copper* or OFHC. The removal of oxygen can also be achieved by the use of P, by forming gas oxides and removing them, in which case is known as *deoxidized phosphorous copper* or DPC (contaminated with Cu_3P).

A considerable amount of Cu alloys in their tough pitch condition are used in architecture, chemistry, mechanics, and electricity. As they are not oxygen-free coppers ($\%O > 0.004$), their use in powerful reductive atmospheres such as acetylene gas welding is not possible, because the oxidules of the eutectic would react with H_2 or with CO of the reductive atmosphere and would produce, in the as-cast grain boundaries, either water vapor or carbon dioxide that, when removed, would originate intergranular disaggregation (hydrogen embrittlement).

EXERCISE 4.5

The binary Cu-O phase diagram has an eutectic at 1066°**C** for 0.39% O (formed by a solid solution of 0.0035% of oxygen in Cu and **Cu_2O**).

(a) Determine if optical microscopy can detect whether an alloy has more than 0.008% oxygen

(b) Explain why alloys with certain amount of oxygen, even if they are TPC, cannot be used in powerful reductive atmospheres.

Solution

(a) Amount of oxygen

For a $x = 0.008\%$ oxygen composition, the fraction of eutectic constituent is obtained by applying the lever rule:

$$f_L = \frac{0.008 - 0.0035}{0.39 - 0.0035} = 0.012\ (1.2\%)$$

On the other hand, this alloy, using an optical microscope, presents a disperse constituent (α rich in Cu) embedded in an eutectic matrix of Cu-Cu_2O, because the fraction of liquid is higher than 1%, all the liquid surrounds the disperse constituent.

(b) Reason why coppers with certain oxygen contents cannot work in reductive atmospheres

When using reductive atmospheres (H_2, CO, etc.), the following reactions take place:

$$H_2 + Cu_2O \rightarrow Cu + H_2O\ (gas)$$

$$CO + Cu_2O \rightarrow 2Cu + CO_2\ (gas)$$

These gases dissolve in the metal, generating high internal pressures and intergranular cracks (embrittlement caused by hydrogen).

4.4 BINARY PERITECTIC REACTION

The eutectic binary system phase diagram of Fig. 4.1, is the result of the interaction of two simple binary diagrams just as shown in Fig. 4.14. At the eutectic temperature there are zero degrees of freedom.

By analogous considerations, in certain cases phase diagrams can be analyzed as the interaction of two total solubility diagrams (Fig. 4.15): peritectic reaction.

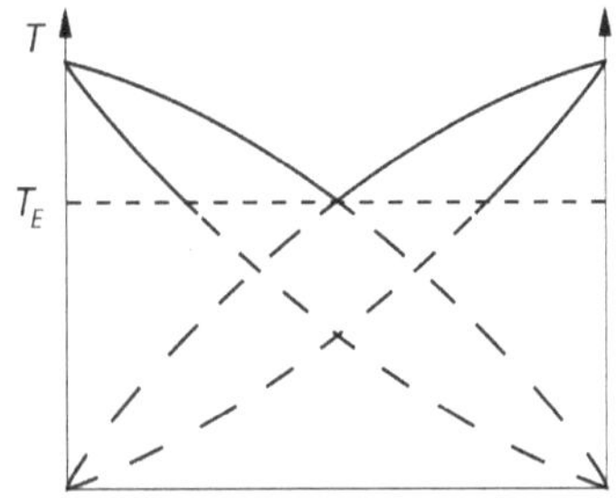

FIG. 4.14 Eutectic as a result of the intersection of two total solubility diagrams.

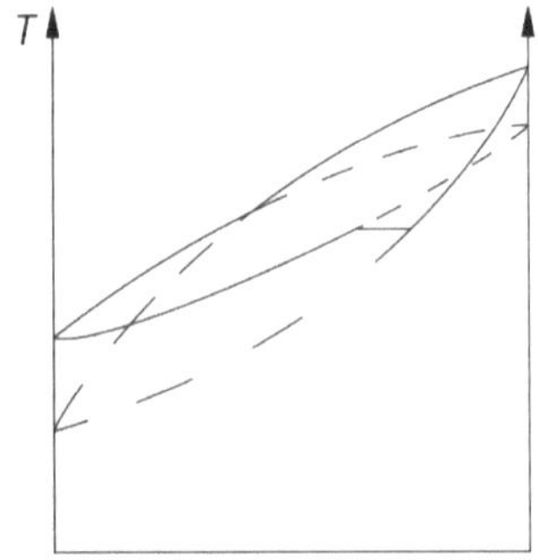

FIG. 4.15 Peritectic as a result of the intersection of two total solubility diagrams.

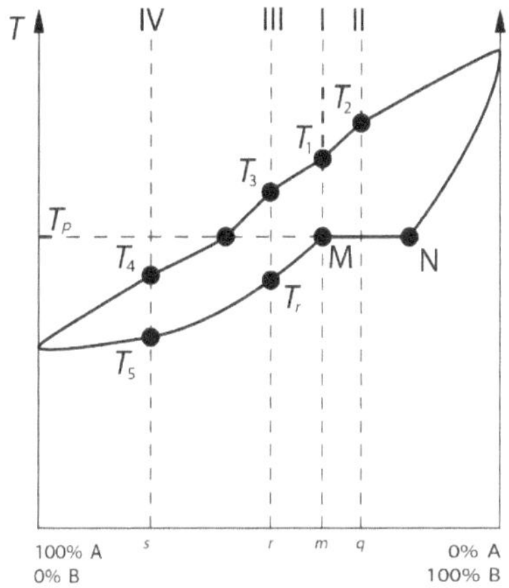

FIG. 4.16 Peritectic diagram in a binary system.

If an alloy with $m\%$ composition in Fig. 4.16 is considered, and whose cooling curve is as indicated in Fig. 4.17, by applying Gibbs' law, T_P temperature will result being constant during the following reaction:

$$\text{Liquid} + \text{Solid } \beta_n \rightleftharpoons \text{Solid } \alpha_m \tag{4.7}$$

When, the reaction of a liquid and a solid at constant temperature results into another solid, the reaction is known as *peritectic* (from the

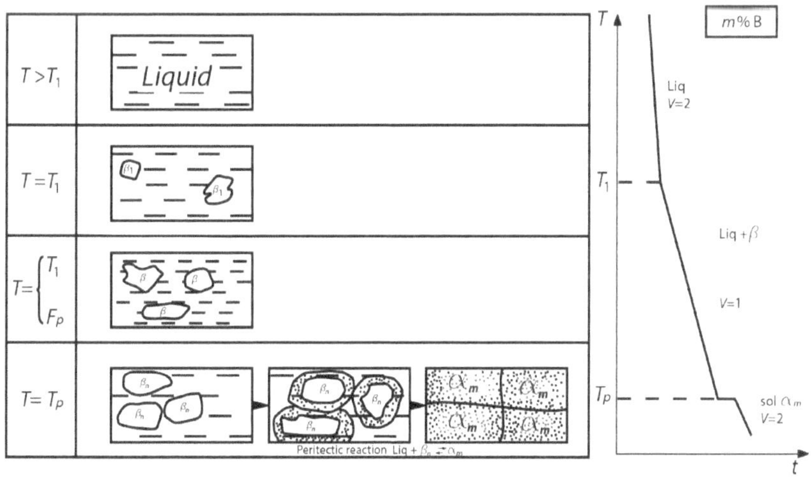

FIG. 4.17 Cooling curve and microstructures of the $m\%$ B alloy at different temperatures.

Greek *peripheral*), assuming that, when freezing the alloy with $m\%$ composition, sufficient time is allowed for all the β_n solid to react with all the liquid. However, in practice, the crystals of β_n solid solution have a defined thickness and their composition can be changed only at the solid/liquid interphase, where the peritectic reaction actually happens. Thus, the periphery of a crystal changes to the $m\%$ composition while the interior remains with $n\%$. This difference in composition between interior and periphery of a crystal can only be leveled out by diffusion, which is time-dependent. Consequently, the alloys in a peritectic system have, after solidification, the first solid solution formed (in this case β_n) and the peritectic solid (α_m).

The micrograph of Fig. 4.18 corresponds to the peritectic reaction between melted Zn and the ε constituent of brasses: at 420°C, liquid $(98.3\%\ \text{Zn}) + \varepsilon(87.5\%\ \text{Zn}) \rightleftharpoons \eta(97.3\%\ \text{Zn})$. By the reaction between liquid and solid, the η solid solution is formed (dark color in the micrograph). In the interior of the η grains the remains of the ε solid (lighter color) can still be seen.

All peritectic alloys correspond to a metastable equilibrium. The formation of the new solid (η) can only occur limitedly: it must be done by diffusion through the first fraction of the solid (η) formed in the contact interface of liquid and ε. Besides, this metastable equilibrium cannot be destroyed by homogenization heat treatments, since it would require very large times.

In the same manner as the one indicated for eutectic systems, the equilibrium microstructure resulting at the end of the solidification can be inferred from the phase diagram of peritectic systems.

As an example, Figs. 4.19–4.21 show the solidification patterns and the cooling curves of the $q\%$ B, $r\%$ B, and $s\%$ B alloys of Fig. 4.16. The

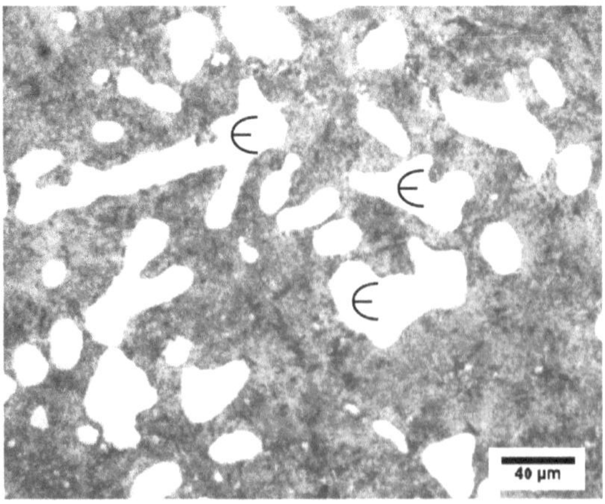

FIG. 4.18 Microstructure of peritectic reaction in the Cu-Zn system.

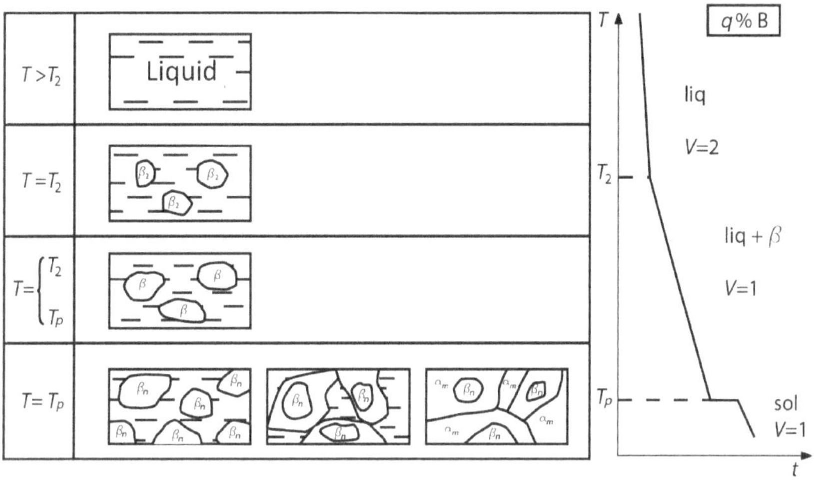

FIG. 4.19 Cooling curve and microstructures of the $q\%$ B alloy at different temperatures.

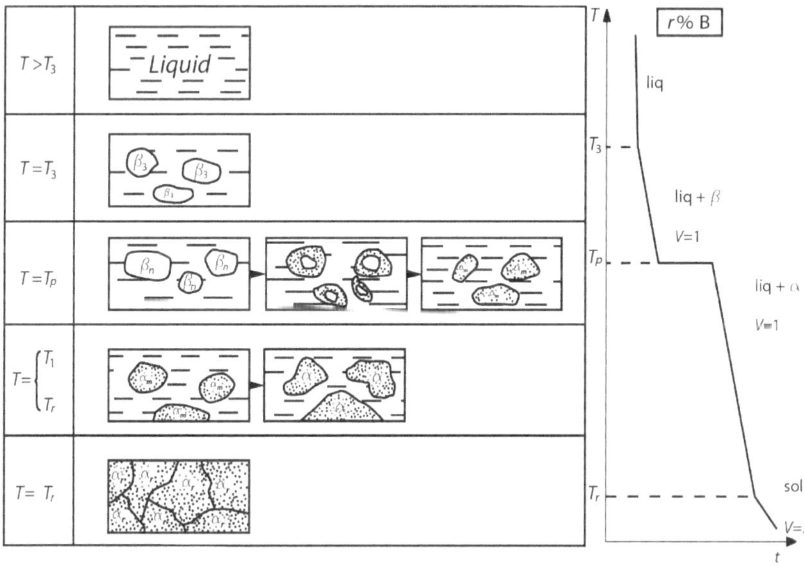

FIG. 4.20 Cooling curve and microstructures of the $r\%$ B alloy (off-peritectic) at different temperatures (solidification does not end at the peritectic temperature).

proportions with which the constituents are presented when the alloy solidifies can be determined, at each case, by applying the lever rule; likewise, the mean compositions of the phases in equilibrium can also be obtained. Peritectic reactions are commonly presented in industrial alloys, such as brasses (Cu-Zn), bronzes (Cu-Sn), steels (Fe-C), etc., and Table 4.2

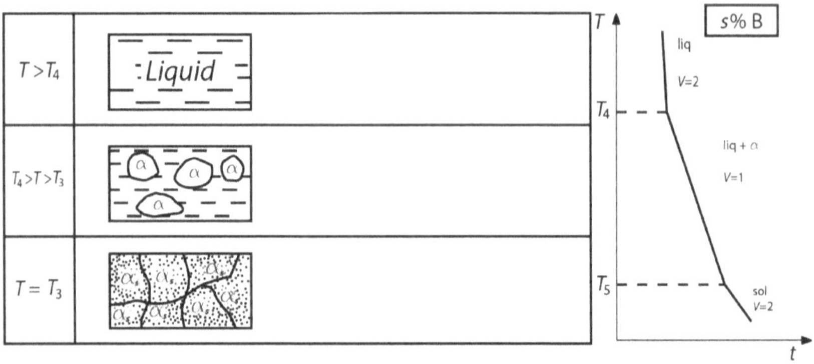

FIG. 4.21 Cooling curve and microstructures of the s% B (off-peritectic) alloy at different temperatures.

TABLE 4.2 Some Binary Peritectic Reactions

Liquid		Solid 1	Solid 2	Temperature (°C)
10.3% Cu	89.7% Fe	SS in Fe δ (6.7% Cu)	SS in Fe γ (8.3% Cu)	1480
97.2% Cu	2.8% Fe	SS in Fe γ (9.5% Cu)	SS in Cu (4.0% Fe)	1093
92.2% Cu	~7.8% Si	SS in Cu (5.3% Si)	γ (~6.7% Cu)	853
74.5% Cu	25.5% Sn	SS in Cu (13.5% Sn)	β (22.0% Sn)	799
81.5% Cu	18.5% Ti	SS in Cu (4.7% Ti)	$\beta(Cu_7Ti_2)$	896
62.5% Cu	37.5% Zn	SS in Cu (32.5% Zn)	β (36.8% Zn)	903
99.47% Fe	0.53% C	SS in Fe δ (0.09% C)	SS in Fe γ (0.17% C)	1495
97.2% Fe	12.8% Mn	SS in Fe δ (9.5% Mn)	SS in Fe γ (10.0% Mn)	1474
94.6% Fe	5.4% Ni	SS in Fe δ (4.0% Ni)	SS in Fe γ (4.3% Ni)	1517
85.5% Fe	~14.5% Si	SS in Fe δ (13.8% Si)	α_2 (14.1% Si)	1275
70.7% Fe	29.3% W	ϵFe_7W_6	SS in Fe δ (35.5% W)	1554

TABLE 4.2 Some Binary Peritectic Reactions—cont'd

Liquid		Solid 1	Solid 2	Temperature (°C)
7.4% Fe	92.6% Zn	SS in Fe δ (46.0% Zn)	$\delta(Fe_3Zn_{10})$	783
3.0% Fe	97.0% Zn	δFe_3Zn_{10}	$\epsilon(FeZn_{10})$	672
0.25% Fe	99.75% Zn	$\epsilon_1 FeZn_7$	$\xi(FeZn_{13})$	530
2.3% Fe	~97.7% Sn	SS in Fe δ (~18.0% Sn)	$Fe_{1,3}Sn$	901
1.0% Fe	~99.0% Sn	Fe_3Sn_2	FeSn	740
0.1% Fe	~99.9% Sn	FeSn	$FeSn_2$	496
39.0% Pb	61.0% Hg	SS in Pb (~27.0% Hg)	β (33.7% Hg)	145
99.59% Al	0.41% Cr	βAlCr	SS in Al (0.77% Cr)	661.4
99.403% Mg	0.597% Zr	βMg_2Zr	SS in Mg (3.6% Zr)	654
50.0% Sn	50.0% Sb	SS in Sb (9.0% Sn)	β (64.0% Sb)	425
3.0% Zn	~97.0% Hg	SS in Zn (~6.0% Hg)	γ (48.0% Hg)	42.9

SS, solid solution.

indicates some of them. The Fe-C diagram also contains a peritectic, however, due to the high diffusivity of carbon at these high temperatures, the peritectic reaction is very fast and is able to transform all the primary (δ) dendrites into austenite, which is the stable phase. This can be considered as an exception of peritectic reactions.

EXERCISE 4.6

Determine, in the Pt-Ag system (Fig. 4.22): (a) the beginning and end of solidification temperatures of the Pt-30% Ag alloy, (b) the weight proportions of the phases present at temperatures slightly higher and lower than the peritectic one, and (c) indicate the matrix and disperse constituents of the solidified alloy at those temperatures.

Solution

(a) Temperatures of beginning and end of transformation

The temperature when solidification begins, T_L, can be read from the liquidus line of the diagram: 1580°C, when the formation of α phase rich in Pt starts, while the end of solidification is at 1186°C, as shown in the diagram.

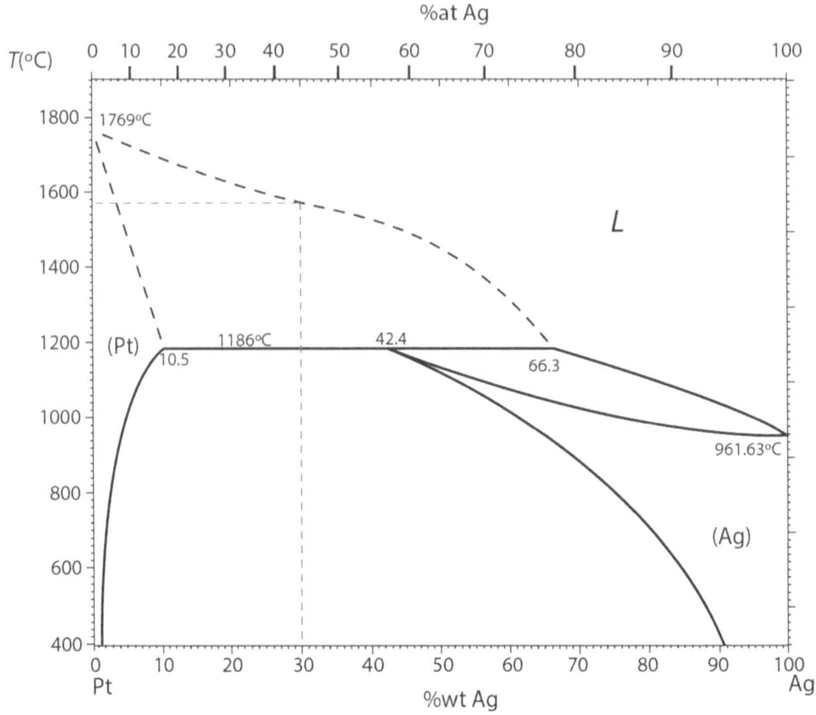

FIG. 4.22 Pt-Ag phase diagram (ASM Handbook, Vol. 3, 1992).

(b) Weight proportions above and below the peritectic temperature

At a temperature slightly above the peritectic one (1186°C), there are two phases present: the already mentioned α phase ($C_\alpha = 10.5\%$ Ag) surrounded by liquid ($C_L = 66.3\%$ Ag), and applying the lever rule:

$$\%L = \frac{30 - 10.5}{66.3 - 10.5} = 0.35\ (35\%)$$

$$\%\alpha_{\text{properitectic}} = 100 - 35 = 65\%$$

Once the peritectic reaction:

$$L_{(66.3\%\ \text{Ag})} + \alpha_{(10.5\%\ \text{Ag})} \leftrightarrow \beta_{(42.4\%\ \text{Ag})}$$

is completed at 1186°C (final solidification temperature for this alloy). At a temperature slightly below, there are two solid phases in equilibrium: a disperse α phase ($C_\alpha = 10.5\%$ Ag) surrounded by a matrix of β ($C_\beta = 42.4\%$ Ag), with the following proportions after applying the lever rule:

$$\%\alpha_{\text{total}} = \frac{42.4 - 30}{42.4 - 10.5} = 0.39\ (39\%)$$

$$\%\beta_{\text{total}} = 100 - 39 = 61\%$$

(c) Matrix and disperse constituents

As mentioned, the disperse constituent would be α crystals (39%) surrounded by a matrix of β (61%).

4.5 BINARY MONOTECTIC REACTION

There are cases, though not very common, where two metals do not form a single phase in liquid state. Between certain composition limits there are two liquid layers with compositions that depend on temperature; and its mix is only complete above a certain critical temperature (T_3 in Fig. 4.23).

Fig. 4.23 presents the Pb-Cu diagram (monotectic type). Applying the rule of phases to an alloy of the system with $m\%$ B, when reaching T_M, an invariant reaction ($V = 0$) takes place (Fig. 4.24):

$$\text{Liquid } L_{II} \leftrightarrow \text{Liquid } L_I + \text{B} \tag{4.8}$$

which is known as a *monotectic reaction*.

Fig. 4.25 shows the solidification scheme for an alloy with $r\%$ B and Fig. 4.26 for alloy with $s\%$ B. The solidification microstructures can be deduced, as well as the proportion of constituents by applying both phase and lever rules.

The existence of a monotectic reaction is particularly interesting in the case of Cu, because it allows for the improvement of its machinability: when using pure Cu for turning or milling operations, it creates very large shavings which make the process very difficult. However, Cu with 5% of Pb presents a large

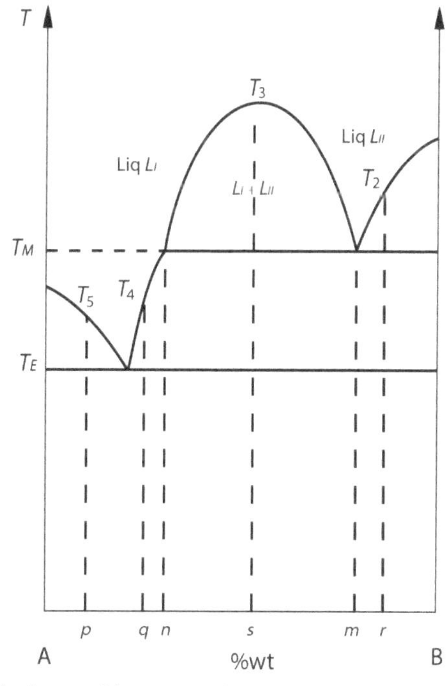

FIG. 4.23 Monotectic diagram (binary system).

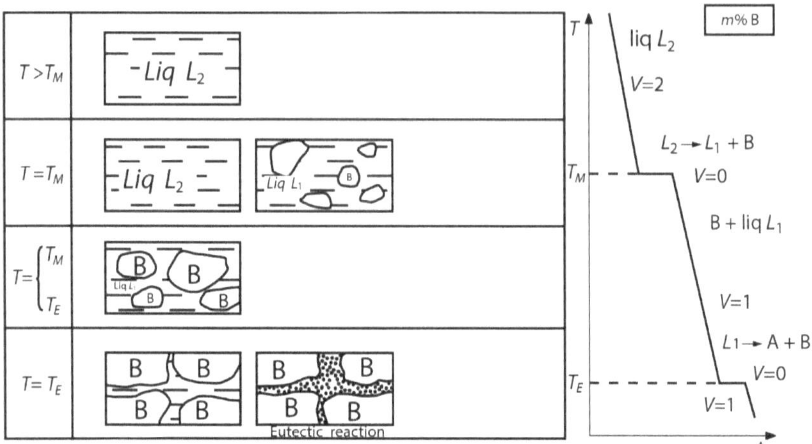

FIG. 4.24 Cooling curve and microstructures of the ***m*** % B alloy at different temperatures.

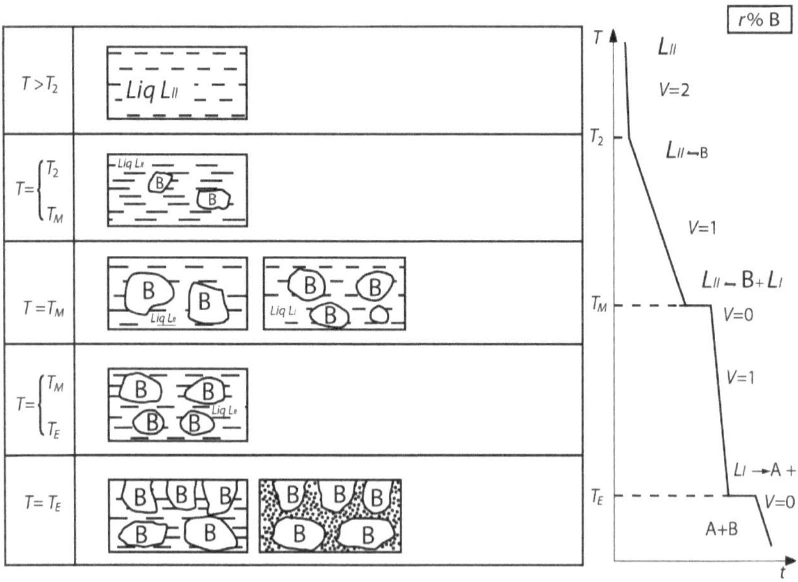

FIG. 4.25 Cooling curve and microstructures of the ***r*** % B alloy at different temperatures.

dispersion of Pb particles in a Cu matrix; when machining this alloy, the shavings will be short since the material breaks as a result of the discontinuity produced by the Pb particles (low melting point, 327°C). The presence of the monotectic reaction in the copper-lead phase diagram is applied in the production of free-cutting coppers, brasses, and bronzes.

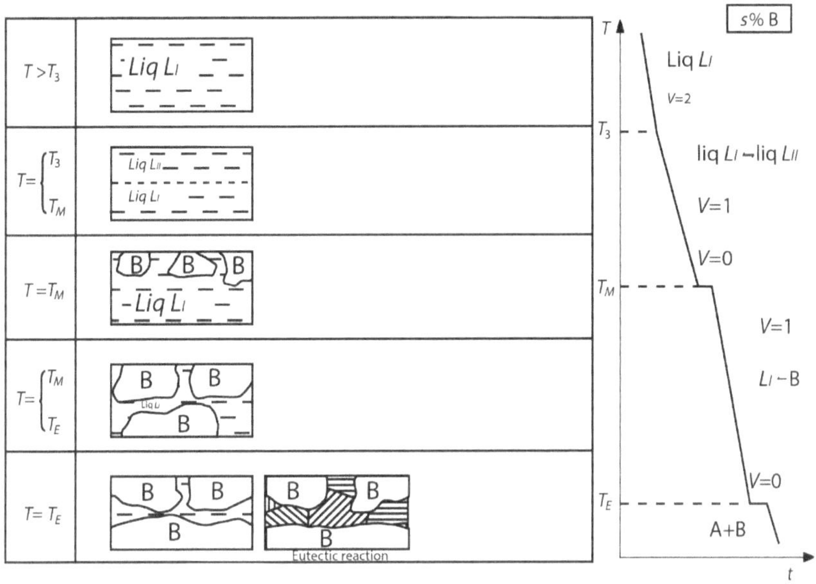

FIG. 4.26 Cooling curve and microstructures of the $s\%$ B alloy at different temperatures.

On the other hand, Pb is sometimes added to Cu and bronzes for another purpose: improving its tribological behavior in a wide range of loads and speeds. In parts such as bearings, Pb is added to prevent risks from faulty lubrication, since direct contact of the shaft and bearing causes friction and a large amount of heat that melts Pb (which does not form an alloy with the steel) and protects the steel shaft and bearing.

Ordinarily, Pb contents between 6% and 30% solve most of the friction problems. In any case, copper-leads, or bronzes with Pb, never have more than 30% Pb: compositions at N and M points of the diagram of Fig. 4.23 are, according to Table 4.3, 36% Pb and 87% Pb, respectively; consequently, an alloy with Pb contents higher than 36% (e.g., the $s\%$ B alloy of Fig. 4.23) would only be uniform in liquid state until reaching temperature T_3, and below this temperature, it would be formed by two immiscible liquids (when reaching the monotectic temperature of 955°C the liquids will have the respective compositions of 36% Pb and 87% Pb) that are separated by gravity and prevent the formation of a copper-lead with homogeneous composition.

Besides, bronzes with high amounts of Pb are highly resistant to corrosion in installations that produce or use sulfuric acid.

The binary Fe-Pb system is very similar to the Cu-Pb one, where the existence of a monotectic reaction is used to obtain steels with good machinability.

TABLE 4.3 Some Binary Monotectic Reactions

Liquid 1		Liquid 2		Solid	Temperature (°C)
89.7% Cu	10.3% O[a]	2.1% O[a]	97.9% Cu	Cu_2O	1218
64.0% Cu	36.0% Pb	87.0% Pb	3.0% Cu	Cu	955
91.7% Cu	8.3% Te	46.3% Te	53.7% Te	Te	1052
>99.95% Fe	<0.05% Pb	99.75% Pb	0.25% Fe	Fe	1532
49.9% Fe	50.1% Sn	82.5% Sn	7.5% Fe	SS Fe δ (16.1% Sn)	1130
93.3% Al	6.7% Cd	~99.0% Cd		SS in Al (0.47% Cd)	649
83.0% Al	17.0% In	96.8% In	SS in Cu	SSin Al (0.17% In)	639
96.6% Al	3.4% Bi	98.1% Bi	1.9% Al	Al	657

SS, solid solution.
[a]At 1 atm.

EXERCISE 4.7

Pb and Zn are practically insoluble between them, in solid state. Their equilibrium diagram (Fig. 4.27) has a monotectic reaction at 418°C:

$$L\,(0.9\%\,\text{Pb}) \leftrightarrow L\,(98\%\,\text{Pb}) + \text{Zn}$$

and an eutectic reaction at 318°C:

$$L\,(99.5\%\,\text{Pb}) \leftrightarrow \text{Pb} + \text{Zn}$$

The melting temperatures of Pb and Zn are 327°C and 420°C, respectively. Explain the possibilities of metal refinement for this system.

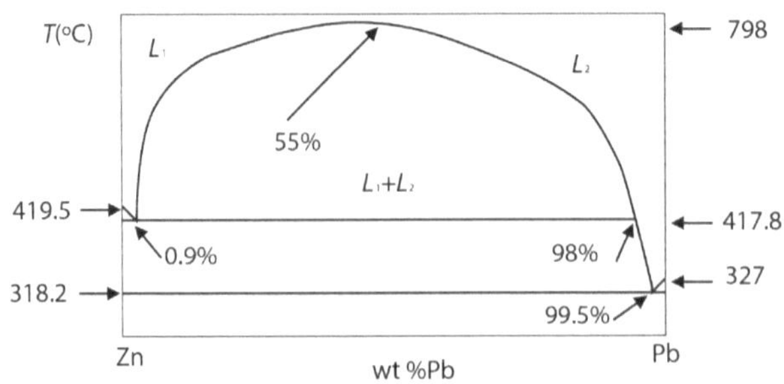

FIG. 4.27 Zn-Pb phase diagram (ASM Handbook, Vol. 3, 1992).

Solution

The points indicated in Fig. 4.27 with arrows on the left side of the diagram and 0.9% for the monotectic reaction, and arrows on the right lower side of the diagram and 99.5% for the eutectic one, are to be considered. If the Zn-10% Pb alloy is selected, the amounts of liquid ($C_L = 98\%$ Pb) and Zn at the monotectic temperature (418°C) would be:

$$f_L \approx \frac{10}{98} \cong 0.10\,(10\%)$$

$$f_{Zn} = 100 - 10 = 90\%$$

and at the 318°C temperature, the liquid rich in Pb ($C_L = 99.5\%$ Pb) forms eutectic crystals of Zn and Pb. Thus, in order to refine the liquid rich in Zn and separate it from Pb, the monotectic reaction can be used; the same would also occur to refine the liquid rich in Pb from the solid Zn (separation by gravity).

Comment: The base of the Parkers process to separate Pb and Ag from silver-rich galena (PbS) is: liquid Pb (rich in Ag) and liquid Zn are placed in contact with each other, Pb becomes saturated in Zn and Ag moves towards Zn (both liquids are insoluble between them); when cooled, the Zn-Ag alloy solidifies (monotectic reaction) and floats in the Pb-Zn alloy, which can later be separated.

4.6 BINARY SINTECTIC REACTION

The sintectic reaction is an invariant one ($V = 0$) involving the decomposition, by heating, of a solid phase into two immiscible liquids:

$$L_I + L_{II} \rightleftharpoons \beta \tag{4.9}$$

as shown in Fig. 4.28. It is important to point out that this reaction is not a very common one. From the equilibrium diagram it is concluded that all melted alloys, whose composition is found in the range of the sintectic line, are

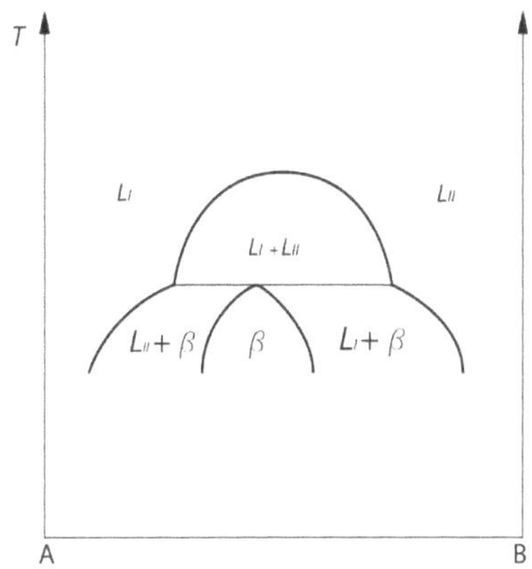

FIG. 4.28 Sintectic reaction (binary system).

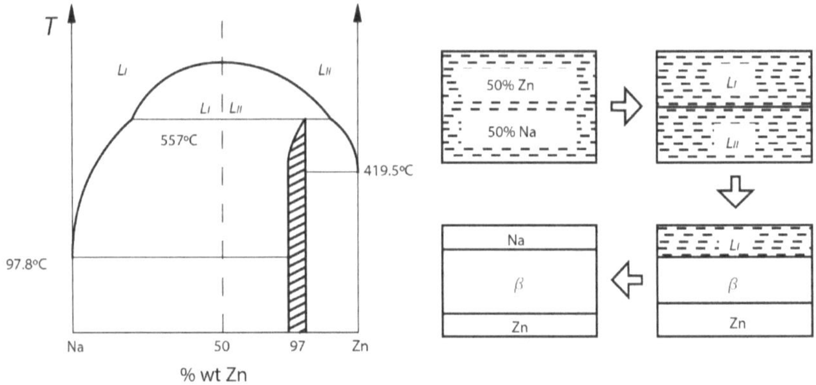

FIG. 4.29 Na-Zn phase diagram and microstructures of the 50% Zn-50% Na alloy.

separated into two liquids. Analogously, when solidifying, the sintectic reaction will take place at the interface of the L_I and L_{II} liquids.

The best known example of the relatively rare sintectic reaction is the Na-Zn system. Above the sintectic temperature, a melt formed by equal parts of Na and Zn exists in the form of two immiscible liquids (L_I and L_{II}). When cooling this melt, it would begin to solidify with the formation of the β constituent in the interphase between both liquids. Afterwards, Zn would solidify and, lastly, Na would solidify from the L_I liquid. The result would have a morphology such as the one of Fig. 4.29.

4.7 OTHER INVARIANT REACTIONS

Invariant reactions between three phases of a binary system can be separated into two groups:

- The first is the transformation of a phase into another two, at constant temperature: Phase $I \leftrightarrow$ Phase II + Phase III. This type of reaction includes eutectics and monotectics.

  The Solid $I \leftrightarrow$ Liquid II + Solid III reaction is not frequent though it occurs in the Cu-Sn system for compositions between 38% and 43% Sn (range not used in industry): a solid previously formed at higher temperatures is liquefied when cooled to 640° C to form another solid (ε) and liquid. An analogous reaction takes place in the Fe-S system, for weight percentages of S between 0.045% and 0.15% (Fig. 4.30).
- The other type corresponds to the reaction of two phases to create a third one: Phase I + Phase $II \leftrightarrow$ Phase III, for example, peritectic and sintectic reactions.

Besides the eutectic, peritectic, monotectic, and sintectic reactions, there are other reactions that can be classified in the types just mentioned (e.g., invariant equilibriums where one of the phases is gas).

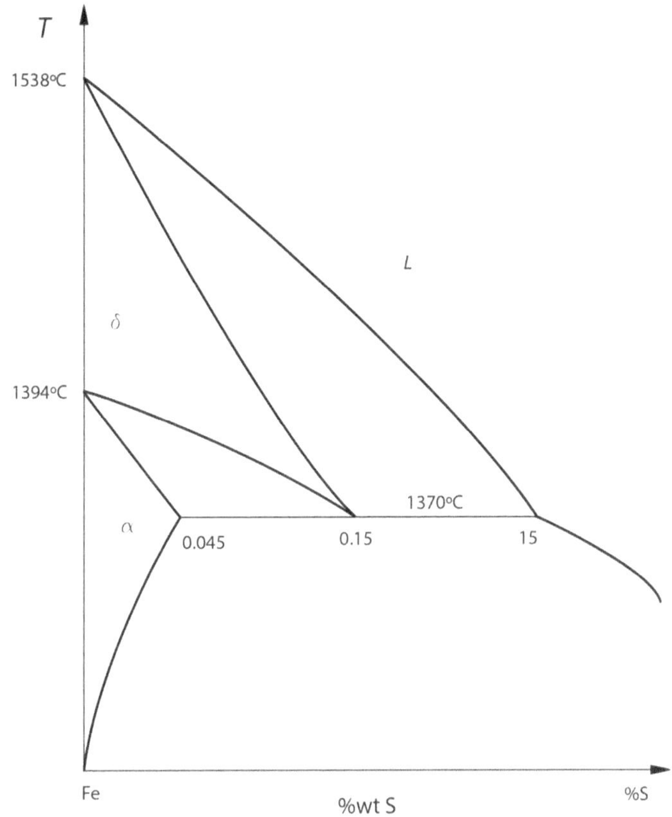

FIG. 4.30 Fe-S system (detail of the left upper part of the diagram) showing invariant reaction.

Among the reactions starting from a gaseous phase, the following are presented in metallic systems only for low pressures or high temperatures:

$$\text{Gas} \leftrightarrow \text{Liquid}\, I + \text{Liquid}\, II$$
$$\text{Gas} \leftrightarrow \text{Liquid}\, I + \text{Solid}\, II$$
$$\text{Gas} \leftrightarrow \text{Solid}\, I + \text{Solid}\, II$$
$$\text{Liquid} \leftrightarrow \text{Gas} + \text{Liquid}\, II$$
$$\text{Liquid}\, I \leftrightarrow \text{Gas} + \text{Solid}\, II$$
$$\text{Solid}\, I \leftrightarrow \text{Gas} + \text{Solid}\, II \quad (4.10)$$

For the reactions of two phases that produce a third (one of them being a gas), the possibilities are: the reaction of the gas + liquid $I \leftrightarrow$ solid II that appears in the Pd-H system; and the reaction of the gas + solid $I \leftrightarrow$ solid II that can occur in some gas-metal systems, though very rarely, at low

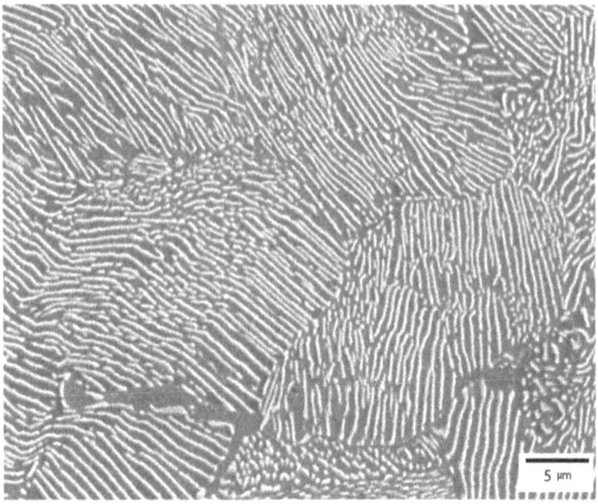

FIG. 4.31 SEM micrograph of eutectoid pearlite.

pressure. There is no knowledge on invariant reactions whose resulting phase is a gas.

Adding to all the reactions mentioned in the previous sections with other types of invariant ones, the reactions with solid phases in equilibrium must be mentioned. For example, the transformation of a solid phase into a complex constituent formed by two other solid phases (solid $I \leftrightarrow$ solid II + solid III). This solid state reaction, at constant temperature, is known as *eutectoid*, for its analogy to the eutectic.

In Fig. 4.31, the morphology of the eutectoid known as *pearlite* can be seen, which is formed by ferrite and Fe_3C, obtained by isothermal transformation at 727°C of the austenite (0.77% C): $\gamma(0.77\%C) \leftrightarrow \alpha(0.02\%\,C) + Fe_3C(6.67\%\,C)$.

In solid state, the reactions of the solid I + solid $II \leftrightarrow$ solid III type can also occur, which are called *peritectoids*.

4.8 COMPUTATIONAL CALCULATION METHODS (CALPHAD)

Metallurgical processes very rarely reach equilibrium from a thermodynamic point of view, yet most processes and operations seek technological solutions to achieve homogeneous mixtures of solid or liquid phases in as shortest periods as possible.

From this perspective, there are three types of processes:

(a) Those that cannot be performed, as thermodynamics prohibit them.
(b) Those systems that are very close to the equilibrium state.
(c) Those that are spontaneous due to a considerable negative variation of free energy.

From a kinetic point of view, the higher the driving force (difference between initial values and those of the thermodynamic equilibrium), the higher the transformation rate will be. As the system approaches equilibrium, transformation rate decreases.

On the other hand, with a system close to equilibrium, nucleation and growth mechanisms are controlled predominantly by superficial free energy phenomena.

Thus, the speed of metallurgical operations and the structures formed are a function of the free energy of the system and the time allowed for their progression.

Software to calculate both phase diagrams and thermodynamic properties have been used for many years to simulate kinetic evolution and estimate properties of materials and processes. The most common example is CALPHAD (calculation of phase diagrams) with the Thermo-Calc system as one of the most widely used software, consisting of a database used to estimate phase equilibrium, phase diagrams, and solid-state transformations.

4.8.1 Fe-C System

The phase diagram calculation for a multicomponent system is based on the free energy of each phase at the pressure, temperature, and composition intervals. For a specific temperature and pressure, equilibrium conditions for each component i is defined by chemical potential μ (also known as partial molar free energy), when reaching the minimum value of free energy ($G=0$), at constant temperature and pressure. Thermo-Calc software considers the specific heat and the activity coefficient (γ) for each component of each phase in equilibrium. In the case of ideal binary solid solutions, the activity is the same as that for the molar fraction (X_i):

$$G = RT\sum X_i(\ln X_i) + \sum G_i X_i \tag{4.11}$$

The Fe-C system is a nonideal regular solid solution, and deviations from the free energy must be defined through activity coefficients ($\triangle G^{ex}$):

$$\triangle G^{ex} = \Omega X_A X_B \tag{4.12}$$

where Ω is a function of heat of dissolution or enthalpy of the mixture, different from zero; X_A and X_B the molar fractions of A and B constituents respectively. Considering the following for the binary system:

$$\mu_A = G_A + RT \ln a_A \tag{4.13}$$

$$\ln\left(\frac{a_A}{X_A}\right) = \ln\gamma_A = \frac{\Omega}{RT}(1-X_A)^2 \tag{4.14}$$

$$\mu_B = G_B + RT \ln a_B \tag{4.15}$$

$$\ln\left(\frac{a_B}{X_B}\right) = \ln\gamma_B = \frac{\Omega}{RT}(1-X_B)^2 \tag{4.16}$$

where a is the activity of the constituent in the mixture. Positive heat of dissolution values indicate that activity coefficients γ_A and γ_B are also positive, indicating A and B will tend to exit the solid solution instead of remaining inside of it. The contrary is also true.

In the case of nonideal ternary systems (e.g., Fe-C-Mn or Fe-C-Si), the excess of free energy may be estimated using expressions such as:

$$\triangle G_{AB}^{ex} = X_A X_B \sum L_{AB}^{(i)} (X_A - X_B)^i \tag{4.17}$$

The interaction energy $L_{AB}^{(i)}$ may be estimated from data of the A-B binary system through linear regression, as $\triangle G_{AB}^{ex}$ depends on the composition of the alloy. Consequently, values for $\triangle G_{ABC}^{ex}$ are the result of the sum of $\triangle G_{AB}^{ex}$, $\triangle G_{AC}^{ex}$, and $\triangle G_{BC}^{ex}$ which can be corrected if deviations are observed and experimental data is available. The same procedure is used in the case of systems with four components such as Fe-C-Mn-Si.

In the case of pure substances (simple chemical elements or compounds), G is calculated through enthalpy (H) and entropy (S) which are obtained using specific heat (c_p):

$$H = H^{ref} + \int_0^T c_p \cdot dT \tag{4.18}$$

$$S = S^{ref} + \int_0^T C_p \cdot \frac{dT}{T} \tag{4.19}$$

$$G = H - T \cdot S \tag{4.20}$$

the super index ref equals the reference values.

Maier and Kelley (1932) proposed a polynomial function to estimate the specific heat:

$$c_p = A_a + A_b \cdot T + A_c \cdot T^2 + \frac{A_d}{T^2} \tag{4.21}$$

with A_a, A_b, A_c, and A_d being constants. Resulting in an expression to calculate free energy under standard conditions:

$$G = B_1 + B_2 \cdot T + B_3 \cdot T \cdot \ln T + B_4 \cdot T^2 + B_5 \cdot T^3 + \frac{B_6}{T} \tag{4.22}$$

with B_1, B_2, B_3, B_4, B_5, and B_6 also being constants which can be found in databases.

For the Fe-Fe_3C system, the software applies the previous equations to calculate the free energy for each phase (austenite, ferrite, and cementite) at a certain temperature. In this fashion, using common tangent lines (Fig. 4.32), compositions and coexistence intervals of the phases in equilibrium can be obtained. As an example Fig. 4.33, obtained at the eutectoid temperature, shows the common tangent for the austenite into pearlite invariant transformation.

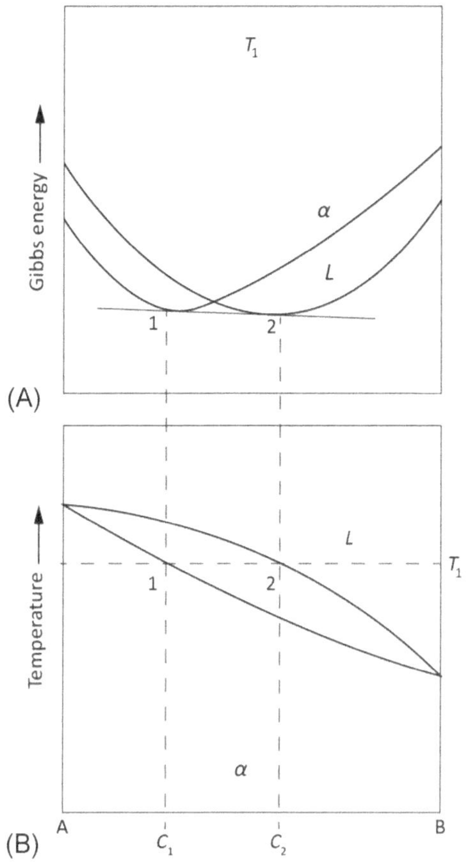

FIG. 4.32 Free energy and common tangent line sketch (A) to estimate binary phase diagrams (B) (ASM Handbook, Vol. 3, 1992).

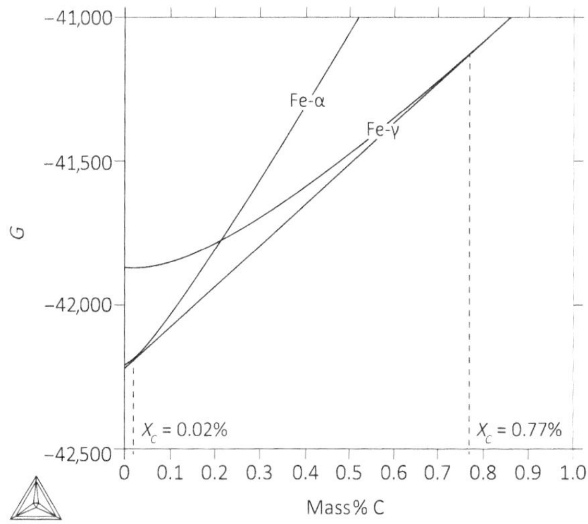

FIG. 4.33 Austenite and ferrite free energy calculations.

4.8.2 Al-Si Pressure Variation

The production of industrial parts using silumin (Al-Si) alloys is attractive due to their low solidification temperature (577°C), good castability, and almost null segregation as the eutectic solidifies at constant temperature, which also results in very low shrinkage and hot tearing. Silumins with 12% Si show a microstructure with acicular Si (polyhedra or needles) dispersed in an Al matrix.

In order to increase toughness and reduce thermal expansion, an addition of 0.1% Na or Sr results in a very fine microstructure of the Si crystals. This modification (Section 4.3.1) lowers the eutectic temperature and increases the eutectic composition (%wt Si) making the alloy slightly hypoeutectic. A further addition of 0.8%–1.1% Fe increases high temperature strength and injection moldability during processes that take advantage of pressure to reduce defects of the final part (pressure molding or pressure die casting).

The displacement of the eutectic point is not only a function of small additions of modifying elements to the original Al-Si alloy, but can also be produced by localized pressure variations: the resulting microstructure will be a combination of needle-like Si crystals and/or eutectic phases of the binary or ternary type. Differences in pressure to produce displacement in composition and temperature of the eutectic point are only relevant for orders of magnitude in Pa values, and even though processes such as Pressure Die Casting only use ~50 MPa, during feeding, the high feeding rate, solidification rate, and complex geometry of the part (in many cases thin cross-sections) result in the entrapment of gasses and very high increments of pressure in localized points. For example, clean-room floor tiles show different microstructures at locations near each other (independently if the alloy includes Na/Sr).

Furthermore, most data for phase diagrams is based on calculations or experiments at atmospheric pressure and in order to estimate relevant modification (eutectic, eutectoid, peritectic points, solubility limits, etc.) of the diagram due to this variable, CALPHAD methods are used since expressions can be added to base free energy values in the databases, which can simulate the increment in pressure, for example, for a certain phase α, the term added to the free energy value would be:

$$dG_\alpha = -S_\alpha \cdot dT + V_\alpha \cdot dP$$

and so on for each phase of the system.

Fig. 4.34A shows the solubility limit calculated at pressures as high as 2000 bar (200 MPa), for the Al-Si system, while Fig. 4.34B indicates the eutectic point at these same pressure values. This type of calculations by the computer model and other similar ones, are the basis for understanding, not only the effect of external factors on the behavior of metal alloys, but also on the development of new materials with enhanced properties or functional applications (bio-chemical, bio-medical, nanostructured, and others).

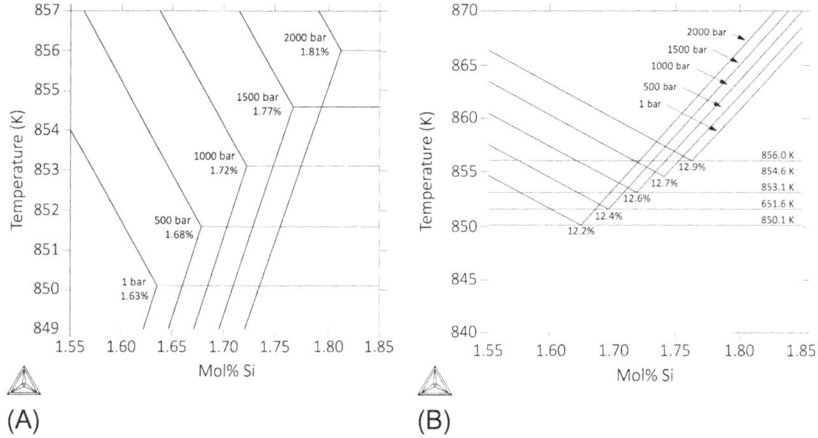

FIG. 4.34 Thermo-Calc calculation for the effect of pressure in the Al-Si system: solubility limit (A) and eutectic point (B).

REFERENCES

Alloy Phase Diagrams. 10th ed. ASM Handbook, vol. 3. ASM International, Metals Park, OH.

Maier, C., Kelley, K., 1932. An equation for the representation of high temperature heat content data. J. Am. Chem. Soc. 54, 3243–3246.

Thermo-Calc Software, n.d. Retrieved September 2016, from http://www.thermocalc.com/.

BIBLIOGRAPHY

Aaronson, H., 2001. Lectures on the Theory of Phase Transformations, second ed. Wiley, New York.

Ballester, A., Verdeja, L., Sancho, J., 2003. Metalurgia Extractiva. Fundamentos, vol. 1. Sintesis, Madrid.

Ferguson, F., Jones, T., 1966. The Phase Rule. Butterworths, London.

Gaskell, D., 1981. Introduction to Metallurgical Thermodynamics, second ed. McGraw-Hill, New York.

Hansen, M., Elliot, R., 1965. Constitution of Binary Alloys, second ed. McGraw-Hill, New York.

Hillert, M., 2008. Phase Diagrams and Phase Transformations. Their Thermodynamic Bases, second ed. Cambridge University Press, Cambridge.

Johnson, W., Smartt, H., 1972. Solidification and Casting of Metals. The Metals Society, London.

Massalski, T., 2001. Binary Alloy Phase Diagrams, second ed. American Society for Metals, Metals Park, OH.

Mesplede, J., 2000. Chimie. Thermodynamique, matériaux inorganiques. Bréal éditions, France.

Porter, D., Easterling, K., 2009. Phase Transformations in Metals and Alloys, third ed. CRC Press, Boca Raton, FL.

Rhines, F., 1956. Phase Diagrams in Metallurgy Their Development and Application. McGraw-Hill, New York.

Sancho, J., Verdeja, L., Ballester, A., 1999. Metalurgia extractiva Vol. II Procesos de obtención. Síntesis, Madrid.

TAPP Database, 1994. Thermochemical and Physical Properties. TAPP, Cincinnati, OH.

Weast, R., 2014. Handbook of Chemistry and Physics, 95th ed. The Chemical Rubber Co., Cleveland, OH.

Winegard, W., 1964. An Introduction to the Solidification of Metals. The Institute of Metals, London.

Chapter 5

Nonequilibrium Solidification and Chemical Heterogeneities

5.1 UNIDIRECTIONAL SOLIDIFICATION AND ZONE MELTING SOLIDIFICATION

The transformation from liquid phase to solid solution, just as it happens with the solidification of a pure metal, always occurs across an interface, independent of the type (equilibrium or nonequilibrium). The perfect equilibrium implies that the composition of both solid and liquid are uniform at all times, which is graphically expressed in Fig. 5.1; furthermore Fig. 5.2, shows the relation between the partition coefficient and the *liquidus* and *solidus* lines.

However, perfect equilibrium never occurs in real conditions; among other reasons, because it is practically impossible to achieve total uniformity of the solid at each temperature during cooling. Considering the solidification of a simple binary alloy with a phase diagram similar to the one in Fig. 5.2 where $K < 1$ (the partition coefficient K is defined in Section 3.2.4, as the ratio of the concentration of solute in the solid to the concentration of solute in the liquid in equilibrium with the solid), and in order to derive a mathematical expression, it is necessary to define the conditions that occur during solidification. These conditions, some of which may practically not occur, are:

- K is a constant (both *liquidus* and *solidus* are straight lines).
- Diffusion in the solid is negligible.
- Diffusion is the only mechanism involved in the mixing of the liquid.
- Equilibrium is maintained at the solid-liquid interface, i.e., the composition of the solid is K times the composition of the liquid at the interface ($C_S = KC_L$).

As an introduction to nonequilibrium solidification, two types of solidification will be analyzed: unidirectional and zone melting. In both cases, it can be stated that the homogeneity of the composition of the solid by diffusion of atoms is negligible during the entire process.

Solidification and Solid-State Transformations of Metals and Alloys. http://dx.doi.org/10.1016/B978-0-12-812607-3.00005-X

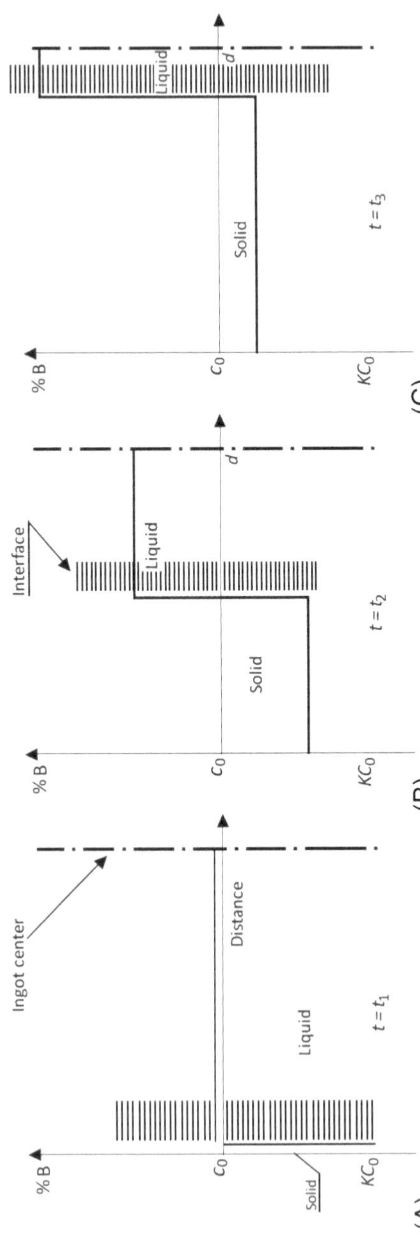

FIG. 5.1 Unidirectional solidification (perfect equilibrium). Composition variation in both solid and liquid for $t_1 < t_2 < t_3$: (A) $t = t_1$, (B) $t = t_2$, and (C) $t = t_3$.

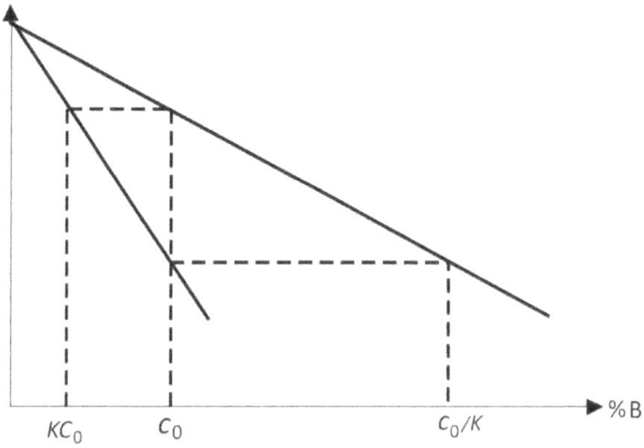

FIG. 5.2 Binary system with constant partition coefficient $K < 1$.

5.1.1 Unidirectional Solidification

For this case, the solid solution formed by two metals A and B is considered. If single-phase diagram (Fig. 5.2) is analyzed: *liquidus* and *solidus* are two straight lines (which can be asserted when the alloy has a composition close to 100% A), and supposing a bar with C_0 mean composition (wt% B) is introduced in a mobile furnace that moves slowly enough to allow, not only that the concentration of the solid in the interface is $C_S = KC_L$, but also the liquid portion of the part inside the furnace always has a uniform composition at all of its points (C_L), as described in Fig. 5.3. Given that the total mass of the solute in the part does not change during the whole process, through balance of matter in B, the following expression can be given:

$$L \cdot C_0 = \int_0^X C_S dx + (L - x) \cdot C_L = \int_0^X C_S\, dx + (L + x) \cdot \frac{C_S}{K} \qquad (5.1)$$

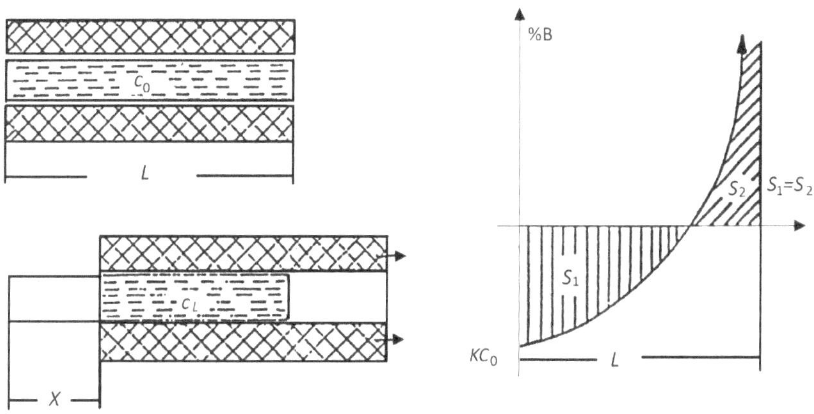

FIG. 5.3 Concentration profile as a function of distance in unidirectional solidification.

where x is the solidified length until a given instant, C_L is the uniform concentration of the remaining liquid, C_S is the solid concentration at the solid-liquid interface, and L is the length of the bar. Differentiating the previous expression, the following is obtained:

$$0 = C_S dx + (L - x)\frac{dC_S}{K} - \frac{C_S}{K}dx \quad (5.2)$$

Integrating Eq. (5.2) and considering the initial conditions as $x = 0$ and $C_S = KC_0$, the following solute distribution law along the part is obtained (that can be seen in Fig. 5.3 for a solid solution with partition coefficient $K < 1$):

$$C_S = KC_0\left(1 - \frac{x}{L}\right)^{K-1} \quad (5.3)$$

And from Eq. (5.3), the Scheil formula for the nonequilibrium solidification (uniform liquid) can be deduced, supposing that $x/L = f_S$ and $(1 - f_S) = f_L$, then:

$$f_L = \left(\frac{C_0}{C_L}\right)^{\frac{1}{1-K}} \quad (5.4)$$

EXERCISE 5.1

A 2000 mm bar of the A-1% B alloy solidifies according to the unidirectional solidification model. Calculate the concentration of solute at the following distances from the first side that solidified: $x = 100$ and $x = 1000$ mm, supposing that (a) $K = 0.5$ and (b) $K = 0.05$.

Solution

For unidirectional solidification (Eq. 5.3) the concentration of the solute will be:

(a) $K = 0.5$

$$C_S(x = 100) = (0.5)(1)\left(1 - \frac{100}{2000}\right)^{0.5-1} = 0.51\%$$

$$C_S(x = 1000) = (0.5)(1)\left(1 - \frac{1000}{2000}\right)^{0.5-1} = 0.71\%$$

(b) $K = 0.05$

$$C_S(x = 100) = (0.05)(1)\left(1 - \frac{100}{2000}\right)^{0.05-1} = 0.05\%$$

$$C_S(x = 1000) = (0.05)(1)\left(1 - \frac{1000}{2000}\right)^{0.05-1} = 0.09\%$$

The graphic representation of the concentration of solute for both cases can be seen in Fig. 5.4.

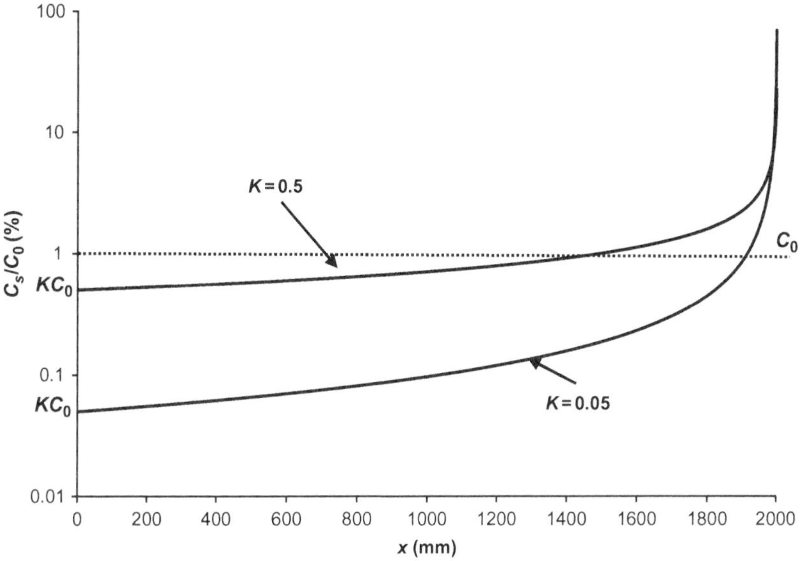

FIG. 5.4 Concentration of solute according to unidirectional solidification model.

EXERCISE 5.2

Applying the Scheil equation, calculate the mean solid concentrations under nonequilibrium freezing conditions for a Ni-50% Cu alloy at the following temperatures: (a) 1319°C, (b) 1240°C, (c) 1217.5°C, (d) 1200°C, (e) 1120°C, and (f) 1080°C.

Solution

In order to use the Scheil equation (5.4), the partition coefficient K must be obtained first, using the concentrations of liquid and solid from Fig. 3.16:

T (°C)	C_S	C_L	$K = C_S/C_L$
1319	26	50	0.52
1300	30	55	0.55
1240	44	71	0.62

which results in a mean partition coefficient $\bar{K} = 0.56$

A balance of matter for B indicates that:

$$f_L \cdot C_L + f_S \cdot \bar{C}_S = C_0 \cdot f_T$$

Then the mean solid concentration, $\bar{C}_S$, can be calculated through:

$$\bar{C}_S = \frac{C_0 - f_L \cdot C_L}{f_S}$$

This expression plus Eq. (5.4) can be used at each temperature, and also data from the phase diagram:

- For the first temperature $T_1 = 1319°C$, $C_L = 55\%$ Cu, then:

$$f_L = \left(\frac{C_0}{C_L}\right)^{\frac{1}{1-K}} = \left(\frac{50}{55}\right)^{\frac{1}{1-0.56}} = 0.81$$

$$f_S = 1 - f_L = 0.19$$

$$\bar{C}_S = \frac{C_0 - f_L \cdot C_L}{f_S} = \frac{50 - (0.81) \cdot 55}{0.19} = 29\% \text{ Cu}$$

- For the second temperature $T_2 = 1240°C$, $C_L = 71\%$Cu, then:

$$f_L = \left(\frac{50}{71}\right)^{\frac{1}{1-0.56}} = 0.45$$

$$f_S = 1 - f_L = 0.55$$

$$\bar{C}_S = \frac{50 - (0.45) \cdot 71}{0.55} = 33\% \text{ Cu}$$

- For the third temperature $T_3 = 1217.5°C$, $C_L = 77\%$Cu, then:

$$f_L = \left(\frac{50}{77}\right)^{\frac{1}{1-0.56}} = 0.38$$

$$f_S = 1 - f_L = 0.62$$

$$\bar{C}_S = \frac{50 - (0.38) \cdot 77}{0.62} = 33.5\% \text{Cu}$$

- For the fourth temperature $T_4 = 1200°C$, $C_L = 80\%$ Cu, then:

$$f_L = \left(\frac{50}{80}\right)^{\frac{1}{1-0.56}} = 0.34$$

$$f_S = 1 - f_L = 0.66$$

$$\bar{C}_S = \frac{50 - (0.34) \cdot 80}{0.66} = 34.5\% \text{Cu}$$

- For the fifth temperature $T_5 = 1120°C$, $C_L = 95\%$Cu, then:

$$f_L = \left(\frac{50}{95}\right)^{\frac{1}{1-0.56}} = 0.23$$

$$f_S = 1 - f_L = 0.77$$

$$\bar{C}_S = \frac{50 - (0.23) \cdot 95}{0.77} = 36.5\% \text{Cu}$$

- And finally, at the Cu melting temperature $T_M = 1080°C$, $C_L = 100\%$Cu, then:

$$f_L = \left(\frac{50}{100}\right)^{\frac{1}{1-0.56}} = 0.21$$

$$f_S = 1 - f_L = 0.79$$

$$\bar{C}_S = \frac{50 - (0.21) \cdot 100}{0.79} = 37\% \text{Cu}$$

All these results modify the solidification interval in the Cu-Ni phase diagram, as seen in Fig. 5.5.

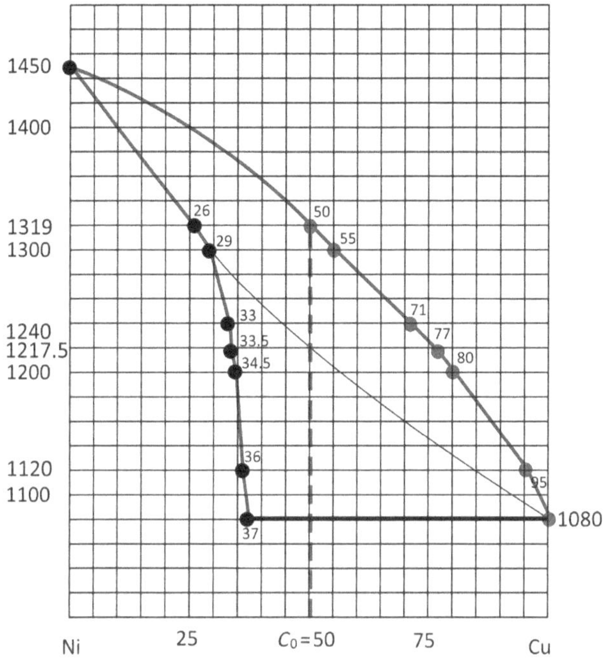

FIG. 5.5 Ni-Cu diagram.

5.1.2 Zone Melting Solidification

Considering the AB alloy of the previous section, and supposing that a bar with uniform C_0 composition is surrounded with a small section by a coil of length a, and according to the zone melting solidification model (Fig. 5.6), the displacement of the coil must be at a rate that allows the $C_S = KC_L$ equilibrium at the solid-liquid interface at all times. This rate can be higher compared to that of the unidirectional model, since only uniform composition of the liquid in the zone a is required.

Taking into account the fact that the total amount of B must be balanced between the origin of the part ($x = 0$) and the right side of liquid-solid interface, then:

$$\int_0^x C_S \cdot dx + a \cdot C_L = (x + a) \cdot C_0 \tag{5.5}$$

where a is the length of the heating coil, x the distance from the origin to the coil at a particular moment, C_L the uniform concentration of liquid inside the coil, and C_S the instantaneous concentration of solid in the left side of the solid-liquid interface. Differentiating expression (5.5), the following is obtained:

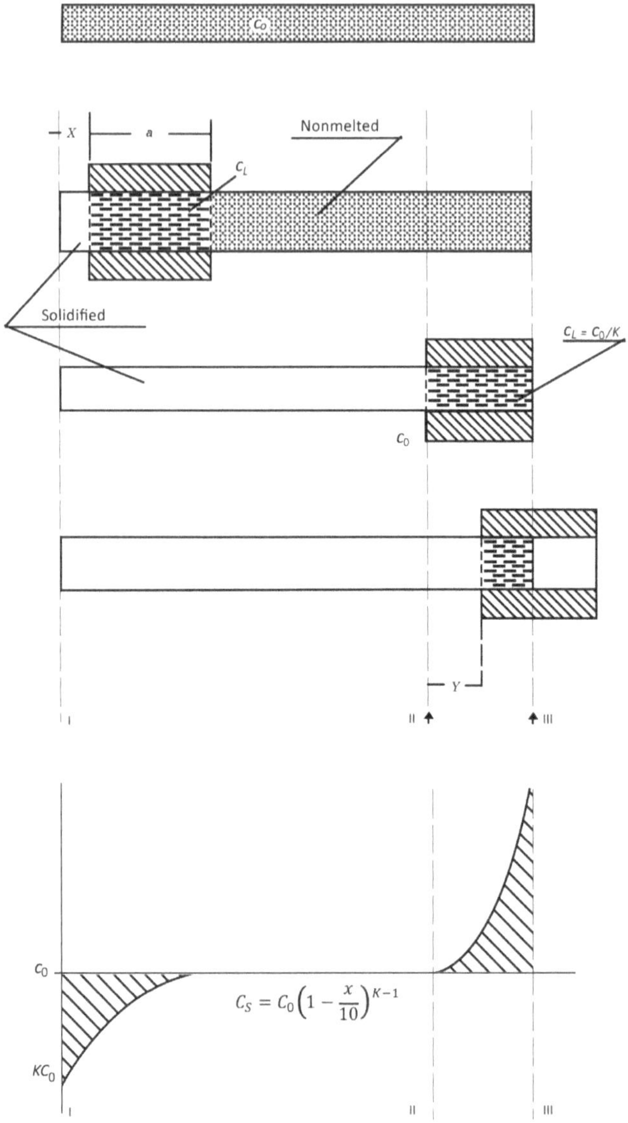

FIG. 5.6 Concentration profile as a function of distance in zone melting solidification.

$$K(C_0 - C_S)dx = a \cdot dC_S \tag{5.6}$$

Integrating Eq. (5.6) and considering that for $x = 0$, $C_S = KC_0$, the solute distribution law is:

$$C_S = C_0\left[1 + (K-1)\exp\left(\frac{-Kx}{a}\right)\right] \tag{5.7}$$

This formula is valid only from $x = 0$ until $x = L - a$, and when sketching this function ($K < 1$), it can be observed that the smaller the length of the heating coil a, the faster the "asymptote" C_0 will be reached; since the slope of the curve for $x = 0$ equals:

$$C_0 K \cdot (1 - K) \cdot \frac{1}{a} \tag{5.8}$$

For the zone between II and III, the distribution of solute B in the solid follows the progressive law deduced by substituting the value of C_0 by C_0/K in Eq. (5.3), which is the composition of the liquid at interface II. Therefore, the solute distribution from II until the end of the part will follow the expression:

$$C_S = C_0 \left(1 - \frac{x}{a}\right)^{K-1} \tag{5.9}$$

EXERCISE 5.3

A bar of the A-1% B alloy with length of 2000 mm solidifies according to zone melting solidification model. Calculate the concentration of solute at the following distances from the first side that solidified $x = 100$ and $x = 1000$mm, supposing that $a = 10$mm and that (a) $K = 0.5$ and (b) $K = 0.05$.

Solution

In zone melting solidification, as shown in Eq. (5.7), the concentration of solute will be:

(a) $K = 0.5$

$$C_S(x = 100) = 1\left[1 + (0.5 - 1)\exp\left(\frac{-0.5 \cdot 100}{10}\right)\right] = 0.999\% \cong 1\%$$

$$C_S(x = 1000) = 1\left[1 + (0.5 - 1)\exp\left(\frac{-0.5 \cdot 1000}{10}\right)\right] = 1\%$$

(b) $K = 0.05$

$$C_S(x = 100) = 1\left[1 + (0.05 - 1)\exp\left(\frac{-0.05 \cdot 100}{10}\right)\right] = 0.424\%$$

$$C_S(x = 1000) = 1\left[1 + (0.05 - 1)\exp\left(\frac{-0.05 \cdot 1000}{10}\right)\right] = 0.994\% \cong 1\%$$

The representation of the concentration of solute in both cases can be seen in Fig. 5.7 (very similar to Fig. 5.6).

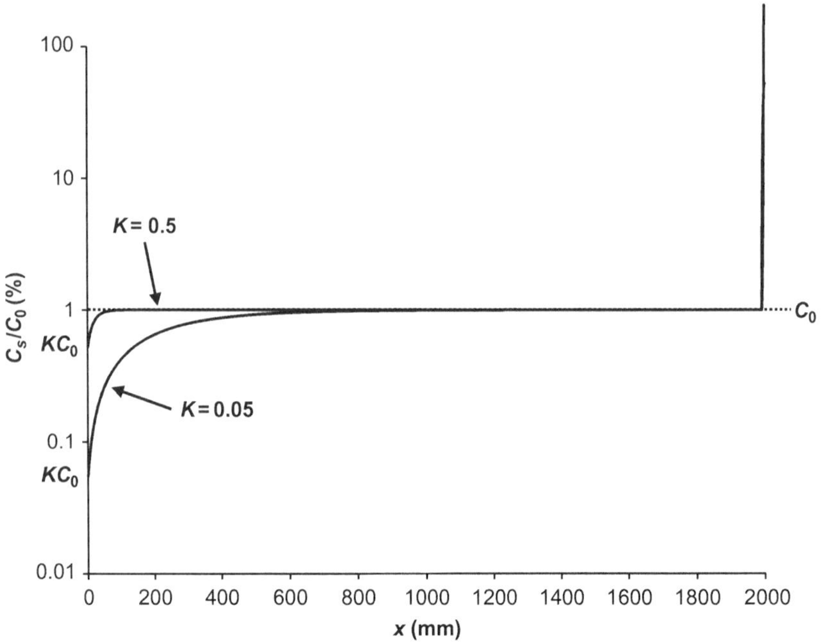

FIG. 5.7 Concentration of solute according to zone melting solidification model.

5.2 MACROSEGREGATION

Industrial freezing conditions of a solid solution are very different from the ones required for equilibrium solidification. With very slow cooling, a perfect homogeneous composition could be reached at all points of the liquid alloy, but uniformity of the solid can never be reached (time required for this would be in the order of years). Consequently, heterogeneities are always present in the chemical composition of a solidified part or ingot.

EXERCISE 5.4

Calculate the time required to eliminate macrosegregation in a cupronickel part of 1 in. thickness solidified under nonequilibrium conditions.

Solution

Selecting $T_{\text{homogeneous}} \cong (T_M)_{\text{Cu}} = 1083 + 273 = 1356\text{K}$, then:

$$D_{\text{Cu}} \cong 1 \times 10^{-8}\, cm^2/\text{s} = 1 \times 10^{-12}\, m^2/\text{s}$$

and knowing that:

$$x \cong \sqrt{Dt}$$

the necessary time will be:

$$t = \frac{x^2}{D} = \frac{(2.54 \times 10^{-2}\,\text{m})^2}{1 \times 10^{-12}\,\text{m}^2/\text{s}} = 6.45 \times 10^8\,\text{s} = 20.46\,\text{years}$$

which makes this evidently unfeasible.

For a nonequilibrium solidification of a solid solution formed by a solvent metal and a solute element with a partition coefficient $K < 1$ (analogously, this is true for the cases of alloys with $K > 1$), the first solid formed is richer in components with high melting points while the last solid formed will present a higher-than-average concentration of the metals with low melting points. This chemical heterogeneity between the fraction that solidifies first and the one that solidifies last, is known as *segregation*. This shows, for example, that the head of the steel ingots shows higher content of C, S, Sn, As, P, etc., than the mean casting composition. This chemical heterogeneity of the ingot or part, at macroscopic scale, is known as *macrosegregation*; while the *microsegregation* (or *coring*) is the one corresponding to microscopic scale (as in the interior of each crystalline grain) which will be analyzed in Section 5.3.

Macrosegregation is the result of a series of macroscopic heterogeneities of different origins which are: normal, inverse, gravity, and band segregations.

5.2.1 Normal Segregation

For a slender (high height/radius ratio) cylindrical ingot mold (with heat transfer preferentially in the radial direction), the distribution of B solute from the periphery towards the center of the ingot mold depends on the composition of the liquid during cooling.

For this type of segregation, two types of solidification can occur:

1. A slow solidification (Section 5.2.1.1) in which the liquid of the ingot mold has uniform composition at all points. Fig. 5.8 presents the variation of the B solute in the solid and in the liquid as a function of time.

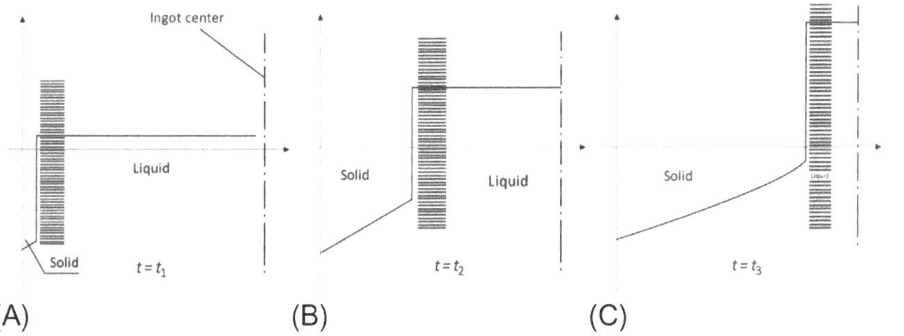

FIG. 5.8 Composition variation in both solid and liquid as a function of distance at different times ($t_1 < t_2 < t_3$): (A) $t = t_1$, (B) $t = t_2$, and (C) $t = t_3$.

2. A faster solidification (Section 5.2.1.2) by diffusion with the formation of a stationary liquid layer in the interface; industrial practices use this model (fast cooling) more frequently than the first one (slow cooling).

5.2.1.1 First Assumption

If the liquid has a perfect and uniform chemical composition at all points, solidification occurs in a similar manner to "unidirectional solidification" and there is a solid-liquid interface where the expression $C_S = KC_L$ is valid.

For an alloy with a composition C_0 of the element B, and $K < 1$ being the partition coefficient, the solute distribution along the solidified mass varies in a progressive way; and by analogy to the case considered in Section 5.1.1, the following can be asserted:

$$C_S = KC_0\left(1 - \frac{x}{L}\right)^{K-1} \tag{5.10}$$

x being the distance to the wall of the ingot mold and L its radius.

Fig. 5.9 shows the composition profile (% B) from the surface towards the center of the ingot and its *macrosegregation*, which includes a *negative*

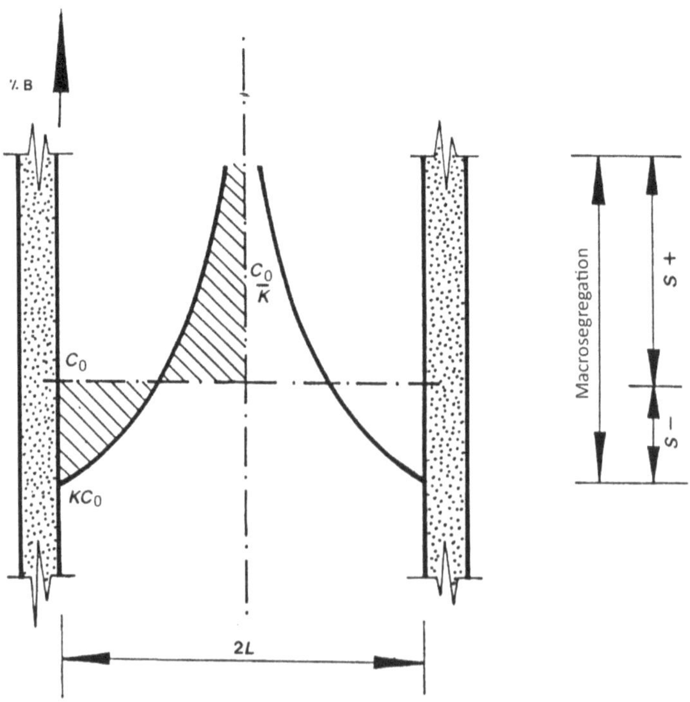

FIG. 5.9 Concentration profile (% B) in the ingot following unidirectional solidification model.

segregation (for compositions lower than C_0) and a *positive segregation*. Due to the constant amount of B in the alloy, the shaded areas are equal.

5.2.1.2 Second Assumption

If conditions are such that solidification is faster than the previous assumption (not allowing perfect homogeneity of the liquid at all times) and, by diffusion, there is a stationary zone (diffusion layer) that can be formed in the liquid after a certain period of time, a different analysis is made.

Initially, the liquid has a composition C_0; the first solid formed has a composition KC_0, and the liquid at the solid-liquid interface will be richer in B than the mean composition of the alloy. As indicated, the solidification is always made either through an interface or interphase; and, given that the concentration of B in the interface is higher than both, the concentration of B in the solid and in the liquid, and an instant later, atoms of B (in the liquid) on the left side of the interphase move towards the solid, enriching it with B. The closer to the composition of the liquid at the interface to C_0/K, the closer to C_0 the composition of the solid (that subsequently freezes) will be.

Due to liquid phase diffusion, after a certain time, a stationary state is reached where the solid in contact with the left limit of the layer has a concentration C_0 and the liquid in contact with the right side of the layer also has a concentration C_0; the solute profile at the interface is shown in Fig. 5.10.

Once that stationary state has been reached, solidification advances by displacement towards the right side of that zone, and the solid formed always has a composition C_0.

The distribution law for solute B in the solid (by considerations deduced from a balance of matter and diffusion theory) is:

$$C_S = C_0\left[1 + (K-1)\exp\left(\frac{-KR}{D}\right)x\right] \tag{5.11}$$

where R (in m/s) is the solidification rate, K the partition coefficient, and D (in m^2/s) is the diffusion coefficient of B atoms in the liquid.

The analogy between this expression and the one deduced for zone melting solidification (Section 5.1.2) is interesting: both are equal if the width D/R of the diffusion layer is considered equal to the length of the coil (a).

Fig. 5.11 shows, for this assumption, the solute concentration in the solid profile from the surface towards the center of the ingot mold, as well as its macrosegregation. The curve will approach the asymptote C_0 at a smaller distance from the wall of the ingot mold when the solidification rate R is faster, since the slope of the curve is $C_0 \cdot (1-K) \cdot K \cdot (R/D)$.

For a theoretical instantaneous solidification ($R=\infty$), the solidified portion next to the wall of the ingot mold would have the same composition as that of the liquid alloy poured into the mold.

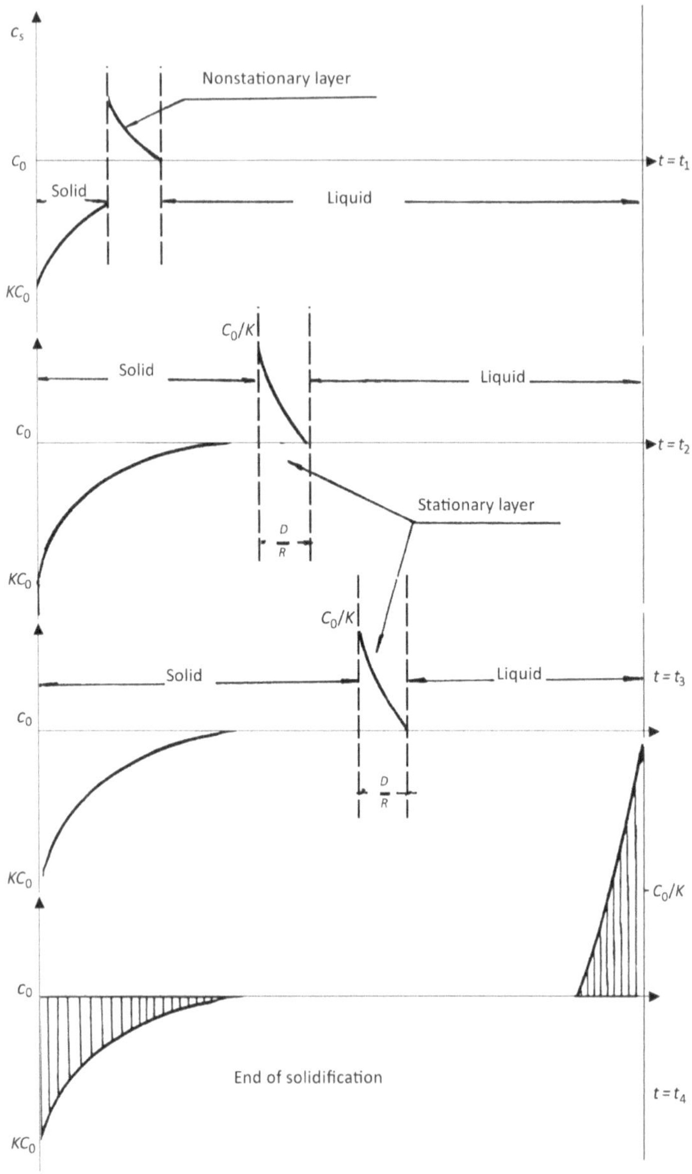

FIG. 5.10 Concentration profile as a function of distance $(K < 1)$: $t_1 < t_2 < t_3 < t_4$. Assuming no stirring in the liquid.

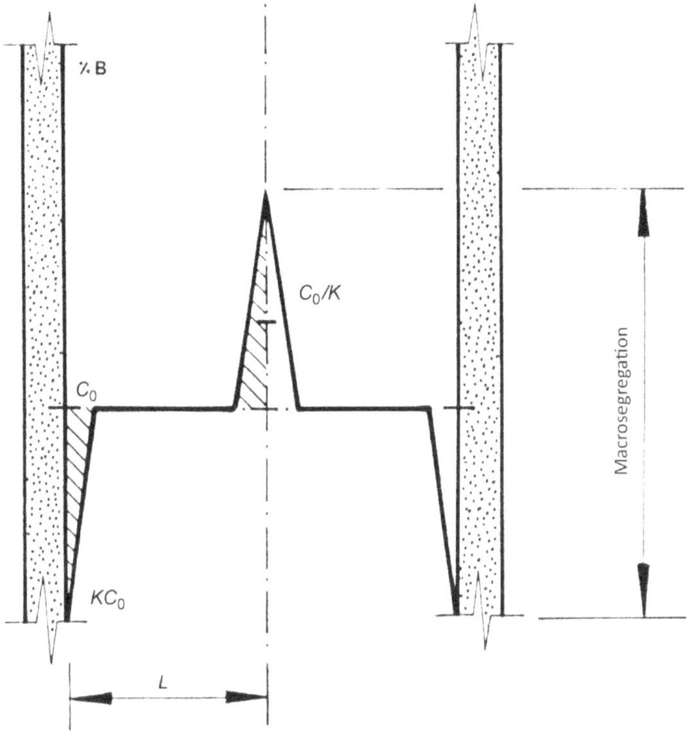

FIG. 5.11 Concentration profile (% B) in the ingot following zone melting solidification model.

EXERCISE 5.5

The concentration profile of solute during a fast solidification is:

$$C_S = C_0[1 + (K - 1)\exp(-\alpha x)]$$

while the concentration of solute at the liquid layer follows the equation:

$$C_L = C_0\left[1 + \left(\frac{1 - K}{K}\right)\exp\left(\frac{-R}{D}x\right)\right]$$

Calculate the value of α. The areas of the depleted solid solute zone and the enriched solute layer zone must be equal.

Solution

Integrating the expressions:

$$C_0\int_0^{\infty}(K - 1)\exp(-\alpha x)dx$$

$$C_0\int_0^{\infty}\left(\frac{1 - K}{K}\right)\exp\left(-\frac{R}{D}x\right)dx$$

And making both areas under the curve equal between them:

$$\left|\frac{1-K}{-\alpha}\right| = \left|\frac{1-K}{-K} \cdot \frac{-D}{R}\right|$$

Then:

$$\alpha = \frac{KR}{D} = \frac{K}{a}$$

where $a = D/R$ is the characteristic distance or width of liquid zone rich in solute, or simply zone width (Eq. 5.7).

EXERCISE 5.6

The fast cooling of an A-2% B alloy follows zone melting solidification model with the following parameters: $\alpha = 820\,\text{m}^{-1}$, $a = 4 \times 10^{-4}\,\text{m}$, and $K = 0.33$. Calculate: (a) the composition of the solid at a 1 mm distance from the mold wall, (b) in stationary state, calculate the amount of solute in the liquid at 60 μm from the interphase, and (c) the value of α.

Solution

(a) Composition of a solid at 1 mm from the wall can be calculated using Eq. (5.7):

$$C_S = 2\left[1 + (0.33 - 1)\exp\left(\frac{-0.33 \cdot 1 \times 10^{-3}}{4 \times 10^{-4}}\right)\right] = 1.413\%$$

(b) Amount of solute in the liquid, 60 μm from the interphase can be obtained through:

$$C_S = C_0\left[1 + \left(\frac{1-K}{K}\right)\exp\left(-\frac{x}{a}\right)\right] = 2\left[1 + \frac{1-0.33}{0.33}\exp\left(-\frac{60 \times 10^{-6}}{4 \times 10^{-4}}\right)\right] = 5.495\%$$

(c) And finally α is:

$$\alpha = \frac{K}{a} = \frac{0.33}{4 \times 10^{-4}} = 825\,\text{m}^{-1}$$

Fig. 5.12 shows the distribution curves for the solute (for a solid solution with a partition coefficient $K < 1$), with unidirectional solidification, and in such conditions that being the solid nonhomogeneous, the composition of the liquid is:

(a) Uniform at all points, with C_S/K value.
(b) Nonuniform, and with a stationary solid-liquid interface.

In the industry, the distribution of solute in the solid usually lies between the *a* and *b* curves of Fig. 5.12 (curve R_2). Type *a* solidifications are slower than that of type *b*. On the other hand, the slope of curve *b*, in its initial part, will increase if higher solidification rates are used (as said, for instantaneous solidifications, or $R = \infty$, curve *b* will match the $Y = C_0$ line).

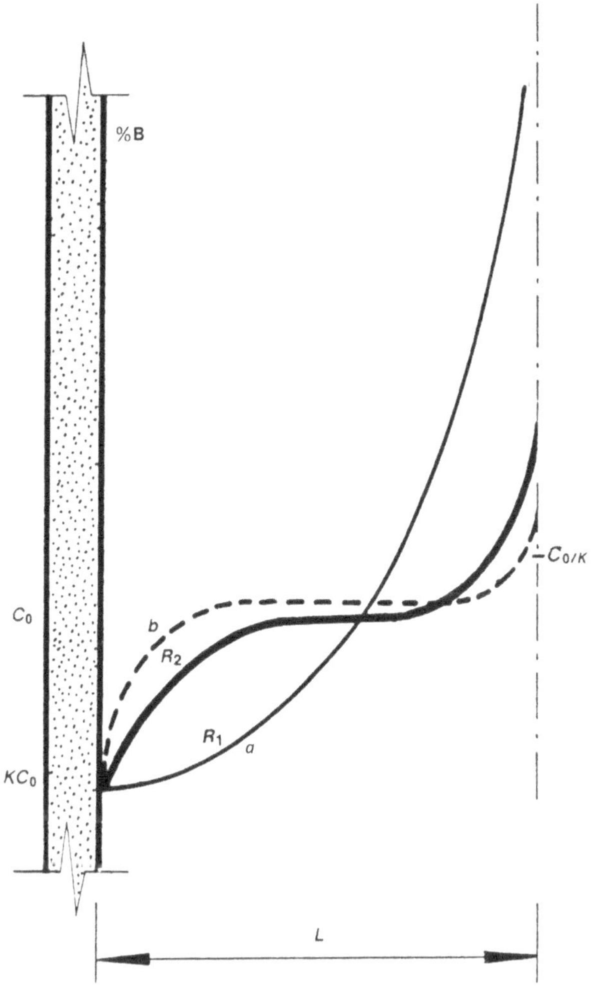

FIG. 5.12 Concentration profile (% B) in the ingot as a function of distance: curve *a* (or R_1) considers unidirectional solidification model, *b* follows the zone melting solidification model, and R_2 is an intermediate case.

The lower the partition coefficient ($K < 1$), the higher the heterogeneity profile of a certain solute B will be.

Iron can form solid solutions with many elements just as with carbon. Their binary diagrams usually follow the shape of those with coefficients $K < 1$. Table 5.1 presents the values of K for different elements that form solid solutions with Fe.

The real shape of the solute distribution profiles in the ingot mold cannot be easily anticipated as a function of the solidification conditions, since, in many cases, the distribution of B can be modified by various extrinsic factors such as:

TABLE 5.1 Partition Coefficient for Elements in Solid Solution With Fe

Element	In Fe-δ	In Fe-γ
Oxygen	0.02	0.02
Sulfur	0.02	0.02
Phosphorus	0.13	0.06
Carbon	0.13	0.36
Nitrogen	0.28	0.54
Copper	0.56	0.88
Silicon	0.66	0.95
Molybdenum	0.80	0.60
Nickel	0.80	0.95
Manganese	0.84	0.95
Cobalt	0.90	0.95
Tungsten	0.95	0.50
Chromium	0.95	0.85

convection vectors in the liquid, release of gasses, differences in the specific volume of the solid and liquid, relative displacements of the crystals in the liquid, etc.

If a steel ingot is sectioned in a transverse direction and etched properly, contour lines are observed which are, in an approximate way, parallel to the surface of the ingot; with each line richer in Fe than the inner surrounding material. The explanation for this pattern is that at the beginning of solidification there is no stationary diffusion ahead of the solid-liquid interface; though, the variation of the solute in the liquid iron is continuous.

The thickness of these lines can reach, in some cases, a few centimeters. The thicknesses of these *solidification contours* are higher if the liquid is mechanically stirred, or if there are currents in the liquid. The currents capable of forming contours in killed steels are caused, mainly, by the liquid stream. When it is done by gravity, the bottom of the ingot begins to solidify before pouring ends and the base is affected by more "contours" which form until mid-height, and then disappear; and can be more evident on one side than the other, for example because of an offset of the liquid stream.

Since the "contours" indicate the positions of the solidification front as a function of time (isotherms), they are very useful in establishing the local solidified thickness during the first instants, and to give an effective orientation or determine the origin of cracks in the ingot.

Despite the limitations just mentioned, both solidification models are useful in the analysis of normal segregation, for example explaining when the liquid is constantly stirred during solidification, macrosegregation, due to the uniformity of the liquid, reaches higher values.

Stirring can be made by mechanical means or produced by the gas evolution in the interior of the ingot mold, as it occurs with *rimmed* steels, whose macrosegregation (for example in sulfur) is, just as it happens with other factors, considerably higher than for killed steels (Fig. 5.13). Normal segregation also increases with the size of the ingot, as the conditions for the uniformity of the liquid are favored by the lower solidification rates (Figs. 5.14 and 5.15).

5.2.1.2.1 Constitutional Undercooling and Columnar Structure

Solidification, according to Section 5.2.1.2 of a solid solution with $K < 1$ partition coefficient, produces a stationary layer with a distribution of solute B as the one presented in Fig. 5.16. Both, in pure metals as shown in Chapter 2, and in alloys, the microstructure can be directly related to undercooling: in pure metals it can only be produced by thermal means, while in alloys it can be indirectly produced by changes in temperature and composition. If it is produced by changes in composition combined with thermal changes, it is known as *constitutional undercooling*.

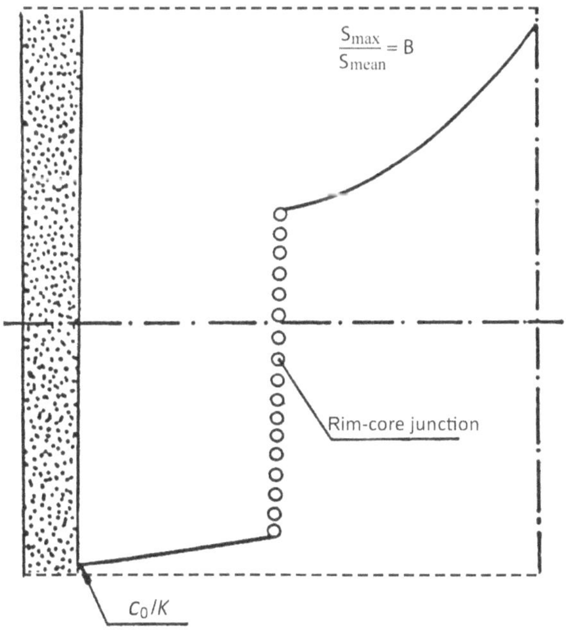

FIG. 5.13 Segregation profile of S from the wall of the ingot mold, of a rimmed steel with 0.030% S mean composition: 60% B in pure zones, 170% B in rim-core junction, and 370% B in internal zone.

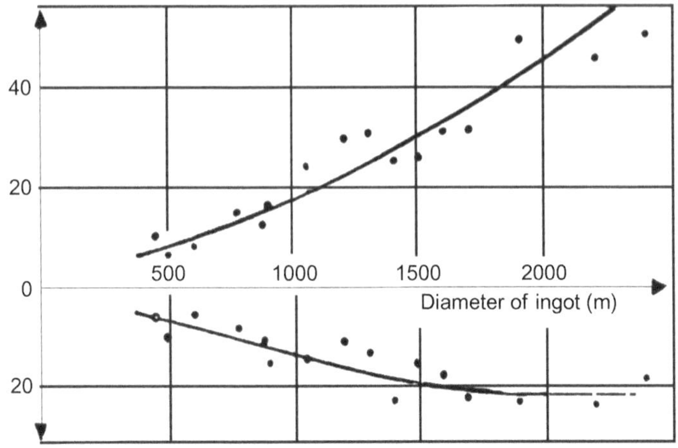

FIG. 5.14 Positive and negative segregations of S in steels as a function of ingot diameter (Bastien, 1960).

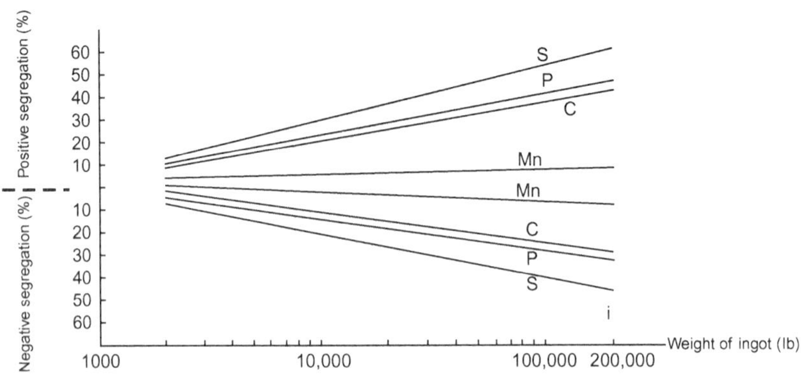

FIG. 5.15 Positive and negative segregation in steels as a function of ingot weight (Derge, 1964).

Equilibrium temperatures in the interface are indicated in Fig. 5.17A, while the amount of solute at these temperatures is shown in Fig. 5.17B.

Considering equal temperature gradients in the liquid, solidification of a solid solution compared to the pure solvent metal one, will present differences in the dendritic columnar structure thickness. The constant solidification temperature T_M of the pure metal is substituted in the case of the solid solution by the curve T_1T_2 (Fig. 5.18) and, consequently, the undercooling in the solid solution layer (constitutional undercooling) is lower than for a pure metal with a solidification point T_M.

The growth of the columnar solid solution dendrites will be slower than in the case of pure metals and, therefore, the nucleation of equiaxed grains will appear sooner, slowing the growth of the columnar grains. In practice, pure

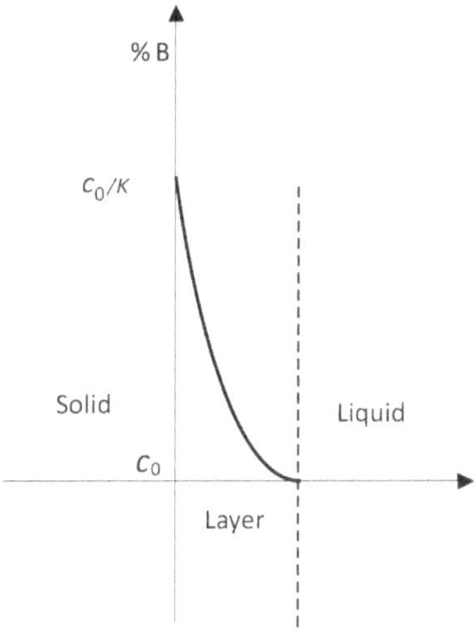

FIG. 5.16 Distribution of solute concentration in the solid-liquid layer during steady-state solidification.

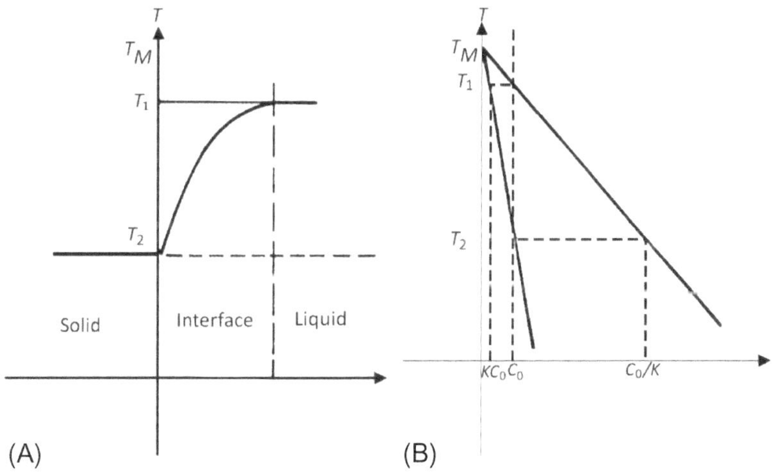

FIG. 5.17 Equilibrium temperature profile (A) in the interface and (B) as a function of solute concentration.

metals present deeper columnar structures compared to their solid solutions; the differences are more evident with smaller K partition coefficient values.

5.2.2 Inverse Segregation

The changes in volume during freezing, specifically for alloys whose specific volume decreases, have important consequences: microshrinkages can occur,

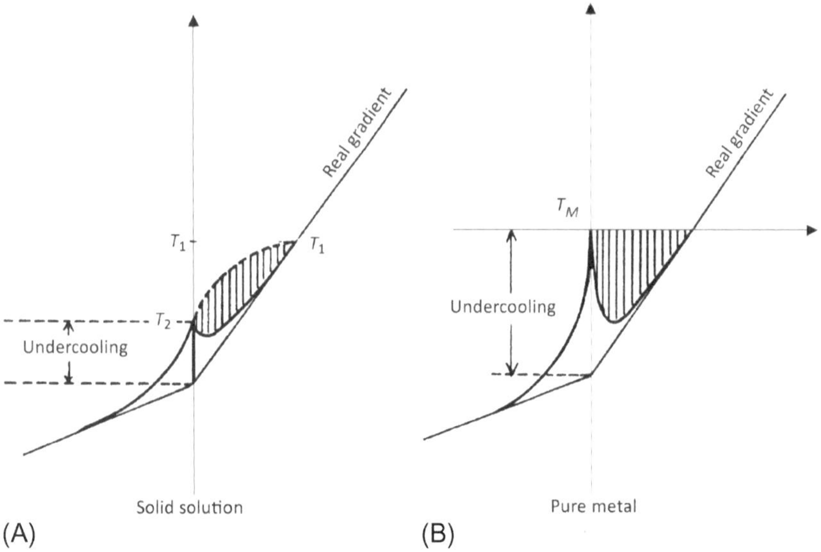

FIG. 5.18 (A) Constitutional undercooling in a solid solution ($K < 1$) and (B) thermal undercooling in a pure metal.

since the contraction in the cortical layer of the dendritic crystals produces fine fissures (microscopically visible), and the created vacuum pumps the liquid through these interdendritic channels; thus, at the end of solidification, outer zones will appear with high content of low melting point solutes. Sometimes, this *inverse segregation* can reach the periphery of the part or ingot, and create zones richer in solutes with low melting point (*exudation*).

Generally speaking, there is contraction by solidification in most alloys, thus producing, in varying degrees, inverse segregation: for example, in Cu-Sn bronzes the defect known as "tin sweat" is typical. In Fig. 5.19, the variation of Sn content along the transverse section of an 8% Sn bronze ingot is presented. The exudation appears as a result of the Sn-enriched liquid (corresponding to the last period of solidification) flowing through the microshrinkages, interconnected between them, until reaching the surface of the part.

Inverse segregation of parts can disappear by turning or milling operations made on the surface, once the parts have solidified.

Metals and alloys that increase their volume when solidifying (for example Bi, Bi-Pb alloys, or gray castings) cannot cause microshrinkages and, thus, cannot present inverse segregation.

5.2.3 Gravity Segregation

The relative displacement of free crystals and liquid modify the behavior of solidification (Section 5.2.1). Liquid, for example, has a tendency to descend

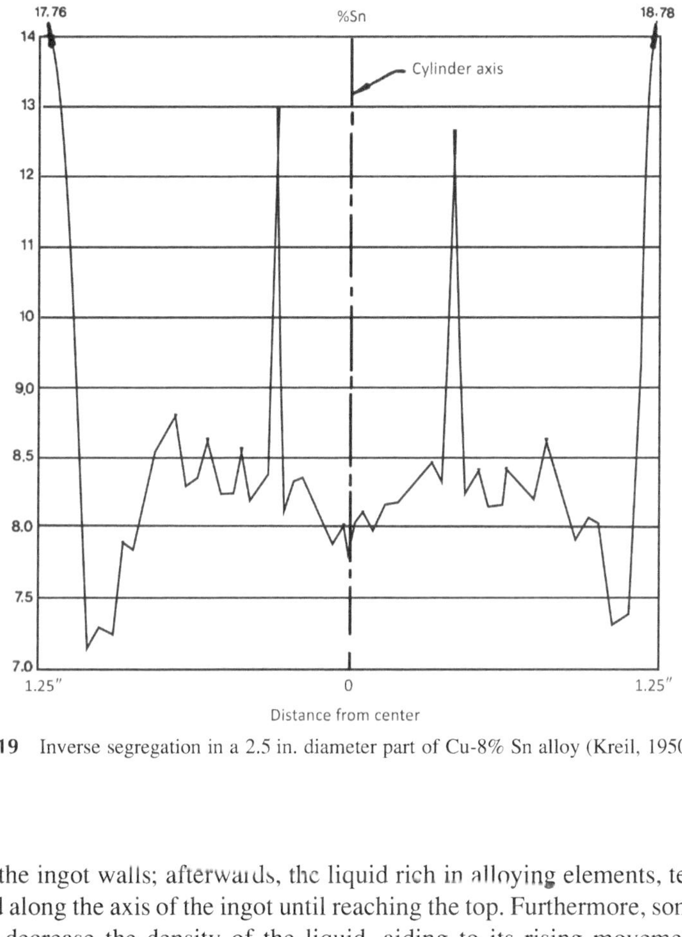

FIG. 5.19 Inverse segregation in a 2.5 in. diameter part of Cu-8% Sn alloy (Kreil, 1950).

along the ingot walls; afterwards, the liquid rich in alloying elements, tends to ascend along the axis of the ingot until reaching the top. Furthermore, some elements decrease the density of the liquid, aiding to its rising movement and increasing segregation. In the case of steels, P and S are found, for this reason, in larger proportions at the top of the ingot.

On the other hand, the first free crystals to be formed, since they are denser than the liquid, usually fall towards the bottom and also drag those crystalline inclusions that solidify at high temperature (used as nucleating agents by the first formed grains). For example, the bottom of steel ingots usually has a large amount of inclusions of the silica-aluminate type, as well as steel crystals with low amounts of carbon and alloying elements.

This type of segregation, known as *vertical segregation* or *gravity segregation*, is characteristic of Cu-Pb alloys and bearings with high content of Pb (Fig. 4.21). To avoid the inconvenience of crystals floating (as soon as they are formed) in the liquid with high Pb content, it is advisable to use fast solidifications and/or other techniques (continuous, centrifugal, or chill casting) that hinder the sink or float of alloying constituents.

In order to avoid vertical segregation in bearings, sintering and powder metallurgy may be used instead of casting techniques (if dimensions of the bearing allow it).

Vertical segregation, analogous to the Cu-Pb system due to high density of the liquid, also occurs in steels with Pb or improved machinability steels.

5.2.4 Local Segregation

5.2.4.1 Segregations Produced by Entrapped Gasses

The gaseous release can modify both, the uniformity of the liquid and the normal segregation; however it can also occur after the first stages of the solidification. For example, hydrogen in bronzes (solid solution of Sn in Cu) remains dissolved in the proportion of 4 cm^3/100 g STP (standard temperature and pressure) of liquid when it has a 97.5% Cu-2.5% Sn composition; however, hydrogen dissolved in a liquid with 80% Cu-20% Sn is 1.5 cm^3/100 g STP of liquid. Therefore, at the end of solidification, when the liquid is richer in tin, the gaseous release is more abundant.

When solidification is in its advanced stages, the residual liquid is aspirated into the microshrinkages where a depression has been formed because of the cooling of undissolved gasses; and that liquid enriches the cavity with low melting point solutes. This type of segregation is known as *segregation in gasses*.

The micrograph of Fig. 5.20 corresponds to a bronze with 15% Sn, showing α grains with coring segregation of Sn inside them (there is more Sn in the periphery of the dendrites with darker color than in the interior of the dendrite).

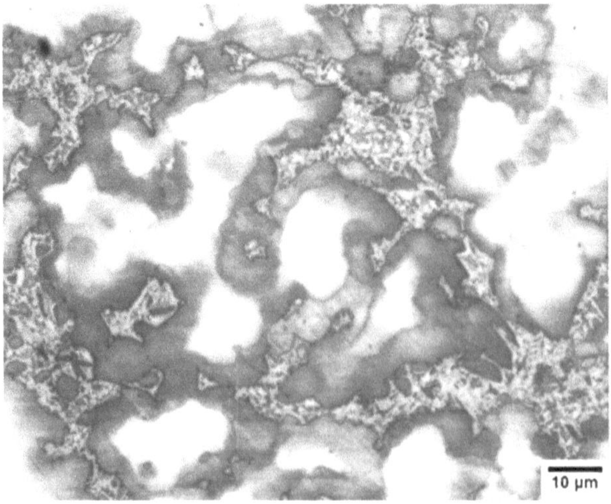

FIG. 5.20 Microsegregation in a Cu-15% Sn alloy.

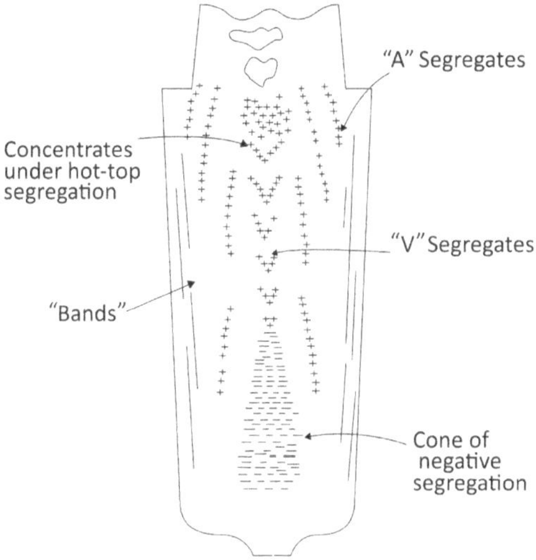

FIG. 5.21 Segregation pattern in a large killed steel ingot. –, negative segregation, +, positive segregation (Derge, 1964).

The presence of 27% Sn eutectoid in the matrix constituent indicates high bubble segregation.

A type of segregation caused by gasses, characteristic of steel ingots, is the one known as *veins in A* or *A segregates* (Fig. 5.21): usually present in both large and small ingots; in killed, rimmed, alloyed, or nonalloyed steels; in both ambient (atmospheric) or vacuum castings. The thread-like shape of the veins is caused by traces left by gas bubbles that rise inside a liquid rich in free dendrites; the path of the bubbles is filled later with segregated liquid. Therefore veins usually have higher amounts of S (in the form of sulfides), P, C, and Cr compared to the mean composition of the steel. In tool steels (high C and Cr content) carbides are usually formed inside the veins.

The local heterogeneity that produces these veins in the steel, once it has been forged, is undesirable since it causes problems in machining, differences in hardenability, tendency to initiate cracks, etc.

5.2.4.2 Segregations by Changes in the Crystallization Structure

A common type of segregation in steel ingots, caused by morphology changes in crystallization, is *veins in* V, V-*shaped veins* or V *segregates* which are clearly evident in ingots weighting more than 20 tons (Fig. 5.21).

The basaltic zone of the bottom of a large ingot has a barrier to the free dendritic columnar growth: fine equiaxic grains fall towards the bottom of the ingot (rain). The basaltic zone can reappear later as there is a liquid, not yet nucleated, in front of it, and in this way, bands of basaltic and equiaxic grains with different grain sizes are formed with a certain frequency, analogous to sedimentation phenomena.

On the other hand, both horizontal and vertical temperature profiles cause convection currents; and the steel, instead of solidifying in the shape of vertical layers with different grain sizes, does so in the shape of superposed conic layers that appear on the upper third of the ingot (or towards the riser, Fig. 5.21): very thin layers of sediments separated by equally conic zones with crystals of larger size.

The fine equiaxic crystals form an almost liquid-proof layer and when the lower layer shrinks, a vacuum is generated between them, causing axial V-shaped porosities; these are dangerous because it is difficult to make them disappear during forging and are located at the axis of the ingot; in order to weld them by recrystallization, it is necessary to use high forging reductions.

In some cases, the vacuum caused in the V-shaped porosities can be large enough for it to aspirate the liquid that exists above the fine equiaxic layer; this liquid may enrich the axial porosities and cause V-shaped veins. Micrographically, it can be seen that V-shaped segregations are surrounded by purer zones.

5.2.4.3 Segregations in Cracks

When a crack is formed by contraction inside the frozen crust and is filled again by liquid (with different composition), it solidifies enclosing a segregation vein.

In the case of steels, this type of local segregation in the interior of previously formed cracks causes potentially brittle zones, sometimes richer in sulfide inclusions that can produce new cracks during cooling or forging.

This type of segregation is formed with some frequency in ingots or in continuous casting billets, following the path where the columnar grains join, especially at the corners of the mold (Fig. 2.23B).

5.2.5 Macrosegregation Indexes

The segregation index (*SI*) of an element at a certain point in the ingot or casting is calculated through:

$$SI = \frac{C - C_0}{C_0} \cdot 100 \tag{5.12}$$

where C is the content of the element at that point and C_0 the mean content of the element in the alloy.

To characterize more precisely the general heterogeneity of an ingot, other indexes must be considered:

- The positive segregation index (*PSI*):

$$PSI = \frac{C_M - C_0}{C_0} \cdot 100 \tag{5.13}$$

- The negative segregation index (*NSI*):

$$NSI = \frac{C_0 - C_m}{C_0} \cdot 100 \tag{5.14}$$

- The global segregation index (*GSI*):

$$GSI = \frac{C_M - C_m}{C_0} \cdot 100 \tag{5.15}$$

where C_M and C_m are the maximum and minimum values respectively of the element being analyzed and considering the casting as a whole.

Fig. 5.22 shows the 3D profile of the local segregation of carbon at different locations in a steel ingot caused by macrosegregation, while Fig. 5.23 presents the usual relationship in steels between the segregation index of carbon and the segregation indexes of other elements.

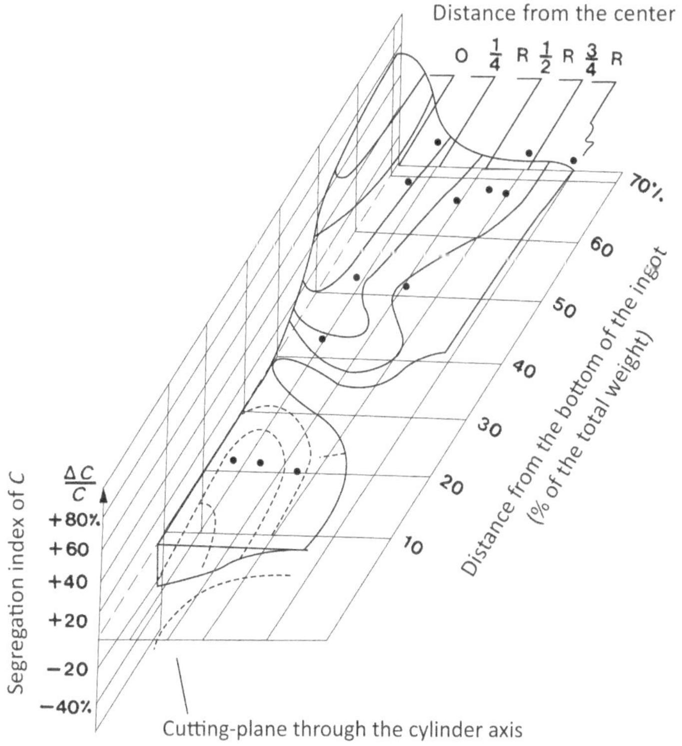

FIG. 5.22 3D profile of the segregation of C in a 135 ton steel ingot (Bastien, 1960).

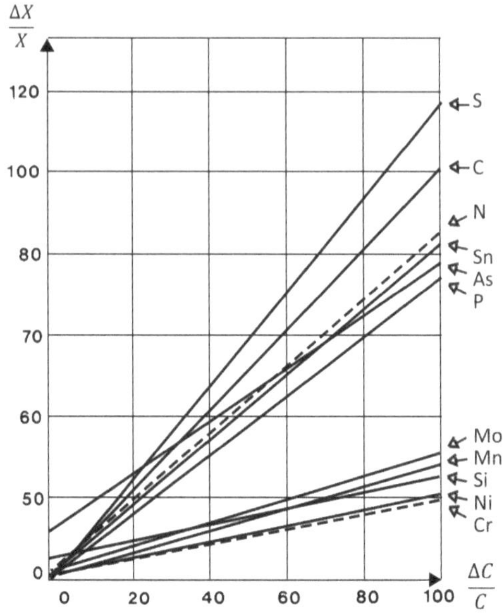

FIG. 5.23 Segregation of various elements with relation to *C* (Bastien, 1960).

EXERCISE 5.7

A 10 ton 0.30% C steel ingot has a carbon positive segregation index of 0.13 and a negative segregation index of 0.3. Calculate (a) the maximum and minimum amounts of C in the ingot, (b) the global segregation index, and (c) the probable location of these points in the longitudinal section of the ingot.

Solution

(a) Maximum and minimum amounts of C in the ingot

The maximum amount of C in the ingot can be obtained through Eq. (5.13):

$$0.13 = \frac{C_M - 0.3}{0.3} \Rightarrow C_M = 0.34$$

while the minimum one can be calculated using expression (5.14):

$$0.3 = \frac{0.3 - C_m}{0.3} \Rightarrow C_m = 0.21$$

(b) Global segregation index

The *GSI* can be determined using Eq. (5.15):

$$GSI = \frac{C_M - C_m}{C_0} = \frac{0.34 - 0.21}{0.3} = 0.43 = 43\%$$

(c) Location

Positive segregation would be found in the upper third section, while the negative one in the lower third section (Figs. 5.19 and 5.21).

5.3 MICROSEGREGATION

Section 5.2.1.2 suggests that the solidification of a ingot with composition C_0 (solid solution) and $K < 1$ causes a concentration vs. distance profile with almost constant composition C_0, a superficial crust with a composition lower than C_0 (specifically, the composition at the wall of the ingot mold would be KC_0) and a composition at the center of the ingot that is higher than C_0/K (Fig. 5.10).

Real solidifications, between *a* and *b* curves of Fig. 5.12, result in wider zones of negative segregation and, therefore, the final C_F composition of the axis of the ingot is higher than C_0/K. The final solidification temperature at the center of the ingot would be T_F (Fig. 5.24).

As a result of nonequilibrium solidification, not only the chemical composition at different zones of the ingot or casting varies, but also the chemical composition of each grain, from its core to its periphery; this heterogeneity is known as *microsegregation*, *minor segregation*, or *dendritic segregation*. The most common case of microsegregation occurs between dendrites and is commonly known as *coring* since the composition of the core of the dendrites is different from their external composition.

An equiaxic grain with mean composition C_0 growing in the liquid, would present a profile with segregation from the core to the periphery of the grain, which is (analogous to the radial macrosegregation) $KC_0 \rightarrow C_F$ (Section 5.2.1).

For the solidus line, the value of C_F, for nonequilibrium freezing, can be obtained by joining the (KC_0, T_{1S}) and (C_0, T_F) points in Fig. 5.24. This line indicates, for each temperature, the mean composition of the solid formed up

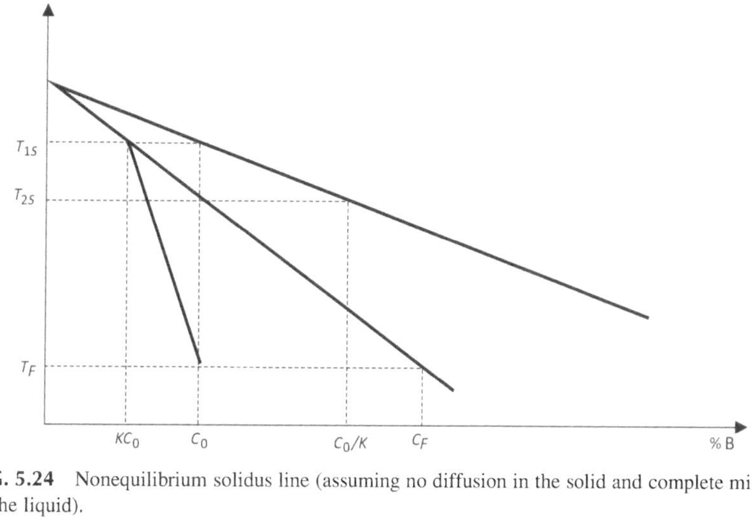

FIG. 5.24 Nonequilibrium solidus line (assuming no diffusion in the solid and complete mixing in the liquid).

to that temperature, unlike the solidus line in equilibrium that indicates, for each temperature, the instantaneous composition.

It can be summed up that, for nonequilibrium temperatures, the solidus line in the binary diagram is displaced to the left for $K < 1$ solid solutions. However, the amount of displacement of this line cannot be quantified since the solute distribution at different distances from the core of the grain will follow a rule (*a*, *b*, or an intermediate one, Fig. 5.12) depending on the conditions at which solidification takes place. Thus, the C_F composition of the periphery of the grain (and therefore, final solidification temperature at grain boundaries) will depend on these conditions.

For limited solid solutions, this heterogeneity at microscopic scale can even result in the formation of nonequilibrium constituents; for example, in the 70/30 brass, the grains formed are α solid-solution crystals when the cooling is of a equilibrium type; however, for nonequilibrium cooling, grain boundaries can be formed by β crystals and the same can be said for an ingot as a whole (macro scale): at zones close to the axis, the solid can be an $\alpha + \beta$ brass.

Microsegregation in a solid solution is visible by optical microscope when etching the polished sample with adequate chemical reagents. Depending on the morphology of crystalline grains, microsegregation will gradually manifest towards the interdendritic spaces or grain boundaries (Fig. 5.20).

Nonequilibrium states imply a risk known as *overheating* or *burning* when the solid alloy is heated again to temperatures close to the solidus line (T_S). If the alloy is solidified at equilibrium conditions, burning would start at T_S, which is a known value for every binary alloy, deduced from the phase diagram. However, if the alloy presents minor segregation, burning, at the grain boundaries or at the interdendritic spaces, burning will start at T_F, which is not a commonly known value.

5.3.1 Homogenization Heat Treatment

The microstructure obtained by solidification (or *as-cast state*) during industrial processes, always presents dendritic segregation and, therefore, anisotropy in its properties as a result of the chemical heterogeneity between the periphery and the core of the grains.

This gradient in the concentration of B can be diminished if the diffusion of B towards the interior of the grain is favored by heating at high temperature for long periods. This treatment, shown in Fig. 5.25, is known as *homogenization annealing*.

The higher the amount of homogenization reached by annealing:

- The faster the diffusion of element B (or its diffusion coefficient) will be. In this case, a high homogenization temperature is convenient, but lower than T_F if burning is to be avoided.
- The smaller the dimensions of the grain will be. A high solidification rate produces grains of smaller sizes and the spaces to be traveled by the B atoms are also small, so homogenization can be achieved faster.

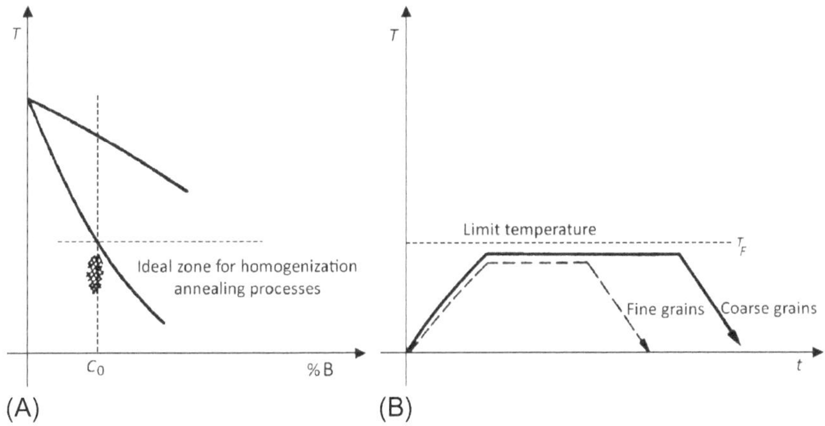

FIG. 5.25 (A) Ideal temperature range and (B) temperature vs. holding time for homogenization annealing.

A process to achieve fast homogenization of the dendrites is mechanical forming (*hot forging*), which is a process that implies (along with metallurgical advantages such as dynamic recrystallization, closing both internal bubbles and microshrinkages, grain refinement, etc.) a previous crushing of the dendrites at high temperature, and with it, making the periphery of the grains closer to their core; the distances to be traveled by B during homogenization are smaller, and diffusion is favored by the fact that it is made at high temperature.

Homogenization annealing eliminates *microsegregation*, but it does not eliminate other types of segregations (Section 5.2).

EXERCISE 5.8

A 0.4% C-2% Ni steel, in its as-cast state has a grain size of 500 μm. Calculate (a) the necessary time to eliminate dendritic segregation (coring) of C and Ni through an homogenization treatment at 1200°C, (b) the necessary time if grain size is refined to 50 μm, and (c) analyze the viability of the treatment.

Data: $D_C \cong 2 \times 10^{-10}\,\text{m}^2/\text{s}$ and $D_{Ni} \cong 7 \times 10^{-15}\,\text{m}^2/\text{s}$

Solution

(a) Time to eliminate dendritic segregation (500 μm)

Considering that the distance atoms have to travel, for diffusion to remove dendrites, is maximum half the grain size (250 μm), and knowing that:

$$x = \sqrt{Dt} \Rightarrow t = \frac{x^2}{D}$$

the necessary time to eliminate C and Ni dendrites is:

$$t_C = \frac{(250 \times 10^{-6}\,\text{m})^2}{2 \times 10^{-10}\,\text{m}^2/\text{s}} = 312.5\,\text{s} \cong 313\,\text{s}$$

$$t_{Ni} = \frac{(250 \times 10^{-6}\,\text{m})^2}{7 \times 10^{-15}\,\text{m}^2/\text{s}} = 8.93 \times 10^6\,\text{s} \cong 103\,\text{days}$$

(b) Time to eliminate the dendritic segregation (50 μm)

$$t_C = \frac{(25 \times 10^{-6}\,\mathrm{m})^2}{2 \times 10^{-10}\,\mathrm{m^2/s}} = 3.13\,\mathrm{s} \cong 3\,\mathrm{s}$$

$$t_{Ni} = \frac{(25 \times 10^{-6}\,\mathrm{m})^2}{7 \times 10^{-15}\,\mathrm{m^2/s}} = 8.93 \times 10^4\,\mathrm{s} \cong 24.8\,\mathrm{h}$$

(c) Viability of the treatment

If the grain size is refined 10 times (from 500 to 50 μm), the required time period is two orders of magnitude shorter. Furthermore, eliminating minor segregation of C is possible by homogenization, while eliminating elements in substitutional solid solution (Ni) is much more difficult.

5.4 NONEQUILIBRIUM BY EUTECTIC AND PERITECTIC REACTIONS

Analogous to solid solutions, both eutectic and peritectic diagrams can also present nonequilibrium states. For example, an alloy that would ideally only form solid solutions, can form eutectic in the interdendritic spaces as a result of the enrichment of one of the components in the liquid during the nonequilibrium solidification.

Fig. 5.26 corresponds to a sand-casted Al-4% Cu alloy: if solidification had been of the equilibrium type, the only constituent would be α with uniform composition at all points, however, in nonequilibrium conditions, the micrograph shows not only *minor segregation* at the interior of the grains (differences in color between core and periphery) but also the presence of a complex matrix constituent (on grain boundaries and at interdendritic spaces) which is the binary eutectic formed by solid solution of Cu in Al and intermetallic compound Al_2Cu.

It is relatively common that alloys of an eutectic system do not freeze under equilibrium conditions which makes the use of compositions lower than *solvus* (for single-phase alloys) advisable in the case of parts that will be forged later.

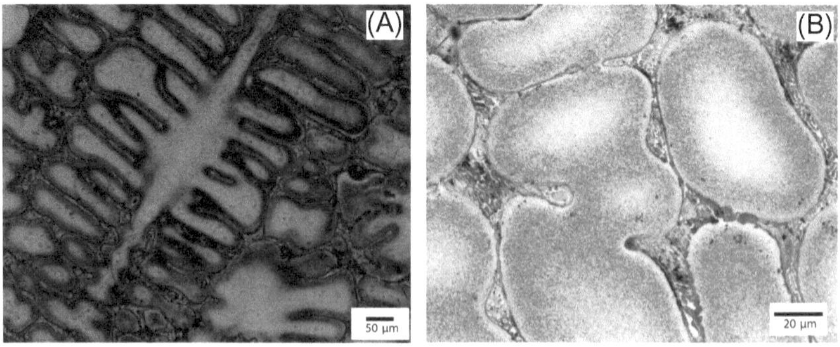

FIG. 5.26 (A) Al-4% Cu (nonequilibrium solidification) and (B) seen at higher magnification.

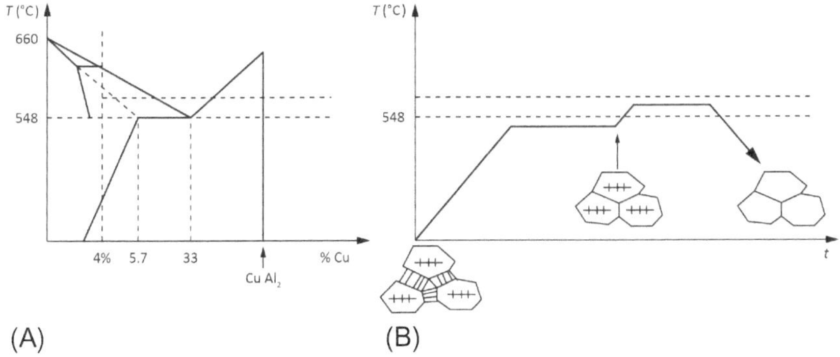

FIG. 5.27 (A) Nonequilibrium Al-4% Cu solidification phase diagram and (B) homogenization thermal treatment.

Another possibility is maintaining the chemical composition and using homogenization heat treatments before forging, in order to recover the equilibrium microstructure (only formed by solid solution grains).

The Al-4% Cu alloy (Fig. 5.27A) cannot be heated above 548°C because, above this temperature, the eutectic melts (known as *burning*). Homogenization treatment of this alloy must follow Fig. 5.27B: remaining at a temperature lower than the eutectic, it disaggregates by solid-state diffusion; once the desired microstructure is reached, temperature can be increased until reaching the homogenization temperature (a few degrees below T_S) and eliminate coring.

Occasionally, forcing the variables to obtain the eutectic microstructure by nonequilibrium conditions can be advantageous to improve casting ability and "hot tearing" resistance (Section 6.4); however, it requires a later homogenization treatment if supplied in a forged state.

Among the Al-based light alloys, the nonequilibrium states are frequent and a result of the wide solidification temperature range. In the Al-Mg system, for example, the eutectic solidifies (Fig. 5.28) at 451°C, while pure Al solidifies almost 200°C above it. The diagram indicates that the maximum solubility of Mg in Al is ~ 15%; in industrial solidifications (because of their nonequilibrium nature) it is common to find eutectic matrixes for Mg contents above 4%; the precautions indicated for Al-Cu also apply.

It is common for single-phase alloys, whose solvus limit corresponds to a peritectic reaction, to present it, even though its chemical composition would only indicate the formation of a simple solid solution. The Cu-Sn bronze with 10% Sn is a distinctive example: when solidification is of the equilibrium type, α solid solution is obtained; however, if it is of the nonequilibrium type, the solidus line is displaced to the left (Fig. 5.29) and β appears by peritectic reaction, in the interdendritic spaces of α.

The properties of a bronze with 10% Sn for bearings, solidified at nonequilibrium conditions (chill cast) differ from the "equilibrium" bronze (sand cast): nonequilibrium bronze cannot be heated above 798°C without the risk of

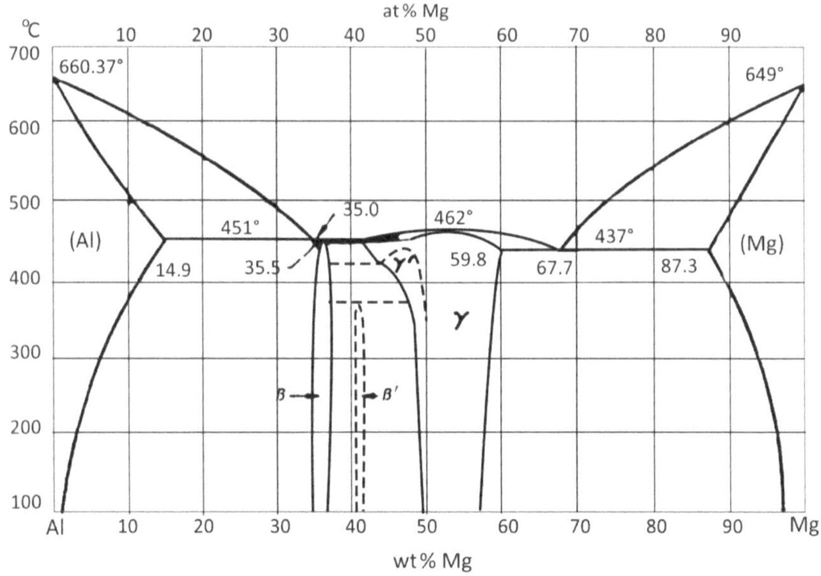

FIG. 5.28 Binary Al-Mg phase diagram (ASM Handbook, vol. 3, 1992).

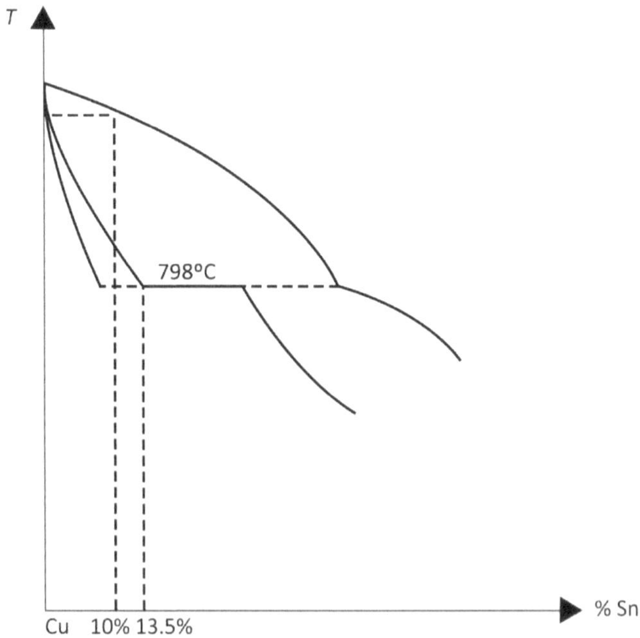

FIG. 5.29 Cu-Sn peritectic reaction (nonequilibrium solidification).

burning (intergranular melting), and since it has phase β in the interdendritic spaces of α (while the equilibrium bronze is formed only by α), it has improved wear resistance. In solid state, during cooling, β phase is transformed in a complex $\alpha+\delta$ aggregate, while the "equilibrium bronze" does not present this reaction.

EXERCISE 5.9

Applying the Scheil equation, calculate the proportion of nonequilibrium constituents that can appear in the solidification of (a) Al-4% Cu, (b) Al-4% Mg, and (c) Cu-10% Sn, using Tables 4.1 and 4.2.

Solution

(a) Al-4% Cu

At 548°C; $L(33\%\text{Cu}) \leftrightarrow \alpha(5.7\%\text{Cu}) + \text{Al}_2\text{Cu}(53\%\text{Cu})$, the partition coefficient is:

$$K = \frac{5.7}{33} \cong 0.17$$

and applying Eq. (5.4):

$$f_L = \left(\frac{C_0}{C_L}\right)^{\frac{1}{1-K}} = \left(\frac{4}{33}\right)^{\frac{1}{1-0.17}} \cdot 100\% = 7.87\%$$

meaning 7.87% of nonequilibrium eutectic.

(b) Al-4% Mg

At 451°C; $L(35\%\text{Mg}) \leftrightarrow \alpha(14.9\%\text{Mg}) + \text{Al}_3\text{Mg}_2(35.5\%\text{Mg})$, the partition coefficient is:

$$K = \frac{14.9}{35} \cong 0.43$$

and applying Eq. (5.4):

$$f_L = \left(\frac{4}{35}\right)^{\frac{1}{1-0.43}} \cdot 100\% = 2.23\%$$

meaning 2.23% of nonequilibrium eutectic.

(c) Cu-10% Sn

At 799°C; $L(25.5\%\text{Sn}) + \alpha(13.5\%\text{Sn}) \leftrightarrow \beta(22\%\text{Sn})$, the partition coefficient is:

$$K = \frac{13.5}{25.5} \cong 0.53$$

and applying Eq. (5.4):

$$f_L = \left(\frac{10}{25.5}\right)^{\frac{1}{1-0.53}} \cdot 100\% = 13.65\%$$

meaning there is 13.65% of liquid that can react with segregated crystals of α (in their boundaries, which have 13.5% Sn) and produce the nonequilibrium β constituent. In bronzes for bearings, it is desirable to have β since it is the precursor for the hard intermetallic δ ($\text{Cu}_{31}\text{Sn}_8$) that increases wear resistance: $\beta \rightarrow \alpha + \gamma$ and $\gamma \rightarrow \alpha + \delta$.

EXERCISE 5.10

The solidus and liquidus of a binary alloy A-10% B are straight lines where $C_L = KC_S$ (where C_S and C_L are the concentration of solid and liquid phases at each temperature). Supposing that, during the solidification process of the alloy, the liquid remains in homogeneous composition at all points, and that diffusion is negligible in solid state, demonstrate that:

$$\ln\left(\frac{m - m_s}{m}\right) = -\int_0^{C_L} \frac{dC_L}{C_L - C_S}$$

where m is the total mass of the alloy, m_s the mass solidified up to a temperature T, C_L the concentration of B in the liquid at T, and C_S the concentration of B in the last layer of solid in equilibrium with C_L at T.

Solution

Making a balance of matter in B, at T:

$$\int_0^{m_s} C_s dm_s + (m - m_s)C_L = mC_0$$

And differentiating:

$$C_S dm_s - dm_s C_L + (m - m_s)dC_L = 0$$

Reordering variables:

$$dm_s(C_S - C_L) + (m - m_s)dC_L = 0$$

$$\frac{dm_s}{m - m_s} = \frac{dC_L}{C_L - C_S}$$

And integrating:

$$\int_0^{m_s} \frac{dm_s}{m - m_s} = \int_{C_0}^{C_L} \frac{dC_L}{C_L - C_S} = -\ln\left(\frac{m - m_s}{m}\right)$$

Q.E.D.

If $C_S = KC_L$, then:

$$\int_{C_0}^{C_L} \frac{dC_L}{C_L - C_S} = \frac{1}{1 - K}\int_{C_0}^{C_L} \frac{dC_L}{C_L} = \frac{1}{1 - K}\ln\frac{C_L}{C_0}$$

$$-\ln\left(\frac{m - m_s}{m}\right) = -\ln f_L = \frac{1}{1 - K}\ln\frac{C_L}{C_0}$$

and

$$f_L = \left(\frac{C_0}{C_L}\right)^{1/(1-K)}$$

which is the Scheil equation or the nonequilibrium lever rule (from the point of view of microsegregation, Section 5.3).

EXERCISE 5.11

If the diagram of the alloy of Exercise 5.10 has an invariant eutectic solidification ($L(50\%B) \leftrightarrow \alpha(25\%B) + \beta(75\%B)$) at 500 °C, determine (a) the end of the nonequilibrium solidification, (b) the microstructure of the as-cast alloy indicating matrix and disperse constituents, and compare it to the microstructure obtained by equilibrium solidification, and (c) its global segregation index.

Data: melting temperature of A is 1000°C.

Solution

The mean solid concentrations of B at each temperature will be $\overline{C_S}$, which define the real solidus line (displaced towards the left of the ideal solidus).

Ideally: $T_L = 900°\text{C}$ and $T_S = 800°\text{C}$.

(a) For nonequilibrium solidification, Scheil equation must be used: at each temperature f_L is calculated, as well as $\overline{C_S}$ through balance of matter in B:

$$T = 900°\text{C};\ C_L = 10;\ \overline{C_S} = 5;\ m_L = 100;\ m_s = 0$$
$$T = 850°\text{C};\ C_L = 15;\ f_L = (10/15)^2 = 0.444$$
$$0.556\overline{C_S} + 0.44 \cdot 15 = 10 \Rightarrow \overline{C_S} \cong 6\%$$
$$T = 800°\text{C};\ C_L = 20;\ f_L = (10/20)^2 = 0.25$$
$$0.75\overline{C_S} + 0.25 \cdot 20 = 10 \Rightarrow \overline{C_S} = 8\%$$
$$\vdots$$
$$T = 500°\text{C (final)};\ C_L = 50;\ f_L = (10/50)^2 = 0.04 = 4\%\ \text{(eutectic liquid)}$$
$$0.96\overline{C_S} + 0.04 \cdot 50 = 10 \Rightarrow \overline{C_S} = 8.333\%$$

(b) The nonequilibrium microstructure is formed by:

Disperse: segregated α with compositions varying from 5% (core) to 25% (periphery).

Matrix: 4% eutectic with mean composition of 50% B ($50\%\alpha(25\%B) + 50\%\beta(75\%B)$).

(c) Global segregation index

Furthermore, its GSI will be:

$$GSI = \frac{25 - 5}{10} \cdot 100 = 200\%$$

The equilibrium microstructure would be monophasic, and formed by crystals of α with 10% B composition.

REFERENCES

ASM International, 1992. ASM Handbook, 10th ed. Alloy Phase Diagrams, vol. 3 ASM International, Metals Park, OH.

Bastien, 1960. Etude de l'heterogénéité des gros lingots de forge. Rev. Met. 57, 1091–1103.

Derge, G. (Ed.), 1964. Basic Open Hearth Steelmaking, third ed. AIME, New York, NY.

Kreil, A., 1950. The continuous casting of copper and its alloys. Met. Rev. 5, 413.

BIBLIOGRAPHY

ASM, 2002. Failure Analysis and Prevention, 10th ed. American Society for Metals, Metals Park, OH.

Cahn, R., Haasen, P., 1996. Physical Metallurgy, fourth ed. North Holland Publishing, Amsterdam.

Chalmers, B., 1977. Principles of Solidification. Krieger Publishing, New York, NY.

Chipman, 1951. Basic Open Hearth Steelmaking. Iron & Steel Division AIME, New York, NY.

Hansen, M., Elliot, R., 1965. Constitution of Binary Alloys, second ed. McGraw-Hill, New York, NY.

Massalski, T., 2001. Binary Alloy Phase Diagrams, second ed. American Society for Metals, Metals Park, OH.

Moore, J.J., 1983. Mechanism of formation of A and V segregation in cast steel. Int. Met. Rev. 28, 6.

Pero-Sanz, J., Verdeja, J., 1976. Estados de inequilibrio. Rev. Tec. Met. 215, 27–40.

Pfann, W., Hollomon, J., 2013. Zone Melting. Literary Licensing LLC, Whitefish, MT.

Chapter 6

Physical Heterogeneities in Solidification

6.1 FACTORS THAT INFLUENCE THE COLUMNAR STRUCTURE

When a liquid metal solidifies inside a cylindrical metallic ingot mold with a high height/diameter ratio, the heat flux is horizontal and radial. At the beginning of the cooling process, this occurs only by heat transfer through the solid, according to model 2 (Fig. 2.22). There will be peripheral-oriented grain growth and the *columnar structure* will appear in the shape of a concentric crust of the ingot which is initially in contact with the ingot mold, because the metallostatic pressure of the liquid mass generates compression against the wall of the mold as long as the crust is not very thick and therefore has plasticity.

With time ($t_1 < t_2 < t_3$) the temperature gradient changes to the shapes shown in Fig. 6.1 until t_3, with cooling following model 1 (Fig. 2.21): almost all latent heat is absorbed by the liquid. The undercooling of the liquid is then, practically null and the directional growth of dendrites is very slow. From this point onwards the remaining liquid will suffer homogeneous and uniform nucleation: nuclei will grow in a thermal isotropic field and promote equiaxic crystals, which during their formation will originate a mushy zone or freezing band, stopping the directional growth of the dendrites that formed the solid crust or columnar structure.

Once solidification ends, the transverse section of the ingot will have three clearly differentiated structural zones (from the surface to the center of the ingot, Fig. 2.24):

- External zone, with very small grains randomly oriented known as *chill grains*.
- Dendrites oriented toward the interior of the ingot (columnar structure).
- Internal zone, with equiaxic grains of globular shape (equiaxed structure).

The columnar structure is interesting in those cases where directional properties are desired, for example, mechanical strength at high temperature. Another example of the creation of columnar structures by solidification through unidirectional cooling is the manufacture of turbine blades for airplanes, since they suffer aerodynamic loads and require good corrosion behavior, they must

Solidification and Solid-State Transformations of Metals and Alloys. http://dx.doi.org/10.1016/B978-0-12-812607-3.00006-1

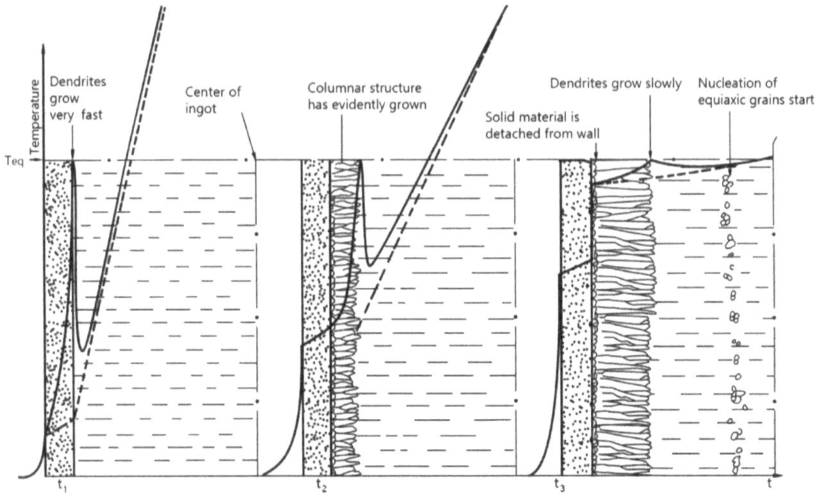

FIG. 6.1 Crystallization zones and corresponding temperature profiles: chill (t_1), columnar (t_2), and equiaxic (t_3) grains.

endure the tendency to gradually elongate (creep) as an effect of centrifugal forces and high working temperatures.

Thermo-creep behavior of Ni alloys (used for the turbine blades) considerably improves when grain boundaries are oriented in the direction of the stress that is causing thermo-creep: compared to "conventional" blades, molded in ceramic molds and with randomly oriented polycrystalline structure, these blades are obtained by solidification with unidirectional cooling. In order to obtain the columnar structure, the mold (with the melted metal inside) is maintained inside a heating hood (Fig. 6.2) that is slowly displaced; while the lower part of the mold is in contact with a refrigerated metallic base.

Creep behavior is improved even more if grain boundaries disappear, thereby making the blade a monocrystal. This is achieved using a technique similar to solidification with unidirectional cooling, but designing an adequate bottle neck in the mold (Fig. 6.2B) in order for solidification of the blade to start with a single crystal.

The columnar structure has advantages just as any manufacturing process that leaves the material with directional properties (magnetic sheets or permanent magnets). In particular, an advantage of molded parts with deep columnar structures is that they turn out more compact since shrinkage is almost null. However, columnar structure is not usually desirable because of the anisotropy of properties it gives to the as-cast metal or alloy, which can cause cracking in subsequent cooling or heating processes, or during forming operations (e.g., rolling).

All the factors that stop heat transfer through the ingot mold as quickly as possible and, instead, direct it towards the liquid metal, are beneficial to obtain a

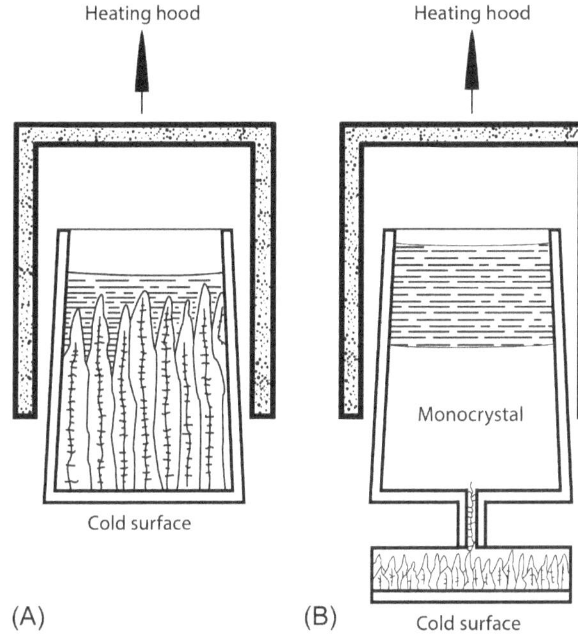

FIG. 6.2 Unidirectional solidification of a turbine blade with (A) dendritic solidification and (B) monocrystalline structure.

thinner columnar structure. On the contrary, all the factors that make heat transfer continuous through the ingot mold will produce deeper columnar structures, for example, with:

- High external refrigeration of the mold.
- High thermal conductivity of the mold material.
- Small dimensions of the part being molded.
- Low stirring of the liquid mass during solidification, since it hinders the homogenization of temperature in the liquid.
- Low amounts of nucleating agents added to the liquid metal, favoring quicker heterogeneous nucleation of equiaxic grains.
- High casting temperature of the alloy, since the center of the ingot will remain at high temperature during longer times. The higher melting (or solidification) temperatures T_M of the metal, the deeper the columnar structure will be.
- Low specific heat of the liquid, since it will be cooled faster.
- High thermal conductivity of the melted metal.

It is important to point out that the melting temperatures of the metals T_M (Table 1.5) and their thermal conductivities (Table 6.1) are not related (the thermal and electric conductivities are, instead, directly related).

Metals without allotropic transformation points (or polymorphic solid-state), such as Al, Cu, Ni, etc. produce deeper columnar structures than those

TABLE 6.1 Thermal Conductivity of Metals at 25°C in (W/m K)

Metal	*K*	Metal	*K*	Metal	*K*	Metal	*K*
Silver	418.00	Sodium	133.53	Iron	75.47	Tantalum	54.57
Copper	392.46	Calcium	125.40	Lithium	70.83	Thallium	39.48
Gold	296.08	Zinc	112.63	Platinum	70.83	Lead	34.83
Aluminum	220.61	Potassium	99.86	Palladium	70.83	Uranium	26.71
Tungsten	200.87	Cadmium	91.73	Cobalt	68.51	Indium	23.22
Magnesium	159.07	Nickel	91.73	Tin	67.34	Antimony	18.58
Beryllium	159.07	Rhodium	88.24	Chromium	67.34	Mercury	8.13
Molybdenum	146.30	Silicon	83.60	Iridium	58.06	Bismuth	8.13

with various allotropic states, such as Fe, since the successive transformations in solid state can erase the columnar structure obtained in solidification. On the other hand, pure metals (Chapter 2) will produce deeper columnar structures than solid solutions.

EXERCISE 6.1

Analyze how variables such as mold material, specific heat of the metal, latent heat of fusion of the metal, and allotropic transformation in the mold material affect the time required to freeze an ingot up to a given distance from the mold wall.

Solution

For this exercise, the Chvorinov formula (2.31) as well its constant C (Eq. 2.32) are used.

(a) Modifying mold material

In this case the following variables will suffer changes: ρ_m, c_m, and k_m:

$$C = (\text{constant})\frac{1}{k_m \rho_m c_m}$$

$$t_f = (\text{constant})\frac{1}{k_m \rho_m c_m}\left(\frac{V}{A}\right)^2$$

Thus, the higher the density of the mold, the higher the specific heat and/or the thermal conductivity of the mold, the lower the time required to freeze the ingot will be. On the other hand if the mold material is able to absorb high amounts of heat, the freezing times will be considerably lower.

(b) Modifying the specific heat of the metal

$$C = \frac{\pi}{4}\left(\frac{\Delta H_f + c_S\left(T_{pour} - T_M\right)}{T_M - T_0}\right)^2 \frac{\rho_S^2}{k_m \rho_m c_m}$$

where c_S is the specific heat of the metal, ρ_S the density of the metal, and T_{pour} the pouring temperature.

$$t_f = (\text{constant})\left(\Delta H_f + c_S\left(T_{pour} - T_M\right)\right)^2\left(\frac{V}{A}\right)^2$$

The higher the specific heat of the metal the higher the necessary time to freeze the ingot will be.

(c) Modifying the melting latent heat of the metal

$$C = (\text{constant})(\Delta H_f)^2$$

$$t_f = (\text{constant})(\Delta H_f)^2\left(\frac{V}{A}\right)^2$$

The higher the melting latent heat, the higher the necessary time to freeze the ingot will be, since a higher amount of heat must be released to change from liquid to solid phase.

(d) Considering an allotropic transformation in the mold material

In this case, if the mold material suffers an allotropic transformation, then theoretically it will be able to absorb a higher amount of heat.

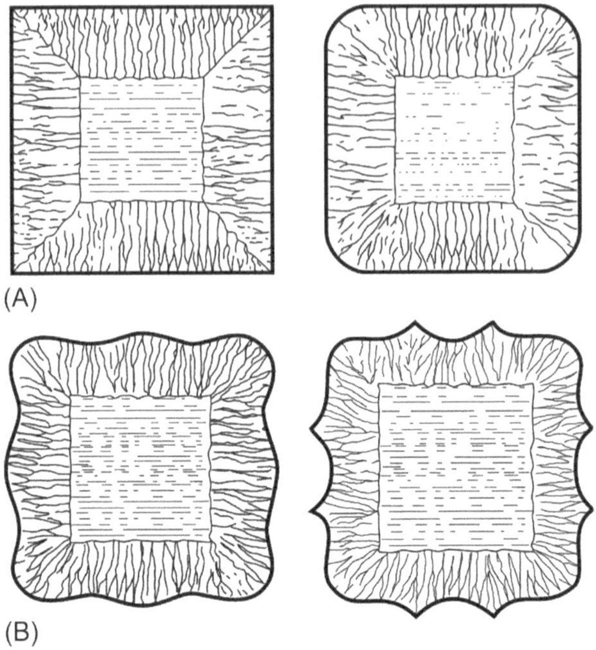

FIG. 6.3 Ingot mold sections: (A) nondesirable and (B) desirable profiles.

Besides the aspects mentioned to avoid the formation of a deep columnar structure (a problem for forging processes), another important factor is the appropriate design of the ingot molds. For example, a mold with a sharp profile as transverse section (Fig. 6.3A) will turn out to favor cracking compared to one with a curved profile (Fig. 6.3B). In an ingot mold of square transverse section, the joining of elongated grains in adjacent faces creates probable fracture paths (Fig. 2.23B).

The columnar structure is not only present in the formation of ingots but in all solidification processes with directional heat flux through a solid (this effect is evident on the microstructure of the filler metal in welding processes).

6.2 CONTRACTION CAUSED BY SOLIDIFICATION

6.2.1 Shrinkage and Solid Contraction

The density of a metal in liquid state is generally lower than the density of the same metal in its solid state. Table 6.2 indicates those density variations for common metals. Therefore, in almost all metals there is a contraction when they solidify; this tendency is not related to the T_M melting temperature.

Fig. 6.4 shows a schematic representation of the total contraction of a metal, Fe, from the liquid phase to room temperature. It is noteworthy that all metals,

TABLE 6.2 Volumetric Contraction Caused by Solidification of Some Metals

Metal	T_M	ρ_L (Measured at Temperature)	ρ_S	$\left(1-\frac{\rho_S}{\rho_L}\right)\times$ 100 (%)
Antimony	630.5	6500 (650°C)	6620	−1.81
Tin	232	6920 (300°C)	7290	−5.07
Uranium	1132.3	17,907 (1150°C)	19,070	−6.09
Lead	327.4	10,570 (340°C)	11,350	−6.87
Molybdenum	2610	9350 (2650°C)	10,220	−8.51
Niobium	2468	7830 (2500°C)	8570	−8.63
Tungsten	3410	17,600 (3450°C)	19,300	−8.81
Titanium	1668	4100 (1700°C)	4510	−9.09
Zinc	419.5	6470 (500°C)	7130	−9.25
Vanadium	1900	5500 (1900°C)	6100	−9.84
Tantalum	2996	15,000 (3000°C)	16,600	−9.64
Magnesium	650	1570 (700°C)	1740	−9.77
Zirconium	1852	5800 (1900°C)	6490	−10.63
Gold	1063	17,240 (1100°C)	19,320	−10.76
Iron	1536.5	7010 (1550°C)	7870	−10.92
Copper	1083	7970 (1100°C)	8960	−11.05
Sodium	97.8	854 (410°C)	970	−11.96
Platinum	1769	18,820 (1800°C)	21,450	−12.26
Aluminum	660	2360 (700°C)	2690	−12.27
Silver	960.8	9200 (1100°C)	10,500	−12.38
Nickel	1453	7710 (1500°C)	8900	−13.37
Cobalt	1495	7660 (1500°C)	8850	−13.45
Chromium	1875	6000 (1950°C)	7190	−16.55
Manganese	1245	5840 (1400°C)	7430	−21.40
Beryllium	1277	1420 (1500°C)	1840	−22.83
Bismuth	271.3	10,030 (300°C)	9800	+2.34
Gallium	29.5	6200 (30)°C	5900	+5.08
Silicon	1404	2510 (1450°C)	2330	+7.72

ρ_L, density of the liquid (kg/m^3); T_M, melting temperature (°C); ρ_S, density of the solid at 20°C (kg/m^3).

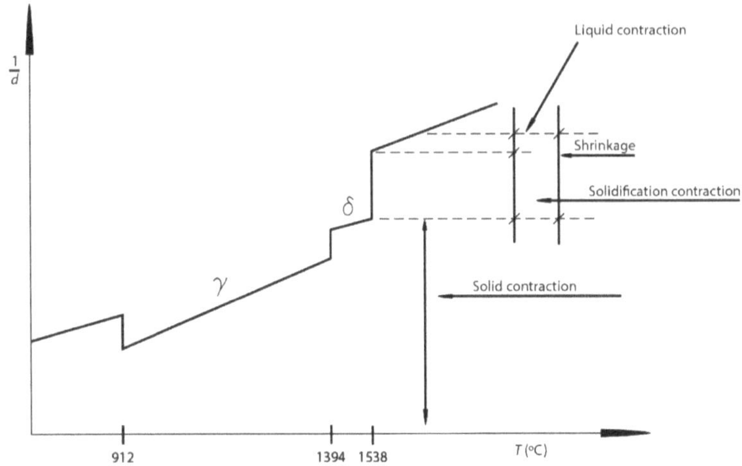

FIG. 6.4 Total contraction of iron starting from liquid state.

once solidified, continue to experiment dimensional variations during cooling. These variations can, sometimes, be expansions, produced by allotropic transformations; which is the case of Fe (Exercise 1.2).

Metals in liquid phase contract beginning from the pouring temperature until reaching the solidification temperature (*liquid contraction*); they also shrink during the transformation from liquid to solid phase, and continue to contract (eventually with partial expansion) in solid state.

The sum of the liquid contraction and the one by solidification is known as *pipe*. The higher the pouring temperature, the most accentuated the difference between the specific volume of liquid and solid will be, and thus the larger the pipe.

However, in alloys, the shrinkage phenomenon is more complex. The density, both in liquid and solid phases, depends on the values of the elements forming the alloy, and also on other factors, such as morphology of crystallization, solidification interval, porosity by gasses, etc.

Shrinkage considers both macroscopic (or *macroshrinkage*, measurable, for example, in mm of linear contraction per meter of melt) or microscopic contraction (or *microshrinkage*, appearing frequently in the interdendritic spaces filled by liquid, not connected to the rest of the liquid metal when dendrites are somewhat developed). When solidifying, and therefore shrinking, these small liquid portions, which are isolated, form internal porosities or *microshrinkage*.

The generalized formation of microshrinkage is usually produced from the moment all the periphery, or crust, of the part, or ingot, solidifies; but firstly, atmospheric pressure and metallostatic pressure force the liquid into filling the interdendritic spaces.

When the solidification structure is columnar, it presents fewer interdendritic spaces and, as a consequence, lesser microshrinkage. This occurs, for example, in parts solidified in metallic molds or chilled casts that, on the contrary, produce higher linear contractions or macroshrinkage compared to parts molded in sand. Actually, as total contraction is divided between these two types of shrinkage, the alloys with high amounts of micro have less macroshrinkage. It is interesting to point out that in industry, macroshrinkages can be eliminated by adding liquid using *risers*, yet this is not possible for microshrinkages, thus it is preferable to avoid them.

Similar to other factors, macroshrinkage will be lower in alloys with a wider solidification range than those where melting takes place at constant temperature (pure metal, intermetallic compounds, eutectics, etc.).

EXERCISE 6.2

The Al-12% Si alloy shrinks 11 mm/m during solidification. Compare this value to that of aluminum (Table 6.2) and determine which alloy would require a larger riser.

Solution

The volumetric shrinkage of Al is:

$$\frac{\Delta V}{V} \cong 12\% = 120‰$$

Furthermore, its linear shrinkage is approximately:

$$120/3 \cong 40\text{mm/m}$$

i.e., the Al-Si alloy has a lower shrinkage, almost four times less, meaning it is more probable to obtain healthy parts compared to pure Al while requiring smaller risers.

When the density of the liquid is lower than that of the solid, as it happens in most cases, shrinkage is an unavoidable phenomenon; the only characteristics that can be changed are position, shape, and dispersion. Macroshrinkage is located at the place where the liquid metal solidifies last, and it depends on the heat flux direction during cooling (Fig. 6.5) and on the massivity of either part or ingot.

In ideal conditions, where the part remains isothermal at all points during solidification and cooling, the external dimensions of the part would vary uniformly without producing any holes or pores. In this case, to obtain a "*sound*" part, it would be enough to have a mold with homothetic ratios to the real part.

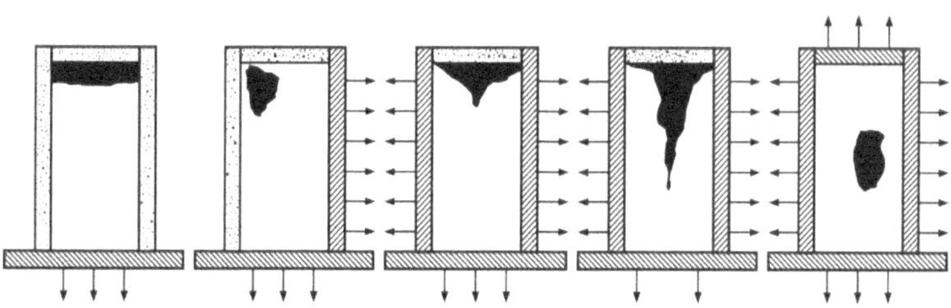

FIG. 6.5 Shrinkage (pipe) shape and location as a function of heat flux directions during cooling.

For that effect, it would be favorable to heat molds and ingots before casting, and a very slow cooling during the casting process.

That ideal solidification, with isothermal cooling at all points of the part, would be possible in metals that have the following characteristics:

- High thermal conductivity in liquid phase (which makes the temperatures of the liquid mass uniform at all times).
- High solidification latent heat (which delays the change into solid state and also contributes to the temperature uniformity).
- High specific heat in solid state (the heat of the crystals favors the thermal equilibrium of the whole mass).

In practice, solidification in the ideal isothermal conditions is almost impossible. The nature of the metal, the differences in mass, and the design of the part and the casting process, are real cooling conditions that are very far from the ideal ones. The problem of shrinkage porosity can be solved by controlling the solidification pattern and thermal gradients so that shrinkage is produced outside the proper casting body. To reduce the negative effects, such as shrinkage, in ingots of simple geometries and molded parts, *risering* techniques are widely used.

Risers are supplementary molds of the part with the function of feeding, at all times, reserve liquid metal to holes created by contraction. The riser must be joined to the main mold by gates with enough transverse section to avoid strangling and it must remain hot for longer times compared to the part or ingot so the unavoidable macroshrinkage is, at the end of the solidification, located at the riser. The riser must be made of a refractory material, and sometimes exothermal compounds are added to maintain the metal liquid in its interior for the longest period possible. Exothermal risers have the advantage that their size can be smaller than conventional ones. It is interesting how the upper part of the riser is maintained liquid during long periods (*open* or *atmospheric riser*) in order to use the atmospheric pressure effect until total solidification of the part. This avoids the formation of a vacuum that, by aspiration, would favor the formation of microshrinkages.

Since the use of open risers, feeding all points susceptible to shrinkage (massive zones, attack inlets, acute angles, etc.), is not always possible; the use of "*blind*" risers (not in contact with the atmosphere) is sometimes necessary. In order for these to cool slowly with the least surface/volume ratio as possible, spherical-type designs are preferred: blind risers sometimes have the shape of an inverted pear.

EXERCISE 6.3

Riser size is determined by two factors: the first is that the freezing time of the riser must exceed that of the casting (at least to some extent) and the second is that the riser must have sufficient feed metal to compensate for the liquid-to-solid shrinkage. The following formula can be established for the riser curve for 0.3% C steel castings:

$$X = \frac{a}{Y - b} + c$$

where X is the freezing ratio of the riser and casting, Y the riser and casting volume ratio, a a constant for freezing characteristics, b the liquid-solid solidification contraction, and c the relative freezing rate of riser and casting. And in order for the riser to provide feed metal to at least compensate the liquid-to-solid shrinkage: $a = 0.12$, $b = 0.05$, and $c = 1.0$.

If shrinkage is approximately 3% of the casting volume, design the adequate riser for the casting of a slab with dimensions $10 \times 10 \times 0.1$ in.

Solution

According to the Chvorinov formula (2.31):

$$X = \frac{V_r/A_r}{V_c/A_c} = \left(\frac{t_r}{t_c}\right)^{1/2}$$

with $X > 1$. Supposing that $X \cong 1.2$, then:

$$\frac{t_r}{t_c} \cong 1.2$$

And using the formula given for this exercise:

$$1.2 = \frac{0.12}{Y - 0.05} + 1$$

which can be graphically seen in Fig. 6.6, being

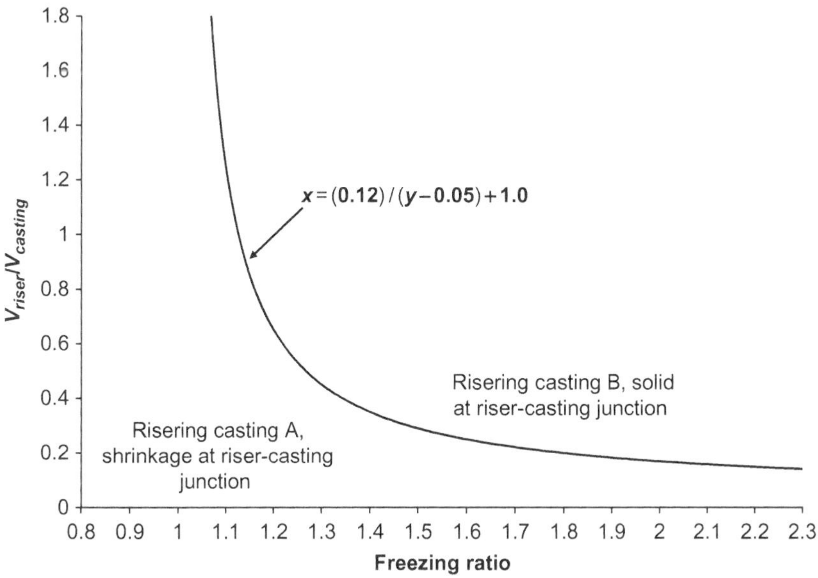

FIG. 6.6 Risering curve to calculate riser dimensions.

$$Y = \frac{V_{riser}}{V_{casting}}$$

Solving for Y:

$$Y \simeq 0.65$$

which is the necessary feed to compensate shrinkage:

$$(10 \cdot 10 \cdot 0.1) \cdot 0.03 \simeq 0.3 < Y$$

EXERCISE 6.4

Using the same method described in Exercise 6.3, design a riser system to produce a sound cubic part (of $4 \times 4 \times 4$ in.) with maximum operation performance and considering riser removal and molding procedure.

Solution

In this case, as casting becomes compact ($L = w = t$), its cooling rate decreases and the riser must be both, more compact and larger than the cast part in all dimensions. When feeding a 4 in.^{3}:

$$\left(\frac{A}{V}\right)_{casting} = \frac{4 \cdot 4 \cdot 6 \text{ faces}}{4 \cdot 4 \cdot 4} = 1.5\frac{\text{in.}^2}{\text{in.}^3} \Longrightarrow \left(\frac{V}{A}\right)_{casting} = 0.67$$

A compact cylindrical riser with a 4.5 in. diameter and a 4.5 in. height will be needed to produce a sound part, as shown by the following:

$$\left(\frac{V}{A}\right)_{riser} = \frac{\frac{\pi}{4}d^2 h}{\pi dh + 2\frac{\pi}{4}d^2} = \frac{d}{6} = 0.75$$

So:

$$\frac{(V/A)_{riser}}{(V/A)_{casting}} = \frac{0.75}{0.67} = 1.12$$

and according to Chvorinov's rule:

$$\frac{V_r/A_r}{V_c/A_c} = \left(\frac{t_r}{t_c}\right)^{1/2} \Rightarrow \frac{t_r}{t_c} = (1.12)^2 = 1.25$$

the volume ratio will be:

$$\frac{V_r}{V_c} = \frac{\pi \cdot (4.5/2)^2 \cdot 4.5}{4 \cdot 4 \cdot 4} = \frac{71.6}{64} = 1.12$$

These coordinates approximately satisfy the equation:

$$X = \frac{0.12}{Y - 0.05} + 1$$

which is the same as the one for Exercise 6.3 (Fig. 6.6) and where $X = 1.12$ and $Y = 1.12$. The performance of the casting operation can be improved by using external chillers, insulating materials in the riser or exothermal materials that decrease the riser volume.

Besides the use of risers, other precautions must be taken into consideration, when obtaining simple geometrical, compact parts, such as ingots. To obtain "*sound*" ingots, the solidification front must be as flat and horizontal as possible, and it is useful to:

- Move the isothermal close to the isobars; therefore contraction can be compensated with a new filling.
- Have similar rates to fill the mold (or *casting speed*, which depends on the diameter of the gates) and to solidify the part (that depends on cooling, among others). If the solidification rate is faster than the one to fill the mold a *cold shut* is formed, a typical ingot defect. On the other hand, if the solidification speed is lower than the casting speed, the mold is "*flooded*."

These two rules are followed with sufficient approximation in the *continuous casting* process (Fig. 6.7). However, since many solidifications do not happen this way, but in molds or ingot molds (Fig. 6.8), it is necessary to modify some factors in order for macroshrinkage, in different molded parts or in ingots, to be flat and horizontal (or at least located outside or in low stress locations). To achieve this flat and horizontal shrinkage (and to avoid it being deep) it is favorable that isothermals are horizontal, which can be obtained by:

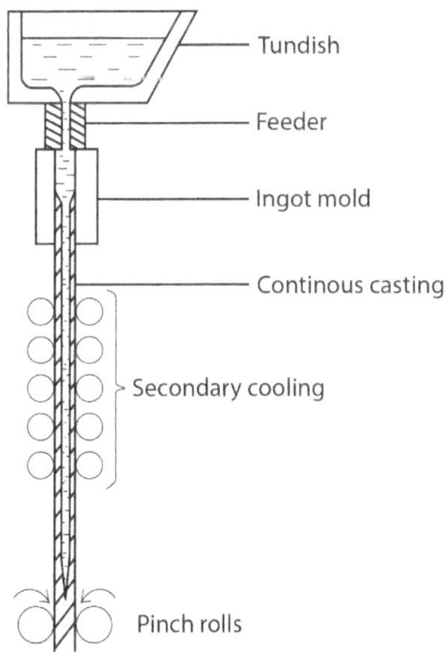

FIG. 6.7 Schematic representation of a vertical continuous casting machine.

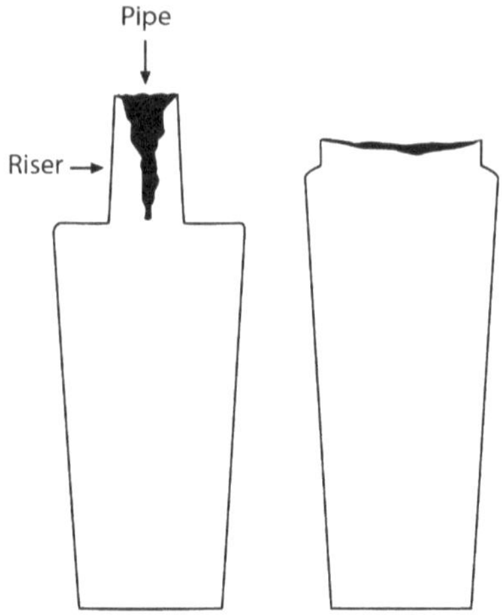

FIG. 6.8 Solidification in an ingot mold.

- Avoiding heat transfer through the lateral walls.
- Pouring by gravity rather than by syphoning. With this last procedure the bottom of the ingot mold is reheated and requires higher pouring temperatures so the liquid has enough fluidity and does not solidify in the gates before being introduced in the ingot. In this case, shrinkage would be higher compared to stream or gravity castings. Casting by syphoning would require more exothermal risers than for the gravity type.
- Constant stirring of the liquid, if possible.
- *Vacuum* casting (Fig. 6.9), so the internal pressure of the gas decreases (gas tends to enlarge the cavities of the porous zones that surround the shrinkage). In fact, the oversaturated gasses, which are released during solidification, tend to be located in the shrinkage which is usually rich in high pressure gas.
- Not opening the mold and removing the part before total freezing. On the contrary, the liquid would cool through the crust of the ingot more quickly than in the interior of the ingot mold, resulting in a deeper shrinkage. Drafting before time is usually made to increase the life of the ingot molds.
- Delaying cooling of the upper part of the ingot and favoring the growth from the bottom. Thus, the design of the ingot molds is done to favor flattening isotherms.

Considering the ingot molds design and the horizontality of the isotherms:

- Among the straight ingot molds (Fig. 6.10B) the ones with low slant are desirable; in other words, those with small height/diameter ratio.
- Analyzing tapered ingot molds, the desirable ones are those with higher section in the upper part (ingot molds with inverse conical shape, Fig. 6.10C), but these ingot molds have difficulties to draft. The easiest ones to draft and

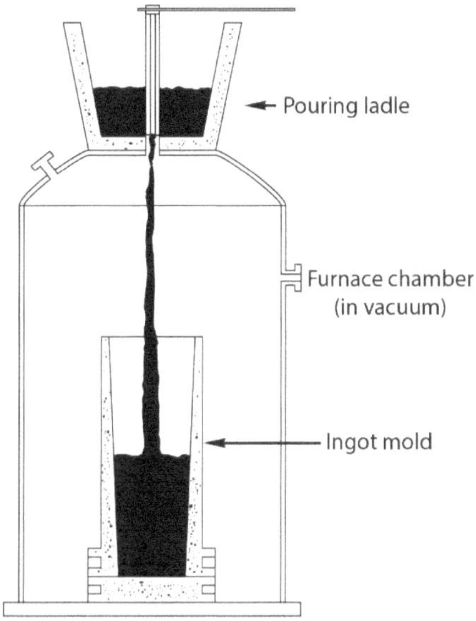

FIG. 6.9 Vacuum casting of an ingot.

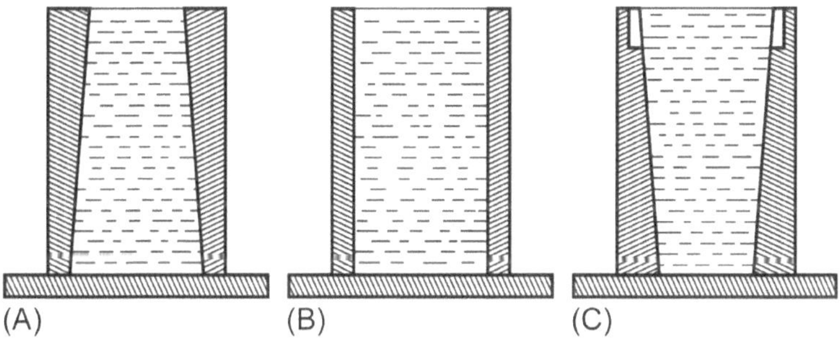

FIG. 6.10 Ingot molds: (A) direct-tapered, (B) straight, and (C) inverse-tapered shapes.

the lowest losses in scraps are obtained with ingot molds with direct conical shapes (Fig. 6.10A), whose higher section corresponds to the bottom and tapered by $\sim 1\%$; shrinkage would be thicker and to solve it, risers are useful. Risers are also helpful because not only do they contain the shrinkage pipe but also because they lodge undesirable gasses and inclusions. In other words, it is the "trash can" of the casting process.

Macroshrinkage can be external (if they are in contact with the atmosphere) or internal. In ingots used in mechanical conformation processes, the detrimental effect of the external shrinkage is evident since its surface, oxidized by air, cannot be welded by recrystallization during rolling or forging processes. If microshrinkage has, instead, smooth, and clean surfaces, it can be closed under the pressure of the rollers; however, if there is slag in the internal shrinkage surfaces, it will prevent, in the same way, a successful welding (seams defect).

In some cases, due to the possibility of eliminating internal voids by recrystallization during forming, bubbles distributed along the metallic mass are not a disadvantage since they compensate the formation of macroshrinkage (e.g., *rimmed steels*). However, for molded parts that will not be mechanically formed later, that internal porosity is not desired since it decreases mechanical characteristics of the part in service.

In molded parts, it is not easy to guide solidification according to the previous considerations, and the ideal isothermal solidification is not feasible at all points. Therefore, another principle, contrary to the ideal isothermal solidification, is followed: solidification must start at those points farthest from the riser and continue towards it, in order for shrinkage to be located at the riser; this is known as *guided solidification* (Fig. 6.11).

An oriented solidification towards only one riser is difficult, so risers must be located at those points susceptible to suffering shrinkage. Also starting points of solidification must be created artificially inside the area of influence of each riser (removing heat by means of internal coolers or *chillers*, for example, rods of the same metal, or external cooler, using more conductive mold materials) while isolating other points so solidification moves towards the closest riser.

When a region of the part is formed by a horizontal cylinder of small diameter in relation to its generatrix, the formation of shrinkage along the axis of the cylinder is probable; to avoid it, the recommendation is to gradually increase the section of the cylinder towards the riser to have a favorable thermal gradient.

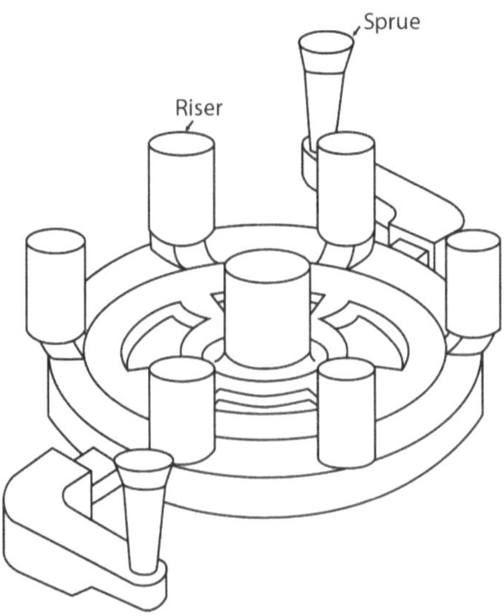

FIG. 6.11 Guided solidification.

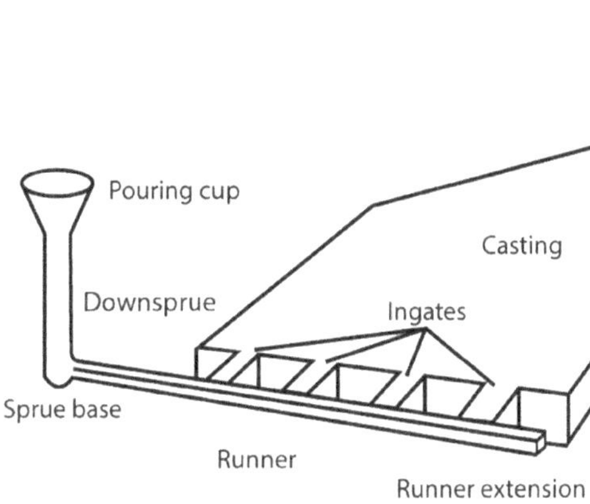

FIG. 6.12 Casting system for a molded part.

Another aspect to consider when sketching and designing the part (in the casting position) is that for each horizontal cross-section, the thickness of the part must be higher than the one for the immediately lower section. Both aspects (regularly growing thicknesses from the bottom upwards as well as horizontally growing towards the risers) are basic principles when designing molded parts.

The design and location of risers must consider that shrinkage is produced at those locations where solidification occurs at later times: zones with higher massivity compared to their neighbors, places reheated by the continuous flux of metal, connection of walls, hot spots, parts with angles, etc. Usually, the solution is not using multiple risers to avoid shrinkages but studying the design of the part (most of the times acute angles can be avoided while effectively reaching the desired tolerances, making the section increments gradual, and avoiding local masses close to others with lower mass, etc.) as well as specifying the way the mold must be fed.

The casting system is usually formed by *pouring cup*, *sprue*, and *gating system*. The sprue holds the metal that was poured from the ladle and guides it to the gating system without dragging air, sand, or slag. The sprue must produce enough metallostatic pressure for the liquid to pass through the remaining casting system, enter the mold, and completely fill it. The metal must enter the mold without turbulences in order to fill it in a laminar manner. The gating system has, precisely, the function of slowing the speed of the liquid, and allowing its distribution through various inlets for a fast filling of the mold (Fig. 6.12), which reduces the temperature differences and diminishes the possibility of cold shuts.

6.3 EVOLUTION OF GASSES DURING SOLIDIFICATION

In liquid phase, metals usually retain large volumes of either dissolved gasses or in the form of unstable liquid compounds. For example, liquid iron dissolves carbon monoxide, carbon dioxide, hydrogen, nitrogen, water vapor, oxygen (which is the main gas dissolved in liquid steel), etc. Liquid copper dissolves hydrogen, oxygen, and sulfur dioxide among others. Hydrogen is also dissolved by aluminum, silver, and tin.

When the temperature of the melt decreases up to the solidification threshold, a fast decrease on solubility is produced and gasses are released from the liquid metal; this release continues along the solidification process. When internal gas pressure becomes high enough, a pore or bubble is formed which will be stable inside the liquid as long as the gas pressure inside the bubble is high enough to balance external forces: liquid-vapor surface energy, metallostatic, and environmental pressures. Thus, for a stable bubble:

$$P_g = P_0 + \rho_L g h + \frac{2\gamma_{GL}}{r} \tag{6.1}$$

where P_g is the total gas pressure inside the bubble, P_0 the environmental pressure, $\rho_L gh$ constitutes the metallostatic pressure, and the term $2\gamma_{GL}/r$ is the pressure over the bubble as a result of the bubble-melt interfacial energy. Assuming the gas inside the bubble is uniform and at equilibrium with the dissolved gasses in the surrounding metal, P_g is the gas pressure calculated through the following thermodynamic expressions for hydrogen:

$$2\underline{H} \rightleftharpoons H_2 \tag{6.2}$$

$$K_H = \frac{P_{H_2}}{\underline{H}^2} \tag{6.3}$$

$$\%H_L = K'_H \sqrt{P_{H_2}} \tag{6.4}$$

$$\%H_S = K''_H \sqrt{P_{H_2}} \tag{6.5}$$

where $\underline{H}$ indicates the dissolved hydrogen in the liquid metal (in cm^3 of gas/100 g STP). The amounts of hydrogen dissolved in the metal (in both solid and liquid) at equilibrium with a partial pressure of hydrogen (P_{H_2}) are given by the Sievert's law (Eqs. 6.4, 6.5).

EXERCISE 6.5

An Al alloy has a solubility of 0.4 cm^3 H_2/100 g STP. Determine the condition (% of liquid left) during solidification for H_2 to start boiling.

Data: Partition coefficient $K = 0.77$, maximum solubility of hydrogen in solid Al is 0.04 cm^3 H_2/100 g STP.

Solution

Making a balance of matter in hydrogen when the alloy has solidified:

$$f_L + f_S = 1$$

$$(1)(0.4) = f_S(0.04) + f_L(0.77)$$

$$f_L = \frac{0.4 - 0.04}{0.77} = 0.4675 = 46.75\%$$

The release of gas starts when the amount of liquid is 46.75%.

EXERCISE 6.6

Calculate the existing porosity, referred to 100 g of solidified Al (a) at atmospheric pressure, (b) in vacuum (85 mm Hg), and finally (c) determine which external pressure would be necessary to completely eliminate hydrogen. Use data from Exercise 6.5.

Solution

(a) At atmospheric pressure and 660°C (933 K)

Calculating the constant of Eq. (6.4):

$$\log K = -2550/T + 2.62$$

$$\log K = -\frac{2550}{933} + 2.62 = -0.113$$

$$K = 0.771$$

Using Sievert's law:

$$S\left(\text{cm}^3/100\text{g}\right) = 0.771\sqrt{P_{H_2}}$$

$$P_{H_2} = 1 \Rightarrow S = 0.771\,\text{cm}^3/100\text{g}$$

Considering the density of Al, 100 g occupy a volume of:

$$V = \frac{m}{\rho} = \frac{100\text{g}}{(2.7\text{g/cm}^3)} = 37.037\,\text{cm}^3$$

The porosity can be calculated through:

$$\text{Porosity} = \frac{S_L - S_S}{V} = \frac{0.771 - 0.04}{37.037} = 0.019 \approx 0.02\,(2\%)$$

(b) If casting is made in vacuum at a pressure of 85 mm Hg then:

$$S = 0.771\sqrt{\frac{85}{760}} = 0.258\,\text{cm}^3/100\text{g}$$

The porosity will now be:

$$\text{Porosity} = \frac{0.258 - 0.04}{37.037} = 0.0058 \approx 0.6\%$$

which is almost four times lower than at atmospheric pressure.

(c) The pressure required to completely eliminate hydrogen would be:

$$0.04 = 0.771\sqrt{P_g} \Rightarrow P_g = \left(\frac{0.04}{0.771}\right)^2 = 2.692 \times 10^{-3}\,\text{atm} = 2.046\,\text{mmHg}$$

making this an extremely high vacuum process which is impossible in industrial operations.

EXERCISE 6.7

A pure aluminum melt contains hydrogen in the amount of 0.5 cm^3 STP/100 g of Al. Assume bubbles form at equilibrium with negligible effect of pressure head and surface tension, and use data from Fig. 6.13.

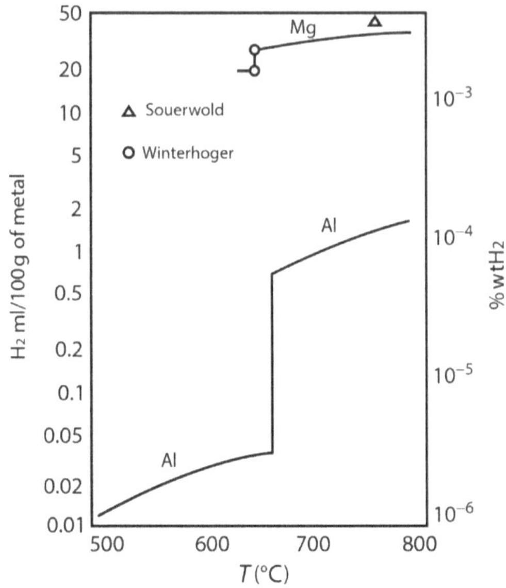

FIG. 6.13 Solubility of hydrogen at atmospheric pressure in aluminum and magnesium.

(a) Calculate the volume of hydrogen in a 100 g casting, solidified at atmospheric pressure.

(b) A widely used quality-control test in aluminum alloys is the "residual pressure tester": a small sample of the Al melt is placed inside a bell jar and allowed to solidify under reduced pressure (vacuum melting), as a result the void volume is increased to an amount visible to the naked eye. Calculate the void volume in a 100 g sample solidified in this manner. The reduced pressure tester is operated at 85 mm Hg absolute pressure.

(c) Calculate minimum pressure to reduce the size of gas pores in order to make them invisible to optical microscopy.

Solution

(a) Hydrogen volume at atmospheric pressure

If aluminum density is $\rho_{Al} \simeq 2700\,\text{kg/m}^3$, then 100 g Al would occupy $3.70 \times 10^{-5}\,\text{m}^3$. As the pores occupy $0.5\,\text{cm}^3 = 5 \times 10^{-7}\,\text{m}^3$, then the porosity would be:

$$\text{Porosity} = \frac{5 \times 10^{-7}\,\text{m}^3}{3.70 \times 10^{-5}\,\text{m}^3} \cdot 100 \simeq 1.35\%$$

And according to the Ideal Gas Law, the volume occupied by pores would be equivalent to:

$$P \cdot V = \frac{a}{M} \cdot R \cdot T$$

$$1 \times 10^5\,\text{Pa} \cdot 5 \times 10^{-7}\,\text{m}^3 = \frac{a}{2\,\text{g/mol}} \cdot 8.314 \frac{\text{Pa}\,\text{m}^3}{\text{mol}\,\text{K}} \cdot 273\,\text{K}$$

$$a = 4.41 \times 10^{-5}\,\text{g}\,H_2/100\,\text{g}\,\text{Al}$$

which means there are $0.441 \times 10^{-6}\,g\,H_2$ or 0.44 ppm of hydrogen dissolved in the liquid melt (Fig. 6.13).

(b) Void volume

Using the Ideal Gas Law, considering the amount of dissolved hydrogen is the same, and knowing that the atmospheric pressure equals 760 mm Hg, the volume occupied by porosity would be:

$$\frac{5 \times 10^{-7}\,m^3}{(85/760)} = 4.47 \times 10^{-6}\,m^3$$

and the amount of porosity would now be:

$$\text{Porosity} = \frac{4.47 \times 10^{-6}\,m^3}{3.70 \times 10^{-5}\,m^3} \cdot 100 \simeq 12.08\%$$

In industrial environments, the solubility of hydrogen in solid state is considered negligible, though it is approximately equal to $0.01\,cm^3/100\,g$ Al at room temperature.

(c) Pressure to eliminate pores

If

$$P_{external} > 2\gamma/r$$

where $\gamma = 0.84\,N/m$ and $d = 1\,\mu m$ (minimum size observed by optical microscopy), then:

$$P_{external} = \frac{2(0.84\,N/m)}{1 \times 10^{-6}\,m} = 3.36\,MPa = 33.6\,atm$$

meaning that any pressure above 34 atm will reduce the bubble size, making them invisible in optical micrographs. Injection casting uses pressures of $300 \sim 500$atm which will completely eliminate porosity: pressure reduces porosity, while vacuum increases it.

The shape and size of pores is related to the amount of solidified metal: large and near spherical shapes with low amounts, and elongated and with the shape of the interdendritic spaces ("worm" shape) with high amounts. In this last case, it is difficult or nearly impossible to distinguish the factor that originated the pore (gas or shrinkage).

All these gas porosities, bubbles, blowholes, etc. decrease both the useful section of molded parts and their resistance. Given that the presence of such cavities is almost always hidden to direct observation, molded parts are less reliable than forged ones (unless they are correctly analyzed with X-ray, ultrasound, gamma-ray, etc. to assure the absence of internal cavities).

Bubbles in ingots subsequently used in thermomechanical forming, are almost harmless if they are located inside the metal and not too close to the surface; its walls are welded and recrystallized during hot-rolling or forging.

As previously mentioned, sometimes the presence of internal bubbles is used to compensate shrinkage. However, if porosity appears on the surface of the ingot or is interconnected to its periphery, oxidation prevents welding in subsequent operations. It is convenient to eliminate superficial bubbles by milling before rolling processes.

Disadvantages caused by gasses are not limited to the possibility of producing bubbles and porosity. Gasses can, for example, produce *solid compounds* with the melt. The formation of Cu_2O at the grain boundaries of deoxidized Cu (Section 4.3.3) is a typical example of this case. The absorption of oxygen in bronzes (CuSn), which later forms SnO and makes the material more brittle, is another example of chemical compound formation.

In other cases, gasses dissolved in the liquid are retained in the solid (*interstitial solid solution*): such is the case of nitrogen in steels, responsible for aging[1] after cold deformation.

Another typical example of interstitial solid solution, with a disadvantageous influence in the properties of the material, is hydrogen in steels. As an effect of the external or internal stresses, hydrogen has a tendency to be released in the molecular state, producing very high tensile stresses and nucleation of cracks (embrittlement).

EXERCISE 6.8

Argon is used to form bubbles (to degasify steel) by injecting it through a ladle containing liquid steel, in order to drag the dissolved N_2 and H_2 in the melt. Gasses above the melt ($P = 1\,\text{atm}$) include 20% in volume of N_2, 10% of H_2, and the rest is Ar being removed progressively. Calculate: (a) the maximum solubility of both N_2 and H_2 in the melt and (b) the vacuum necessary to reduce the H_2 content to 1 ppm.

Data: The Sievert law constant for N_2 is 32 and for H_2 is 27, and the curve of solubility of nitrogen in iron is shown in Fig. 6.14.

Solution

(a) Maximum solubility of nitrogen and hydrogen

From Fig. 6.14, the solubility of nitrogen is 0.04% per 100 g Fe or:

$$400\,\text{ppm} \simeq 0.04\,\text{g}\,N_2/100\,\text{g}\,\text{Fe}$$

And applying the Ideal Gas law considering atmospheric pressure and room temperature, the volume would be:

$$P \cdot V = n \cdot R \cdot T$$

$$V = \frac{n \cdot R \cdot T}{P} = \frac{\frac{0.04\,\text{g}}{28\,\text{g/mol}} \cdot 8.314 \frac{\text{Pa}\,\text{m}^3}{\text{mol}\,\text{K}} \cdot 273\,\text{K}}{1 \times 10^5\,\text{Pa}} = 3.24 \times 10^{-5}\,\text{m}^3\,N_2/100\,\text{g}\,\text{Fe}$$

1. An aged material has higher yield and fracture stresses compared to the unaged one and its stress-strain curve has a plateau (Luders' strain) after tensile stress as well as smaller elongations.

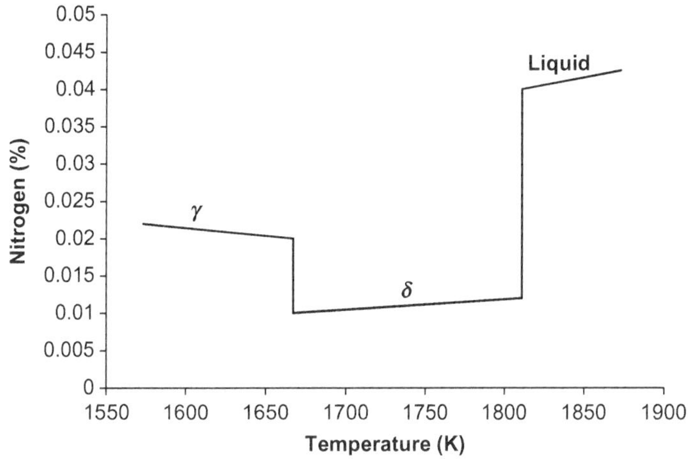

FIG. 6.14 Solubility of nitrogen in iron as a function of temperature.

For a pressure of:

$$P = \frac{0.2\,N_2}{(0.2\,N_2 + 0.1\,H_2 + 0.7\,Ar)} = 0.2\,\text{atm}$$

$$(S_{N_2})_L = 32\sqrt{0.2} = 14.31\,\text{cm}^3/100\,\text{g} \simeq 180\,\text{ppm}$$

For a pressure of $H_2 = 0.1\,\text{atm}$:

$$(S_{H_2})_L = 27\sqrt{0.1} = 8.53\,\text{cm}^3/100\,\text{g} \simeq 7.5\,\text{ppm}$$

(b) Vacuum to reduce hydrogen content

The vacuum necessary to reduce the H_2 content to 1 ppm ($1 \times 10^{-4}\,\text{g}H_2/100\,\text{g}Fe$) can be determined through:

$$V = \frac{n \cdot R \cdot T}{P} = \frac{\dfrac{1 \times 10^{-4}\,\text{g}}{2\,\text{g/mol}} \cdot 8.314 \dfrac{\text{Pa}\,\text{m}^3}{\text{mol}\,\text{K}} \cdot 273\,\text{K}}{1 \times 10^5\,\text{Pa}} = 1.13 \times 10^{-6}\,\text{m}^3\,H_2/100\,\text{g}Fe$$

And applying the Sievert's law:

$$(S_{H_2})_L = 27\sqrt{P_{H_2}} \Rightarrow P_{H_2} = \left(\frac{1.13\,\text{cm}^3}{27}\right)^2 = 1.75 \times 10^{-3}\,\text{atm} = 1.33\,\text{mmHg}$$

which is evidently lower than the atmospheric pressure and almost impossible to achieve under industrial conditions.

Irrespective of the origin of gasses inside an alloy (bubbles, porosity, chemical compounds, solid solutions, etc.), the general recommendations to avoid their disadvantageous effect can be summed up in two: reduce the amount of gasses dissolved in the liquid and aid the exit of the gasses from the mold.

The higher the casting temperature, the higher the amount of gasses dissolved in liquid state will be. Therefore, measures to eliminate dissolved gasses are particularly interesting in the case of alloys with high melting temperature (even small turbulences may cause a high release of gasses during solidification and the liquid would appear to be boiling, as is the case of rimmed steels).

The first step to effectively diminish the amount of gasses dissolved in the melt, is to know the chemical-physical characteristics of the reactions and equilibrium in liquid state in order to avoid oxidized states (using the right deoxidizing agent, for example, Si, Al, and Ti in steels) or sulfurized states (e.g., using alkaline carbonates in bronzes), etc. *Chemical correctors* must be used in the right amounts; bubbles or porosities can be less harmful than shrinkage (caused by an excess of deoxidizing agents, inclusions, or chemical segregations).

In order to reduce the gas evolution during solidification of metals and alloys, as much as possible, it is desirable to:

- Use low casting temperatures (copper-aluminums, for example, have high melting temperatures between 1083°C and 1037°C and it is necessary that their pouring temperatures do not exceed the *liquidus* line by more than 100°C).
- Maintain the metal for a considerable time in the ladle, so gasses can escape.
- Cast in vacuum.
- Reduce the moisture in sand molds.
- Decrease the proportion of organic binders in nonmetallic molds.
- Avoid the drag of air into the mold.

To ease the release of gasses in the mold it is useful to put air outlets (*taps*) in the upper part of the mold and have holes that penetrate up to the main cavity (*vent*), to allow the escape of gasses.

It is also useful to increase the mold permeability to gasses with thicker sand (and reducing the content of fine grains in the sand, since they fill the holes of the larger grains and reduce permeability) or with a lower packing of the sand.

Although it is a recommendation to ease the release of gasses through the mold, sometimes, it is useful to ignore this aspect by a fast solidification, with an early formation of a metallic crust, so the solid crust protects the internal cavities from contact with air.

To eliminate the gaseous elements present in the form of *solid solutions*, it is necessary, when possible (as in the case of hydrogen in steels), to heat treat the material for long periods to improve diffusion of the gas towards the exterior. To remove free hydrogen from steel, Al is used as a chemical corrector.

6.3.1 Boiling Produced by Carbon Monoxide in Steels

Most metals are cast with the lowest gas amount possible in order to obtain a dense casting or ingot. Exceptions to this are *semikilled* and *rimmed* steels, in which boiling produced by CO during solidification is used to compensate the solidification shrinkage and thus avoid *pipes* at the top of the ingot. Blow holes formed by CO are subsequently closed by mechanical processing such as rolling or forging.

As a result of the manufacturing process, liquid iron has dissolved gasses, from which oxygen, always present, is the main one. When a steel has been refined in the converter or in a ladle, until reaching the specific chemical composition, the liquid always contains considerable amounts of dissolved oxygen.

Even though the carbon of the steel is not used as a deoxidizing agent (more powerful deoxidants such as Ti, Al, Mn, Si, etc. are used), it plays an important role in relation to the amount of dissolved oxygen due to the chemical equilibriums established between FeO, Fe, C, and O. At each temperature, the lower the %C, the higher the oxygen content in the liquid iron will be. For example, at 1600°C, for steels with C higher than 0.3%, the oxygen percentage in equilibrium is only 0.005%; meanwhile for steels with 0.2% C it increases to 0.02% and even more for steels with $<0.15\%$ C $(0.05 \sim 0.1\%)$.

In liquid state, oxygen and carbon in solution react to form carbon monoxide that tends to escape. The saturation of CO in the liquid is higher when temperature drops, as the chemical equilibrium is displaced towards the formation of this compound.

Formation of gas bubbles in the melt is not spontaneous: if there are no substrates to promote nucleation, the melt is simply oversaturated in CO but no bubbles are formed. On the other hand, in the presence of substrates or heterogeneous nuclei (such as coarse ingot mold walls, solid inclusions, etc.) bubbles appear even with low or null oversaturation.

Spontaneous nucleation in the interior of a liquid, of a bubble (formed by molecules of CO) should exceed a high external pressure, combined result of superficial tension of the liquid, ferrostatic pressure head, and atmospheric pressure. If only surface tension is considered, a spherical gas pore would be under the following external pressure of the liquid:

$$p = \frac{2\gamma_{LG}}{r} \tag{6.6}$$

where γ_{LG} is the liquid-gas surface tension and r the radius of the bubble. The pressure of the liquid on the bubble is inversely proportional to its diameter, and it is impossible for a few molecules of CO to form bubbles by homogenous nucleation. Thus, bubble-nucleation is inhibited, such as nucleation of a solid from a liquid. Bubbles are formed, therefore, heterogeneously over a solid substrate, and afterwards will grow to such an extent that they overcome ferrostatic head and atmospheric pressures. Furthermore, ferrostatic pressure head is important, especially on large ingots: for example, a steel column with a height of 1470 mm (approximately that of an ordinary ingot) is equivalent to one atmosphere of pressure; the external pressure at the bottom of the ingot mold would then be two atmospheres.

EXERCISE 6.9

Calculate the gas pressure required to form a stable bubble, with a diameter of 1 μm, near the surface of (a) an aluminum melt and (b) a steel melt, at ambient pressure (1 atm).

Data: $\gamma_{LG_{Al}} = 0.84\,N/m$ and $\gamma_{LG_{Steel}} = 1.5\,N/m$.

Solution

Using Eq. (6.6):

(a) Aluminum

$$P = \frac{2\gamma_{LG}}{r} = \frac{(2)(0.84\,N/m)}{0.5 \times 10^{-6}\,m} = 3.36\,MPa = 33.6\,atm$$

(b) Steel

$$P = \frac{2\gamma_{LG}}{r} = \frac{2(1.5\,N/m)}{0.5 \times 10^{-6}\,m} = 6\,MPa = 60\,atm$$

Comment: Both pressure values are, evidently, extremely high, though the zone affected by the bubble is small.

The instant the liquid is poured into the mold, it reacts violently (with considerable carbon boil) according to the following reaction:

$$C + O \leftrightarrow CO \tag{6.7}$$

Fig. 6.15 illustrates the formation and morphology of the first bubbles or *skin bubbles* (blowholes): the movement of the solidification front compresses the bubbles and gives them an elongated or horn/worm shape. In a transverse section of the ingot, these skin bubbles, either touch the wall of the ingot mold (external bubbles) or are surrounded by a peripheral zone (outside crust) free of bubbles that is known as *ingot skin* (Fig. 6.16).

When the head of the ingot solidifies and the new gas bubbles formed next to the already solidified crust cannot escape and are trapped, sometimes a crown of bubbles is formed, generally with rounded shapes, homothetic to the surface of the ingot, and known as *rim-core junction* because it results from the blocking or closing of the head of the ingot (Fig. 6.17). The region of the ingot, including skin and rim blowholes, that solidifies before the rim-core junction is formed, is known as *pure zone* (resulting from an oriented solidification model, Section 5.1).

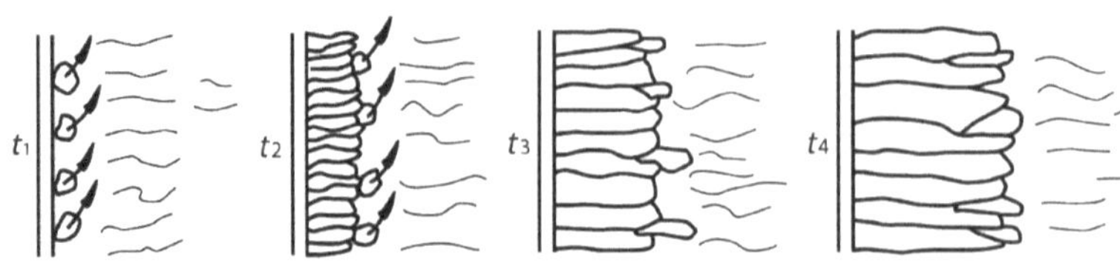

FIG. 6.15 Mechanism of formation of blowholes caused by gas evolution.

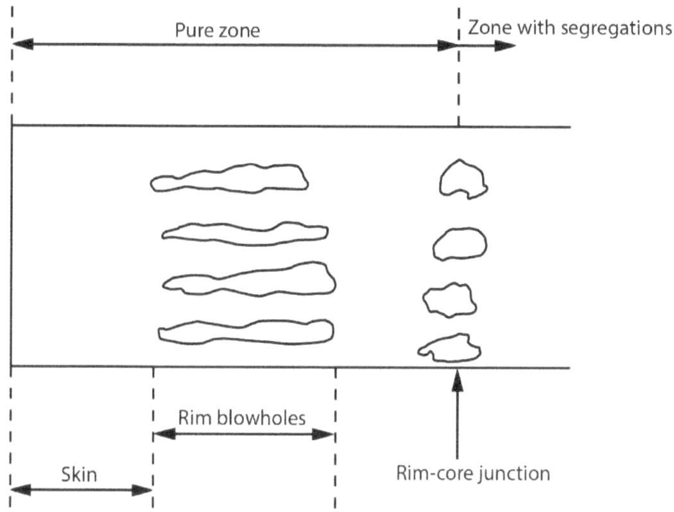

FIG. 6.16 Structure of the pure zone in rimmed steels.

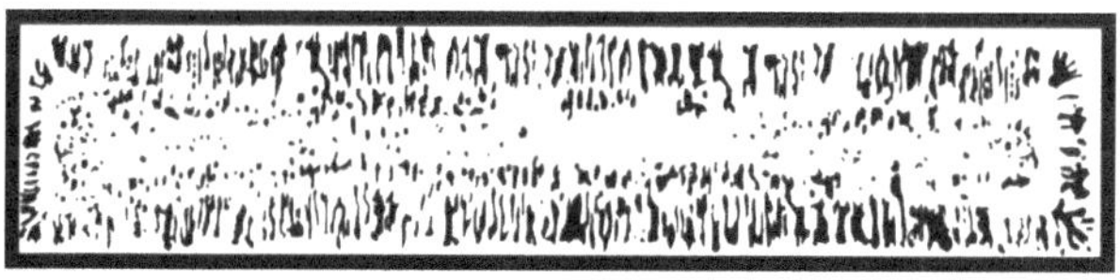

FIG. 6.17 Cross-section of a rimmed steel slab showing elongated and densely distributed blowholes and rim-core junction.

In the upper half of the ingot some isolated bubbles usually appear since the segregated carbon concentration (and saturation of CO) is high; furthermore, the weak ferrostatic pressure head and the nucleation substrate of macro and microshrinkages, promotes the formation of these bubbles.

EXERCISE 6.10

A plain carbon steel is melted from a charge of pig-iron and scrap steel, and iron ore is added to obtain a carbon boil and reduce the carbon content to 0.02%.

(a) When the carbon content reaches 0.02%, calculate the oxygen content in the steel.

(b) Calculate the critical oxygen concentration to avoid porosity.

(c) Determine the amount of silicon to be added in order to reduce carbon.

Note: Both melting and solidification take place at 1550°C, and at this temperature the equilibrium constant for the $CO(g) \leftrightarrow C + O$ reaction is:

$$K = \frac{(\%C)(\%O)}{P_{CO}} = 1.97 \times 10^{-3}$$

where %C and %O are in wt% and P_{CO} in atm.

When the oxygen solubility in the solid is negligible, the equilibrium constant for the $Si + 2O \leftrightarrow SiO_2$ reaction at 1550°C is:

$$K = (\%Si) \cdot (\%O)^2 = 1.14 \times 10^{-5}$$

For this exercise, ignore the effects of surface energy and metallostatic head.

Solution

(a) Oxygen content in the steel

From the C, O, and CO equilibrium:

$$(\%O) = \frac{1.97 \times 10^{-3}}{0.02} \simeq 0.1\% (1000\text{ppm})$$

(b) Critical oxygen concentration to avoid porosity

C and O burn with their respective atomic weights:

$$\frac{O}{C} = \frac{16}{12} = \frac{4}{3} = 1.33$$

A straight line can be constructed with a slope of 1.33, and knowing the (0.02, 0.1) point, the equation of this line would be $y = 1.33x + 0.073$. Both the concentration of C and O as well as this line is shown in Fig. 6.18, with the intersection between curve and straight line at (0.02, 0.1).

With this oxygen content (0.073), all the C has been burnt and boiling stops.

(c) Amount of silicon to reduce carbon

Si has a higher affinity for O than for C:

$$Si + 2O \leftrightarrow SiO_2 \text{ (inclusions)}$$

$$[Si] \cdot [O]^2 = 1.14 \cdot 10^{-5} \Rightarrow [Si] = 0.0011\% (11\,\text{ppm})$$

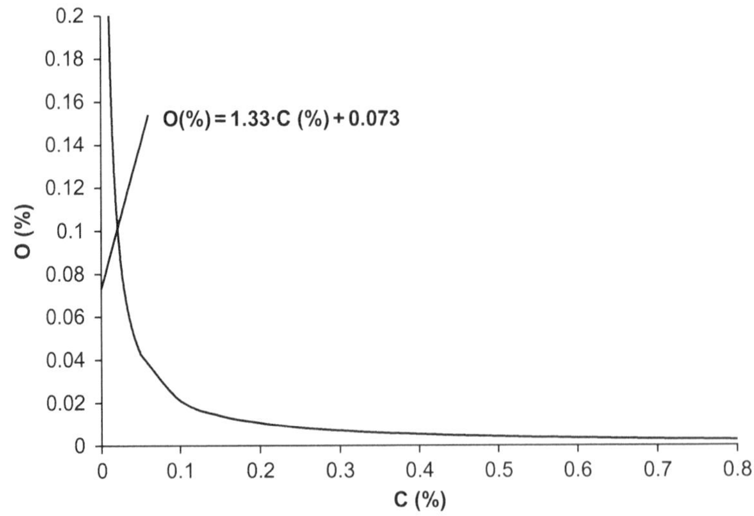

FIG. 6.18 Carbon-oxygen relationship under equilibrium conditions at 1550°C.

This amount of Si would correspond to the equilibrium with O (0.1%) in the liquid. If this equilibrium is broken (or temperature decreases), silica (SiO_2) inclusions are formed. The minimum amount of Si necessary to neutralize oxygen and avoid boiling will be:

$$0.1 \cdot \frac{28}{32} \simeq 0.090\%$$

6.3.2 Macrostructure of Steel Ingots According to Their Degree of Deoxidation

Killed steels are deoxidized in order to avoid release of gas during solidification, which can be achieved by the addition of Fe-Si and/or Al to the ladle, or by vacuum degassing (exposing the melted metal to a low environmental pressure to remove all the dissolved oxygen, without adding deoxidizing agents).

The head of the ingot killed with deoxidants is slightly concave and presents macroshrinkage in the upper zone (Fig. 6.19A). Usually these defects are avoided by using risers that contain the shrinkage, though continuous casting

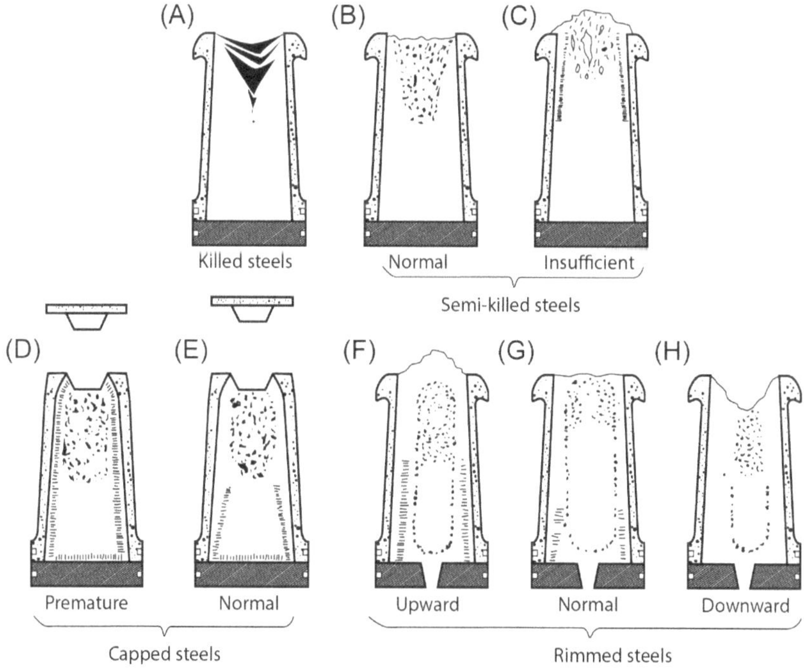

FIG. 6.19 Morphology of ingots with different states of deoxidation: (A) killed, (B) normal semikilled, (C) insufficient semikilled, (D) prematurely capped, (E) normal capped, (F) upward rimmed, (G) normal rimmed, and (H) downward rimmed steels.

is also used; the progress of this technique and the absence of gaseous release during solidification has displaced, in a large part, the casting of killed steels in ingot molds.

Killed steels can be produced with any amount of carbon content and other alloying elements, in order to hinder the rimming action, resulting in homogeneous, and sound parts. All steels with more than 0.30% C are of the killed type, since the amount of oxygen dissolved in the liquid, given its equilibrium with C, is small. Thus, the gaseous release in solidification does not reach the levels required for the semikilled, capped, or rimmed types. Furthermore, low carbon steels can also be of the killed type, except those with large amounts of dissolved oxygen requiring deoxidant (raising the costs of the process and producing a large number of inclusions). As previously mentioned, by vacuum casting, it is possible to obtain inclusion-free (0.5% volumetric fraction) low carbon killed steels.

Steels are *semikilled* when they have more dissolved oxygen than killed steels and are able to produce, by reaction with the C of the steel, a slight release of CO (that approximately compensates for shrinkage). The bubbles appear at the end of solidification, when temperature is lower and at those places where the ferrostatic pressure head is lower, i.e., the upper half of the ingot.

Carbon content in semikilled steels is $0.15 \sim 0.30\%$C for structural applications, and their deoxidizing level is lower than in killed steels. Deoxidation with Fe-Si and some Al is usually made either in the ladle or, sometimes, in both ladle and mold. If deoxidation is not enough, the ingot can adopt the morphology shown in Fig. 6.19C: the pressure of the gas in the region that solidifies last can sometimes break the already solidified head of the ingot and force the residual liquid out, producing bleeding and skin bubbles.

When there are very large gas emissions, as in the case of steels with high oxygen content in the liquid and carbon percentages lower than 0.15% C, a very pure skin is formed in the outer zone of the ingot (Section 5.2.1.1).

Gas evolution, in the boiling mode, is not produced instantaneously when pouring the ingot, but when solidification begins, and with it, the abrupt drop in solubility of gasses dissolved in the liquid. Furthermore, when CO is formed, oxygen and carbon are eliminated from the melt in the proportion corresponding to their atomic weights. This is why the concentration of oxygen in the remaining liquid decreases faster than the carbon content, which influences the rate of subsequent gaseous release.

If boiling is considerable, a lot of bubbles are formed in the top and bottom of the ingot. Bubbles at the top are released from the solid front and escape the liquid mass, while at the bottom of the ingot, starting from a certain solidification time, bubbles cannot overcome the ferrostatic pressure head and become trapped, with worm-like (ellipsoidal) shapes, located at the bottom half (the closer to the bottom of the ingot, the closer to wall they will be, Fig. 6.19F and G). The amount of entrapped bubbles can be higher than the magnitude of shrinkage and the top of the ingot can be convex.

As boiling increases, a higher number of bubbles are formed at the top and bottom of the ingot, and therefore they can escape more easily, while skin bubbles only appear on the bottom of the ingot and farther away from the periphery. High boiling diminishes the amount of blowholes and approximately compensates shrinkage. Fig. 6.19F–H schematizes the morphology of a *typical rimmed steel*.

In steels with too much boiling, because of excessive amount of oxygen, bubbles are also formed at the top and bottom, though they escape easily. Excessive boiling frees almost all of the gas, and bubbles are barely present: the already formed ones cannot compensate shrinkage, similar to the behavior of bubbles in killed steels (Fig. 6.19H).

If the manufacturing process is considered, it is evident that rimmed steels are not obtained by continuous casting but always made in molds. The skin of rimmed ingots is very pure and has excellent behavior in drawing processes, as well as very good corrosion resistance after galvanizing or painting, and other advantageous properties. For certain applications, such as manufacture of galvanized steel sheet, the presence of a skin with those characteristics makes rimmed steels preferable to killed ones. Wider skins are desirable because if skin bubbles reach the periphery, they oxidize and cause peripheral defects during rolling. The use of chill molds promotes an energetic gaseous release, solidification with columnar grains and, as a consequence, thick "skins."

To favor the maximum thickness of the metal before external pressure (ferrostatic and atmospheric) prevents the exit of gasses towards the atmosphere and favors adherence of bubbles, slow castings (small diameter of the inlet) are favored, as well as by bottom syphoning; the characteristics of slow castings delay the increase in ferrostatic pressure head and favor the release of gasses.

For a quick start of boiling, it is convenient to use low casting temperatures. However, this would require casting at high speeds to avoid the formation of solidified material at the bottom of the ladle. Therefore, the selection of temperature and pouring rate imply a compromise: to achieve the adequate effervescence and, at the same time, prevent the formation of solidified crust at the bottom of the ladle, increasing the performance of the casting.

The skin can be thicker, by adequately increasing the boiling degree of the steel; for example, adding oxygen or powders that favor oxygen release in the liquid bath. In this way, a higher pressure of CO is achieved and, just as it happens with other factors, a higher release of bubbles next to the solidification front is produced. On the other hand, when boiling is stopped by mechanically blocking the top of the ingot (*capped steels*, Fig. 6.19D and E), the skin will be thinner, and skin bubbles closer to the periphery.

These *capped* steels are a consequence of a variation in the technique used to obtain rimmed steels. Oxygen content of the steel before casting should be low, and ideally somewhat lower than for rimmed steels. The carbon content must be $0.15 \leq \%C < 0.3$; they are cast in bottle-type ingots, covering them a little over 1 min after boiling has started.

At first, there is a high gaseous release with convectional flow that drags the gasses formed at the wall and allows their exit to the atmosphere. In the lower half of the ingot, bubbles do not appear until boiling is interrupted by covering the ingot mold, which increases the pressure and subsequently, formed bubbles cannot be removed from the wall (Fig. 6.19E).

The interruption of the boiling can also be achieved by chemical means such as massive additions of Al (or other deoxidants like Ti) at the top of the ingot (by killing the upper zone) to obtain a fast solidification of the top and premature closing of the ingot. Thus, once the desired thickness of the skin is reached, the release of CO is stopped by mechanical or chemical means, and the rest of the liquid steel solidifies in a way similar to semikilled steels.

A very premature interruption of the boiling results in the morphology of Fig. 6.19D, with skin bubbles and negative effect since they would almost be in contact with the mold. The higher the tendency of a rimmed ingot to "ascend" (morphologies of Fig. 6.19 corresponding to the top of the ingot), the higher the "sound" thickness of the skin in the lower zone of the ingot will be (samples are usually taken at 30 cm from the base).

Generally speaking, during solidification boiling increases the chemical segregation caused by stirring the liquid. Thus, the segregation at the center of the ingots of capped steels is lower than in rimmed steels, but higher than semikilled ones.

Between the four types considered (killed, semikilled, capped, and rimmed), killed steels are the ones with the lowest degree of segregation (Chapter 5).

6.4 MOLDABILITY

A metal or alloy is known to be apt for molding, when (starting from the liquid state) sound and compact parts can be obtained by solidification, they reproduce the mold perfectly and do not crack during freezing.

6.4.1 Castability

The castability measures how well a metallic liquid fills the space at the interior of a mold, implying access of the liquid, both to the thinnest zones of the mold and the dendritic spaces that are being created in the interior of the mass as solidification takes place.

During the solidification of a metal, three successive stages can be identified:

- At the first one, the proportion of solid phase crystals inside the liquid is very low. These crystals, not in contact with each other, allow free movement of the continuous liquid phase.
- At the second one, crystals (solid phase) are big enough to form a porous mass through which the liquid circulates, but with greater difficulty.

The liquid and solid phases are continuous, but only the liquid phase suffers relative displacements.

- The third one appears as the amount of solid phase increases, and it opposes (like a barrier) the displacement of the liquid. The solid phase is continuous, while the liquid one is discontinuous (entrapped between dendrites) and its relative displacement is not possible.

In general, feeding becomes more difficult as soon as there are solid crystals at the center line. The higher the total solidification time, during which these centerline crystals are present in a casting, the more difficult the feeding will be. The centerline feeding resistance (CFR) equals the time during which crystals form at the centerline divided by the total solidification time of casting:

$$CFR = \frac{\text{time interval between start and end of freezing at centerline}}{\text{total solidification time of casting}} \cdot 100\% \quad (6.8)$$

It is important to point out that CFR is an index of the width of the freezing band of the alloy (the liquid-solid mushy zone), i.e., the region of the part at which solid and liquid coexist at a given time during solidification.

In the case of metallic melts that do not present the third stage when solidifying, its casting ability would be excellent. All the intrinsic factors that contribute to delaying the stage of discontinuity of the liquid until the end of the solidification, improve castability. Among those factors is, for example, morphology of the crystals: when they are convex (as it occurs with intermetallic compounds or intermediate phases of equilibrium diagrams), castability is better than in the case of equiaxed dendrites (typical for solid solutions).

Other factors that contribute to obtaining compact parts, such as a small solidification temperature range, also contribute to a good castability. In general, alloys with smaller liquidus to solidus intervals are the easiest to feed, compared to those with long cooling ranges. As a general foundry experience, it is difficult to satisfactorily feed alloys with $CFR > 70\%$.

Apart from the examples mentioned, it is not simple to list in a general manner the intrinsic factors for a good casting ability. It is true that fluidity and surface tension of the liquid alloy contribute to this property, but their role is basically secondary, since they only consider the liquid state, and castability is the result of the properties of liquid state and the crystallization that occurs during solidification. Likewise, theory on fluid statics and dynamics gives valuable information but it does not define in a general way the castability of an alloy and neither can be compared to other metals.

In practice, castability is tested by pouring the liquid alloy in a testing mold, in the shape of a flat and horizontal spiral (Fig. 6.20), and the solidified length of the spiral is reported, though this is a merely comparative analysis. If a high-purity Al (99.9%) is compared to a commercial-purity Al (99.0%), the spiral length during this type of testing would be, for example, 27 in for high-purity, while only 20 in for commercial-purity, as this last one has a small solidification

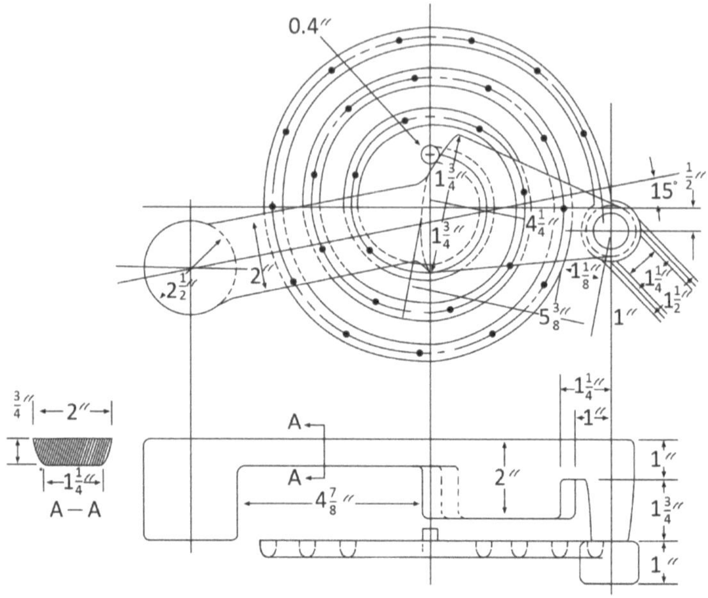

FIG. 6.20 Mold for castability testing (Saeger and Krynitsky, 1931).

interval (primary dendrites increase viscosity) before solidifying at constant temperature. The high-purity Al has an almost null interval and a lower viscosity (or higher fluidity).

6.4.2 Soundness

Fast solidifications with adequate risers are favorable in order to obtain a compact solid; if the columnar structure reaches the axis of the ingot or the center of the part, it would have enough time for the shrinkage to be located at the riser. The effect of chilling in alloys reduces *CFR*, both in steels and nonferrous alloys.

Instead, when the columnar structure is very narrow or nonexistent (as it occurs in very slow solidifications), shrinkage is distributed amongst microshrinkages with diameters usually lower than 1 mm: the equiaxic dendrites can intertwine their branches during their growth and enclose small isolated portions of the liquid alloy, that when solidifying result, by contraction, in internal pores.

If the relation between soundness and columnar structure is considered, the wider the solidification interval, the more porous the alloy will be; increasing the solidification range, the constitutional undercooling also increases, with the columnar structure less deep (Chapter 5).

In order for a part with equiaxic grains to be compact, it is necessary that the residual liquid (flowing through the interdendritic channels) fills the microshrinkages. This will be easier when the residual liquid has higher fluidity.

However, if there are inclusions that prevent the free capillary circulation of the liquid, more microshrinkages or more porosity can appear (e.g., in badly manufactured Al parts that contain alumina in suspension).

An interesting case of improving soundness of Cu-based alloys (such as bronzes) is adding small amounts of Pb (<7%). Lead, insoluble in solid phase in Cu and in solid solutions of Cu, allows (Pb-Cu has a phase diagram similar to Fig. 4.23) the liquid L_1 to fill microshrinkages produced during solidification (more specifically, solidification of α solid solution of Sn in Cu). That easiness to fill microshrinkages is consequence of the low temperature at which the liquid L_1 (rich in Pb) ends its solidification: lower than the melting point of Pb (327.4° C). The maximum content that can be added to Cu alloys is 30% Pb (Chapter 4), however, when Pb is added with the only goal of improving soundness, the maximum Pb content is 7% because of the high density of L_1 rich in Pb; the amount of this liquid is important because the α crystals will float in the liquid before making contact between them and, thus, before microshrinkages appear.

Gasses dissolved in the liquid also influence the formation of microshrinkages and, therefore, diminish soundness. The gas concentration necessary for the nucleation of a bubble is lower if negative metallostatic pressure is applied to the liquid. These negative pressures are created at zones where liquid is trapped by the solid; consequently, it can be expected that dissolved gasses will nucleate and a bubble will grow at isolated liquid portions (interdendritic), forming microshrinkages.

All recommendations from Section 6.3.1 to reduce the amount of gasses dissolved in the liquid, are also efficient to improve soundness.

Desulfuring bronzes through the use of alkaline carbonates (magnesium or calcium carbide) before pouring eliminates sulfur, otherwise concentrations higher than 0.1% S result in a large number of bubbles and a decrease in the soundness of the bronze.

On the other hand, given that alloys with wide ranges of solidification usually present important boiling during this process (by solubility variation in the segregated liquid), consequently those alloys have a higher tendency to form pores.

Summarizing, just like with other factors and considering alloys of the same system, the deeper the columnar structure, the sounder the solid part will be, the amount of gasses dissolved in the liquid will be lower, the liquid will be more fluid and the solidification interval will be smaller.

It is usually said that for a system in equilibrium, the casting ability is maximum when melting is congruent at constant temperature, as it happens with pure metals and eutectics; and minimum for limit solid solutions.

In any case, if an alloy does not shrink during solidification (e.g., with Bi-Pb alloys and gray castings), it will not present porosity due to microshrinkages and, therefore, will be more compact.

Besides the use of nondestructive tests such as X-ray, gamma-ray, ultrasound, etc., soundness is frequently evaluated by analyzing the soundproofing ability of the alloy.

6.4.3 Hot Tearing

When designing a part for casting, it is useful to have adequate geometric shapes to avoid stresses during solidification. The presence of tensile stresses at the last stage of solidification can produce tears by hot deformation that cannot be filled by the limited residual liquid. This risk of fracture is higher in those alloys with a wider solidification range (in particular in the case of solid solutions without eutectic matrix) due to the mushy nature of the solid when approaching the final solidification temperature. Hot tearing takes place between the temperature at which the metal becomes coherent and the temperature at the end of cooling (solidus line): hot tearing occurs at or slightly above the solidus line in castings under tensile stresses. Hot tearing is the result of the combined effect of both metals in coherence with existing liquid films, and sufficient tensile stresses at all times. Thus, an alloy sensitive to tearing may be cast without tears if the mold is designed to avoid tensile stresses during the coherent range. If a metal has extensive final solidification at *constant temperature* it will not present hot tears in uniform cross-sectioned castings; there will not be tensile stresses due to solid contractions when the metal is at this stage since temperature is constant. Therefore, eutectics and pure metals are less susceptible to hot tearing than alloys with wide solidification range (Fig. 6.21).

Consequently, hot tears are extremely rare in gray castings as solidification in its final stage, takes place at an almost constant eutectic temperature. Furthermore, hot tears are also prevented by expansion ($\sim 2.5\%$ in volume) during final solidification product of the reaction:

$$\text{Liquid} \rightarrow \text{graphite} + \text{austenite}$$

To avoid this tendency to crack, it is desirable to prevent the formation of a cold peripheral crust (with low plasticity), using fast cooling. When this cortical solid contracts, it tends to reduce the volume of the internal liquid mass that, by being incompressible, produces more surface cracks by tensile stresses. The

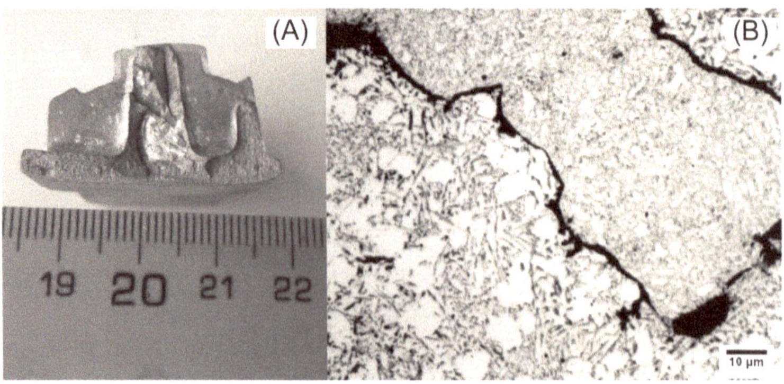

FIG. 6.21 (A) Macro and (B) micrograph of a cast part that suffered hot tearing.

fracture is more probable if the alloy is less plastic, since it is a function, not only of temperature, but also of thickness, depth of columnar structure, crystalline system of the solvent, etc.

To improve the plasticity of the periphery of the ingots, molds with noncylindrical shape are desirable; in this way, deformations or stresses without straining are more likely than fractures.

When casting is poured by gravity, sometimes peripheral cracks are produced by the off-center of the liquid *stream* that results in differences in the thickness of the solidified crust and, therefore, zones with less resistance.

During cooling, when the alloy has fully solidified, tension stresses, deformations, and cracks can also be produced. Evidently, the design of the part is a key factor; another one is solid contraction (Table 6.2) that, among others, depends on the thermal expansion coefficient. Alloys with high thermal expansion coefficients and low thermal conductivity produce larger temperature differences, and thus tensile stresses and cracks. It is also important to consider volume variations by allotropic transformation in solid state (Exercise 1.2). In some cases a stress relief heat treatment may be carried out to remove these stresses and turn elastic strain into plastic.

EXERCISE 6.11

To analyze hot tearing of an 87% Cu-13% Sn alloy, a part is cast in the shape of a bar with square transverse section and a length of 200 mm ending on both sides in an U shape (to avoid its free shrinkage during cooling), as shown in Fig. 6.22. Shrinkage takes place between 800° C and 300° C. Calculate the nature and amount of residual stresses, and determine if the bar deforms plastically.

Data: $E = 105\,\text{GPa}$, $\alpha = 18 \times 10^{-6}\,\text{K}^{-1}$, $\sigma_y = 150\,\text{MPa}$, $R_m = 300\,\text{MPa}$, and $A = 15\%$.

Solution

The Cu-13% Sn alloy has a wide solidification interval (very close to the solubility limit of Sn in Cu), thus it can present tearing.

The strain at the yield stress value following Hooke's law is:

$$\varepsilon = \frac{\sigma_y}{E} = \frac{150\,\text{MPa}}{105 \times 10^3\,\text{MPa}} \cdot 100 = 0.14\%$$

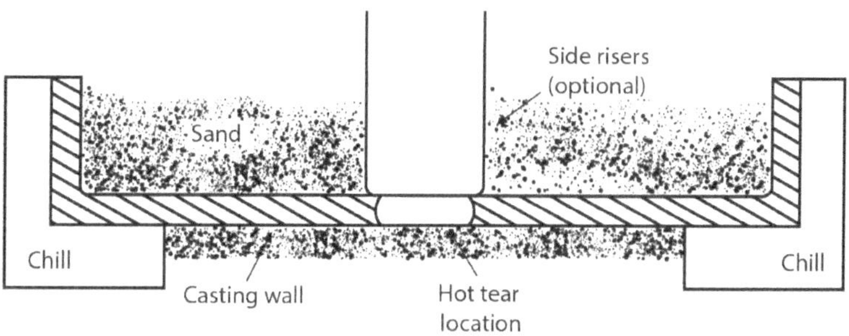

FIG. 6.22 Schematic cast shape to determine hot tears.

Furthermore, in solid state, the free shrinkage of the bar would be:

$$\delta = l_0 \alpha \Delta T = (200\text{mm})(18 \times 10^{-6}\text{K}^{-1})(800 - 300)\text{K} \simeq 1.8\text{mm}$$

equivalent to a 0.9% strain, but since it is constrained on both sides (dilate or shrink), there will be tensile stresses that plastically deform the bar and reduce the risk of hot tearing.

To reduce the internal cracks produced in an ingot or a part after solidification, during cooling, it is desirable that each section is homothetically compressed by zones that surround it. Furthermore, a fast heat release through the walls of the ingot or chill molds would be better (though, as previously mentioned, this can cause cracks by temperature gradients in the solid).

Finally, alloys with wide solidification ranges, considering not only their mushy nature but also their tendency to have pores, have higher risks of developing tensile stresses, deformations, or cracks in solid state, compared to alloys with congruent melting point.

In conclusion, it is important for the engineer and scientist to develop solid metals that have homogeneous chemical compositions (no segregations), without gross physical defects (shrinkage, porosity, tears), and uniform mechanical properties for room-temperature applications. Fine-grained materials improve both strength and toughness. For high temperature applications, coarse or monocrystalline materials improve creep resistance.

REFERENCE

Saeger, C.M., Krynitsky, A.I., 1931. A practical method for studying the running quality of a metal cast into foundry molds. Trans. A.F.S. 39, 513–540.

BIBLIOGRAPHY

A.S.M., 2002. Failure Analysis and Prevention, 10th ed. American Society for Metals, Metals Park, OH.

Flinn, R., 1963. Fundamentals of Metal Casting. Addison-Wesley Publishing Company, Reading, MA.

Kear, B., 1986. Nuevos metales. Investigación y Ciencia 123 (December), 123–132.

Le Breton, H., 1975. Defectos de las piezas de fundición. URMO, Bilbao, Spain.

Lecompte, 1962. Cours d'Aciérie. Revue Metállurgie, Paris.

Massalski, T., 2001. Binary Alloy Phase Diagrams, second ed. American Society for Metals, Metals Park, OH.

Pokorny, A., Pokorny, J., 1967. De Ferri Metallographia, vol. 3 Berger-Larrault, Paris, France.

Roesch, K., Zimmermann, K., 1969. Acero Moldeado. Editecnia, Madrid.

Sancho, J., Verdeja, L., Ballester, A., 1999. Metalurgia Extractiva II. Procesos de Obtención. Editorial Síntesis, Madrid.

Weast, R., 2014. Handbook of Chemistry and Physics, 95th ed. The Chemical Rubber Co., Cleveland, OH.

Chapter 7

Equilibrium Transformations

The yield strength of an alloy can be obtained considering properties such as the yield strength of a monocrystal formed by the metal in its pure form and solid solution, dispersion or precipitate, and grain size strengthening, as follows:

$$\sigma_y = \frac{G}{10,000\,\text{MPa}} + G\sum(\delta_i \cdot c_i) + \frac{G \cdot \mathbf{b}}{L} + G \cdot \mathbf{b} \cdot \sqrt{\rho} + K_{gb} \cdot d^{-1/2} \tag{7.1}$$

where G is the shear modulus, δ_i is the relative difference between the atomic diameters of solute and solvent, c_i is the amount of solvent, $\mathbf{b}$ is the Burgers vector, L is the tangent distance between precipitates, ρ is the dislocation density[1] $(m_\perp/m^3)$, K_{gb} is the grain boundary hardening, and d is the grain size.

EXERCISE 7.1

Calculate the yield strength of a binary solid solution with the following properties: $G = 50\text{GPa}$, $c = 1$ at. %, $\delta = 0.10$, $d = 30\mu\text{m}$, $L \approx 1\,\mu\text{m}$, $K_{gb} = 10\text{MNmm}^{-3/2}$, $\mathbf{b} \approx 3\text{Å}$, $\rho = 1 \times 10^6\,\text{cm}_\perp/\text{cm}^3$

Solution

Using Eq. (7.1) with adequate units:

$$\sigma_y = \frac{G}{10,000\text{ MPa}} + G\sum(\delta_i \cdot c_i) + \frac{G \cdot \mathbf{b}}{L} + G \cdot \mathbf{b} \cdot \sqrt{\rho} + K_{gb} \cdot d^{-\frac{1}{2}}$$

$$= \frac{50,000}{10,000} + (50,000)(0.10)(0.01) + \frac{(50,000)(3)}{10,000} + (50,000)(3 \times 10^{-8})\sqrt{1 \times 10^6}$$

$$+ (10)(0.03)^{-\frac{1}{2}} = 130\text{MPa}$$

Phase diagrams are formed by both solidification (transformation of solid into liquid and vice versa) and solid-state transformation lines, when solidification has ended and the alloy suffers equilibrium transformations during cooling. These last ones are generally caused by variations in solubility, allotropic transformation, or a combination of both.

1. The density of recrystallized state is $1 \times 10^{10}\,m_\perp/m^3$, and of the cold work one is $1 \times 10^9\,m_\perp/m^3$.

Solidification and Solid-State Transformations of Metals and Alloys. http://dx.doi.org/10.1016/B978-0-12-812607-3.00007-3

7.1 TRANSFORMATIONS CAUSED BY SOLUBILITY VARIATIONS

7.1.1 Precipitation Process

If an eutectic binary phase diagram is considered (Fig. 7.1), below T_E there are two solvus lines separating the saturated α_s and β_s solid solutions; during cooling, solubility decreases for both, B in the crystalline lattice of A (α_s) and A in the crystalline lattice of B (β_s).

Fig. 7.2 shows the cooling curves and microstructures of alloy I (of Fig. 7.1), with $n\%$ B: at T_p (known as *critical* temperature) a precipitate of α is formed and equilibrium is broken because a new phase appears. According to the Le Chatelier rule, there is a heat release manifested as a change in the slope of the cooling curve. This amount of heat is so small that in practice, T_p is very difficult to be determined by direct thermal analysis; and therefore, to obtain the critical points, other techniques, such as X-ray diffraction, dilatometry, microscopy, etc. are used.

The precipitation equilibrium process occurs, generally, *by nucleation and growth*. The mechanism is analogous to the one indicated for solidification with thermally activated diffusion. Fig. 7.3 schematizes the precipitation of alloy I, isothermally transformed at different temperatures. The proportion of α and β for each temperature lower than T_p can be determined by both the lever and phase rules applied to the phase diagram. Fig. 7.3 also indicates that the lower the isothermal precipitation temperature, the lower the mean size of the precipitates will be.

Precipitation is continuous when particles appear to be almost uniformly distributed in the interior of the crystalline grains. Usually, continuous precipitation is produced when certain crystallographic planes and directions of the

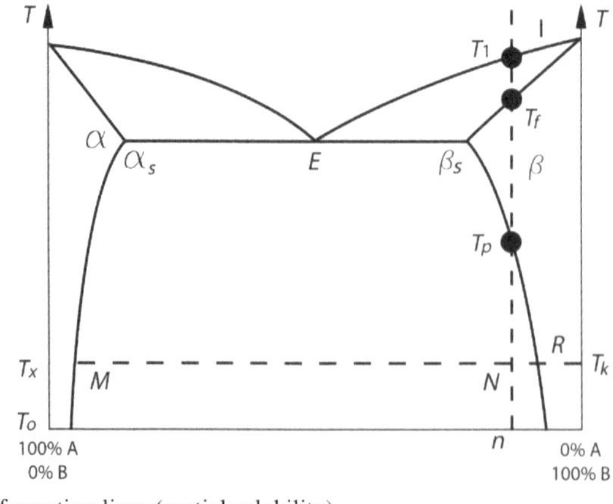

FIG. 7.1 Transformation lines (partial solubility).

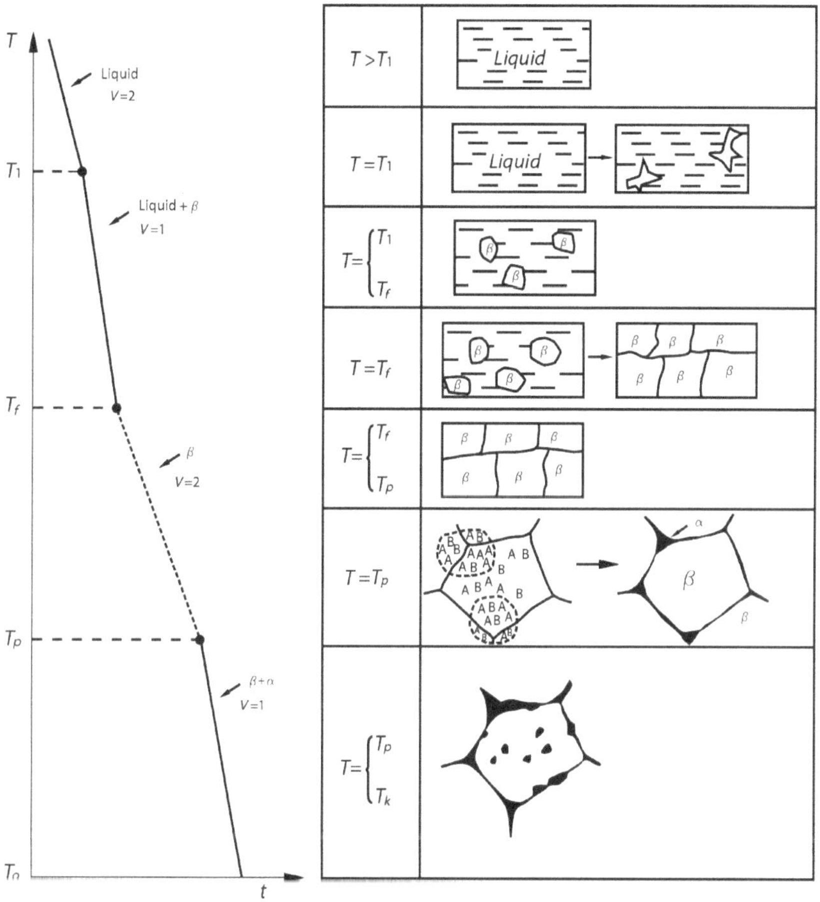

FIG. 7.2 Cooling curves and microstructures of alloy I (n% B) of Fig. 7.1.

precipitated particle match with specific planes and directions of the crystalline grain. Table 7.1 shows some preferential orientations (or epitaxial relations) between precipitates and crystalline grain matrix. In general, these relationships minimize distortions caused by volume variations produced in the interior of the crystalline grains during precipitation: the formation of the new equilibrium phase reduces free energy, but, at the same time, the increase in energy caused by volumetric distortion to create the new phase tends to balance that reduction. Therefore, preferential orientations decrease the total free energy and make the continuity of the reaction possible.

This continuous precipitation (caused by orientation relationships) must not be confused with the *discontinuous* or *localized* one: since precipitation is a nucleation and growth process, there can be regions of the crystal that, being

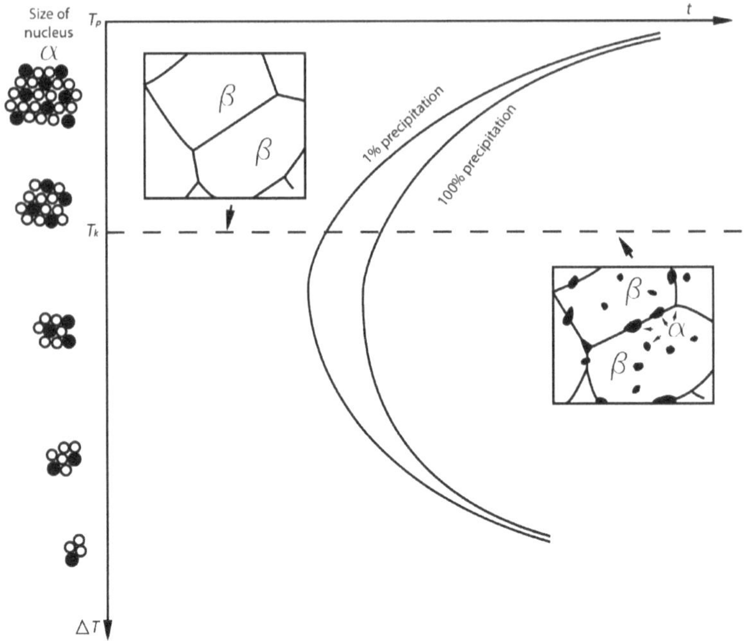

FIG. 7.3 Precipitation by nucleation and growth of alloy I (n% B) of Fig. 7.1.

TABLE 7.1 Crystallographic Relationships Between Matrix and Precipitated Phases Produced by Cooling

Binary System	Matrix Phase and Its Crystalline System	Precipitated Phase and Its Crystalline System	Crystallographic Relationships (the Precipitated Phase is Indicated First)
Ag-Al	Solid solution in Al; fcc	γ (Ag_2-Al); hcp	(0001) // (111), $[11\bar{2}0]$ // $[1\bar{1}0]$
	Solid solution in Al; fcc	γ′ (transitional); hcp	(0001) // (111), [1120] // [110]
Ag-Cu	Solid solution in Ag; fcc	Solid solution in Cu; fcc	Plates // (100); all directions //
	Solid solution in Cu; fcc	Solid solution in Ag; fcc	Plates // (111) or (100); all directions
Ag-Zn	β (AgZn); bcc	Solid solution in Ag; fcc	(111) // (110), $[1\overline{10}]$ // $[\overline{111}]$
	β (AgZn); bcc	γ (Ag_5Zn_8); bcc	(100) // (100), [010] // [010]

TABLE 7.1 Crystallographic Relationships Between Matrix and Precipitated Phases Produced by Cooling—cont'd

Binary System	Matrix Phase and Its Crystalline System	Precipitated Phase and Its Crystalline System	Crystallographic Relationships (the Precipitated Phase is Indicated First)
Al-Cu	Solid solution in Al; fcc	θ ($CuAl_2$); bct	Plates // (100); (100) // (100), [011] // [120]
	Solid solution in Al; fcc	θ′ (transitional); tet	(001) // (100), [010] // [011]
Al-Mg	Solid solution in Al; fcc	β (Al_3Mg_2); fcc	Plates firstly // (110); then // (120)
Al-Mg-Si	Solid solution in Al; fcc	Mg_2Si; fcc	Plates // (100)
Al-Zn	Solid solution in Al; fcc	Zn almost pure; hcp	Plates // (111); (0001) // (111), $[11\bar{2}0]$ // <100>
Au-Cu (b)	Unlimited solid solution; fcc	a_1' (AuCu I); ord fct	(100) // (100), [010] // [010]
Be-Cu	Solid solution in Cu; fcc	γ_2 (BeCu); ord bcc	GP Zones // (100); then γ_2 with [100] // [100], [010] // [011]
0.4C-Fe	Austenite (γ-Fe); fcc	Ferrite (α-Fe) (proeutectoid); bcc	(110) // (111), $[1\bar{1}1]$ // $[1\bar{1}0]$
0.8C-Fe	Austenite (γ-Fe); fcc	Ferrite in perlite; bcc	(011) // (001), [100] // [100], $[0\bar{1}1]$ // [010]
	Austenite (γ-Fe); fcc	Ferrite in upper bainite; bcc	(110) // (111), $[1\bar{1}0]$ // $[\bar{2}11]$
		Ferrite in lower bainite; bcc	(110) // (111), $[1\bar{1}1]$ // $[1\bar{1}0]$
1.3C-Fe	Austenite (γ-Fe); fcc	Cementite (Fe_3C); ortho	No plates // (111); (001) Fe_3C // to plate's plane
Co-Cu	Solid solution in Cu; fcc	α-solid solution in Co; fcc	Plates // (100); same orientation of precipitate and matrix phases
Co-Pt (b)	Solid solution in Pt-Co; fcc	α″ (CoPt); ord fct	Plates // (100); all directions are parallel

Continued

TABLE 7.1 Crystallographic Relationships Between Matrix and Precipitated Phases Produced by Cooling—cont'd

Binary System	Matrix Phase and Its Crystalline System	Precipitated Phase and Its Crystalline System	Crystallographic Relationships (the Precipitated Phase is Indicated First)
Cu-Fe	Solid solution in Cu; fcc	γ-Fe (transitional); fcc	Cubes (100); same orientation of precipitate and matrix phases
		α-Fe; bcc	Plates // (111); random lattice orientation
Cu-Ni-Co	Solid solution in Cu; fcc	Solid solution in Coα; fcc	Plates // (100); same orientation of precipitate and matrix phases
Cu-Ni-Fe	Solid solution in Cu; fcc	Solid solution in Feα; fcc	Plates // (100); same orientation of precipitate and matrix phases
Cu-Si	Solid solution in Cu; fcc	β (Cu-Si); hcp	Plates // (111); (0001) // (111), $[11\bar{2}0]$ // $[1\bar{1}0]$
Cu-Sn	β phase; bcc	Solid solution in Cu; fcc	(111) // (110), $[1\bar{1}0]$ // $[1\bar{1}1]$
Cu-Zn	β (Cu-Zn); bcc	Solid solution in Cu; fcc	(111) // (110), $[1\bar{1}0]$ // $[\bar{1}11]$; variable plane // [556] plates or needles
	β (Cu-Zn); bcc	γ (Cu_5 Zn_8); ord bcc	(100) // (100), [010] // [010]
	τ (Cu-Zn); hcp	Solid solution in Zn; hcp	$(10\bar{1}4)$ // $(10\bar{1}4)$, $[11\bar{2}0]$ // $[11\bar{2}0]$
Fe-N	Ferrite (α-Fe); bcc	γ_1 (Fe_4N); fcc	(112) // (210)
Fe-P	Ferrite (α-Fe); bcc	δ (Fe_3P); bct	Plates // (21, 1, 4)
Pb-Sb	Solid solution in Pb; fcc	Solid solution in Sb; rhom	(001) // (111), [100] // $[1\bar{1}0]$

bcc, body-centered cubic; bct, body-centered tetragonal; fcc, face-centered cubic; hcp, hexagonal closed-packed; ortho, orthorhombic; rhom, rhombohedral; tet, tetragonal; ord, ordered.

favorable for heterogeneous nucleation or aiding the diffusion of atoms (such as inclusions, imperfection in the crystalline lattice, fields of microtension by mechanical deformation, grain boundaries, etc.), are more adequate to form precipitates. This localized precipitation is usually abundant with small undercooling (heterogeneous precipitation, Fig. 7.3).

EXERCISE 7.2

Obtain the solubility limit curve of Sn in Pb supposing that it follows an Arrhenius-type law.

Data: the solubility of Sn in Pb at 183°C is 19% and at 100°C is 5% (Table 3.1).

Solution

Arrhenius expressions are of the type:

$$C = Ke^{-Q/RT}$$

$$\ln C = \ln K - \frac{Q}{RT}$$

At 183°C:

$$\ln 19 = \ln K - \frac{Q}{R(456)}$$

At 100°C:

$$\ln 5 = \ln K - \frac{Q}{R(373)}$$

Subtracting these last two expressions:

$$\ln 19 - \ln 5 = \ln K - \ln K - \frac{Q}{R(456)} + \frac{Q}{R(373)}$$

$$\ln \frac{19}{5} = \frac{Q}{R} \cdot \frac{83}{(456)(373)}$$

solving for Q:

$$Q = \frac{(8.314)(456)(373)}{83} \ln \frac{19}{5} = 22{,}745.07 \frac{\mathrm{J}}{\mathrm{mol}}$$

and the K value, substituting Q in both equations, is:

$$K = Ce^{Q/RT} = 19e^{22{,}745.07/8.314(456)} = 7661.07$$

$$K = Ce^{Q/RT} = 5e^{22{,}745.07/8.314(373)} = 7661.08$$

At room temperature, the concentration will be:

$$C = 7661.07e^{-22{,}745.07/8.314(300)} = 0.84\%$$

The solubility curve with temperature can be seen in Fig. 7.4.

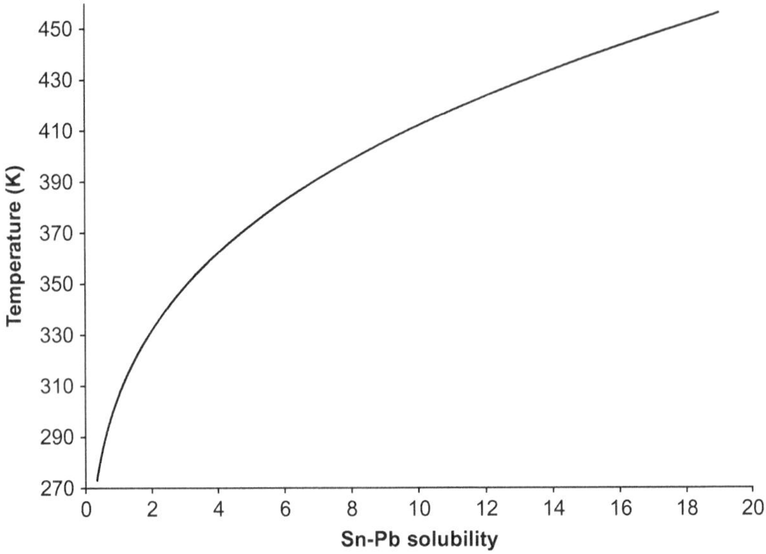

FIG. 7.4 Solubility curve of Sn-Pb.

EXERCISE 7.3

Applying the general model for transformation kinetics by nucleation and growth (Chapter 2), plot the TTT curves for precipitation by solubility loss of the Pb-15% Sn alloy.

Data: $\gamma_{\alpha\beta} = 500\,\text{mJ/m}^2$ and $Q_{Sn} = 106\,\text{kJ/mol}$

Solution

Usually, the driving force of a precipitation process ΔG_v is a function of the chemical potential difference of the precipitating element (in this case Sn):

$$\Delta G_v \cong RT \ln \frac{C_0}{C}$$

where C_0 is the solubility at T_p and C is the solubility at the desired temperature. Using data from Table 1.2:

$$V_m(\text{Sn}) = \left(\frac{1}{7310}\frac{\text{m}^3}{\text{kg}}\right)\left(0.1187\frac{\text{g}}{\text{mol}}\right) = 1.62 \times 10^{-5}\,\frac{\text{m}^3}{\text{mol}}$$

The nucleation barrier ΔG^* can be obtained by:

$$\Delta G^* = \frac{16}{3}\cdot\frac{\pi\gamma^3}{\Delta G_v^2} = \frac{16\pi}{3}\frac{\left(0.5\,\text{J/m}^2\right)^3}{\left(RT \ln \frac{C_0}{C}\frac{\text{J}}{\text{mol}}\right)^2\left(\frac{1\,\text{mol}}{1.62\times 10^{-5}\,\text{m}^3}\right)^2} = \frac{7.95\times 10^{-12}}{\left(T \ln \frac{C_0}{C}\right)^2}\,\text{J}$$

$$\frac{\Delta G^*}{kT} = \frac{7.95 \times 10^{-12}\,\text{J}}{\left(13.8 \times 10^{-24}\frac{\text{J}}{\text{K}}\right) T \left(T \ln \frac{C_0}{C}\right)^2} \simeq \frac{5.76 \times 10^{11}}{\left(\ln \frac{C_0}{C}\right)^2 \cdot T^3}$$

And now the nucleation rate and speed can be obtained as a function of temperature, considering the following values for the constants: $n \simeq 1 \times 10^{29}\,\text{atoms/m}^3$, $\mathbf{b} \simeq 10\text{Å}$, $\Omega \simeq 10\text{Å}^3$, $n_s \simeq 1 \times 10^{19}\,\text{atoms/m}^3$, $\omega^* \simeq 1$, $D_0 \simeq 1 \times 10^{-4}\,\text{m}^2/\text{s}$, which results in:

$$I = n^{\omega^*} \cdot \exp\left(\frac{-\Delta G^*}{kT}\right) \cdot \frac{D_0}{b^2} \cdot \exp\left(-\frac{Q_{dif}}{RT}\right)$$

$$= 10^{44} \cdot \exp\left(\frac{-\Delta G^*}{kT}\right) \cdot \exp\left(-\frac{Q_{dif}}{RT}\right)\left[\frac{\text{nuclei}}{\text{m}^3\text{s}}\right]$$

$$I \simeq 1 \times 10^{44} \cdot \exp\left[\frac{5.76 \times 10^{11}}{\left(\ln \frac{C_0}{C}\right)^2 \cdot T^3}\right] \cdot \exp \frac{-106{,}000\,\text{J}}{8.31 \frac{\text{J}}{\text{mol K}} \cdot T}\left[\frac{\text{nuclei}}{\text{m}^3\text{s}}\right]$$

$$v = \Omega \cdot n_s \cdot \frac{D_0}{\mathbf{b}^2} \cdot \exp\left(\frac{-Q_{dif}}{RT}\right) \cdot \frac{\Delta G_v}{RT} = 10^5 \cdot \exp\left(\frac{-Q_{dif}}{RT}\right) \cdot \frac{\Delta G_v}{RT}\left[\frac{\text{m}}{\text{s}}\right]$$

$$v = 1 \times 10^5 \cdot \ln \frac{C_0}{C} \cdot \exp\left(\frac{-106{,}000}{8.314 \cdot T}\right)\left[\frac{\text{m}}{\text{s}}\right]$$

The Johnson-Mehl-Avrami law is used to calculate the precipitated fractions at a certain temperature T (as a function of time):

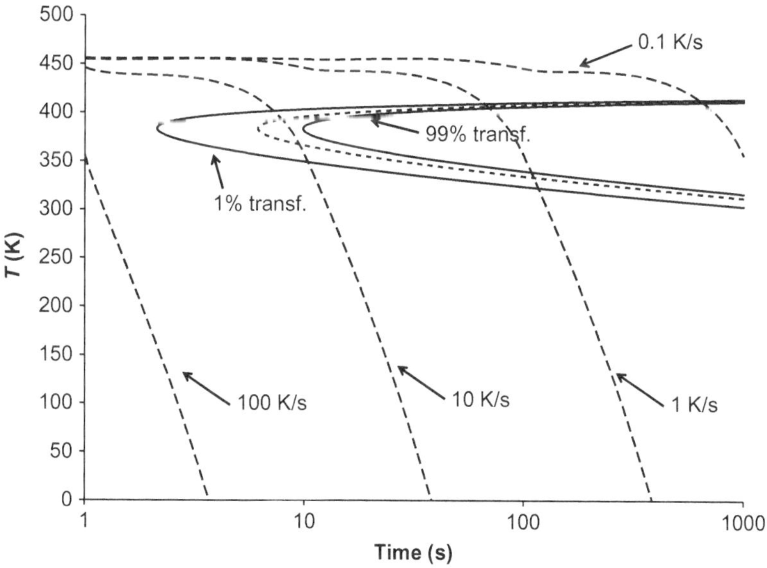

FIG. 7.5 TTT curves for a shape factor of 10 degrees in a Pb-Sn alloy.

$$t_x = \sqrt[4]{\frac{-4 \cdot \ln(1-x)}{I \cdot v^3}}$$

For heterogeneous conditions it can be calculated using various shape factors:

$$f(\theta) = (1 - \cos\theta)^2 \cdot (2 + \cos\theta)/4$$

The TTT curves considering 1% and 99% of transformation, as well as different cooling rates are found in Fig. 7.5 (for $\theta = 10°$) and Fig. 7.6 (for 15°).

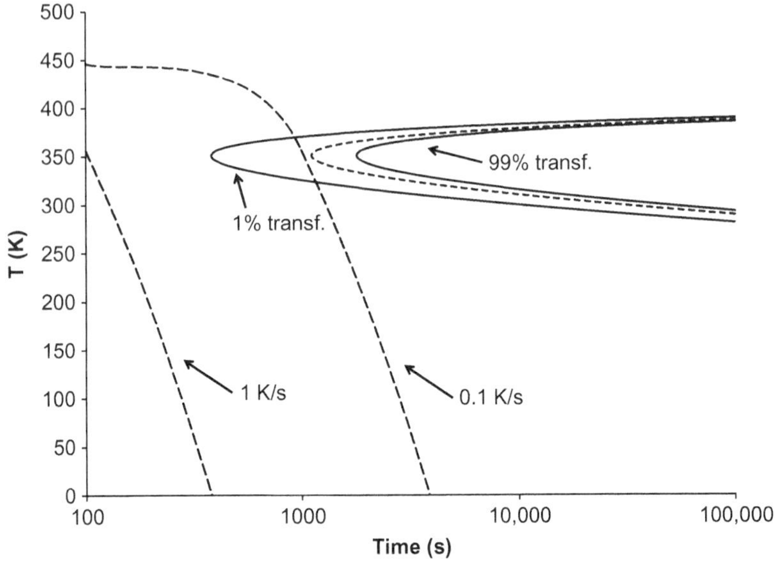

FIG. 7.6 TTT curves for a shape factor of 15 degrees in a Pb-Sn alloy.

EXERCISE 7.4

For the Pb-15% Sn alloy, used for bearings (Fig. 7.7), calculate the approximate value of its elastic limit.

Data: solubility of Sn in Pb at room temperature is 1.5% (Fig. 4.1), $G_{Pb} = 7.3\,\text{GPa}$, $G_{Sn} = 15.6\,\text{GPa}$, $\mathbf{b}_{Pb} = 3.5\,\text{Å}$ and $\mathbf{b}_{Sn} = 3.02\,\text{Å}$.

Solution

Calculating the ratio of phases present in the alloy at room temperature:

$$\frac{P_\beta}{P_\alpha} = \frac{(15 - 1.5)}{(100 - 15)} \approx 0.159(16\%)$$

$$\frac{\rho_\beta \cdot V_\beta}{\rho_\alpha \cdot V_\alpha} = \frac{7.3 \cdot V_\beta}{11.3 \cdot V_\alpha} \approx 0.16 \rightarrow \frac{V_\beta}{V_\alpha} \approx 0.25 \rightarrow \frac{V_\beta}{V_\alpha + V_\beta} = \frac{0.25}{1.25} = 0.2(20\%)$$

There is a relationship between grain size ($\bar{X}$) and fraction of the phases:

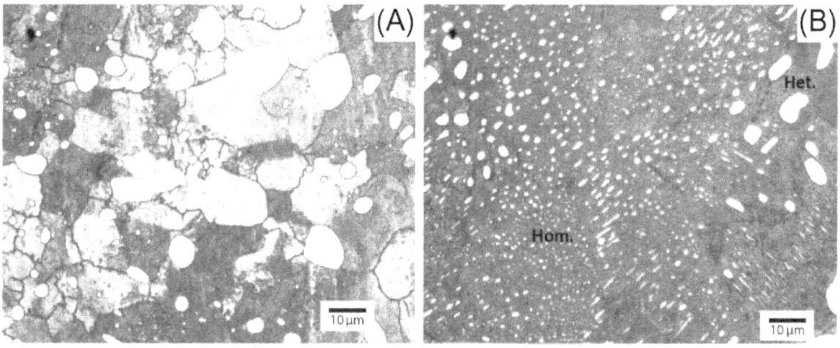

FIG. 7.7 (A) Micrograph of Pb-15% Sn and (B) showing homogeneous and heterogeneous solidification.

$$f_p \approx f_a \approx f_v \approx \frac{\bar{X}^2}{L^2} = \left(\frac{\bar{X}}{L}\right)^2 \rightarrow L \approx \frac{\bar{X}}{\sqrt{f_v}} - \bar{X}$$

Considering a grain size of 1 µm:

$$L \approx \frac{1\,\mu\text{m}}{\sqrt{0.2}} - 1\,\mu\text{m} = 1.24\,\mu\text{m}$$

and the solute composition at room temperature, considering the maximum solubility of Sn in Pb at this temperature as 1.5% wt Sn, is:

$$c_i = \frac{\dfrac{w_{Sn}}{P_{Sn}}}{\dfrac{w_{Sn}}{P_{Sn}} + \dfrac{w_{Pb}}{P_{Sn}}} = \frac{\dfrac{1.5}{118.7}}{\dfrac{1.5}{118.7} + \dfrac{98.5}{207.2}} 0.02589 \approx 0.026$$

The elastic limit can now be obtained using Eq. (7.1):

$$\sigma_y\left(\frac{P_\beta}{P_\alpha}\%\text{Sn}\right) \cong \frac{G}{10{,}000\,\text{MPa}} + G\sum(\delta_i \cdot c_i) + \frac{G \cdot \mathbf{b}}{L}$$

$$\cong \frac{7300}{10{,}000} + (7300)\left(\frac{3.5 - 3.02}{3.5}\right)(0.026) + \frac{(7300)(3.5 \times 10^{-4}\,\mu\text{m})}{1.24\,\mu\text{m}}$$

$$\cong 28.56\,\text{MPa}$$

Comment: This value indicates that the solid solution hardening of Sn in Pb is more important than the precipitation of Sn (β) in Pb (α).

Equilibrium precipitation is generally produced by nucleation and growth, though other types of precipitation reactions can happen without the formation of nuclei and even requiring holding at temperatures below the solubility limit and therefore inducing the diffusion of atoms. These reactions are very rare and are usually produced when a solid solution presents a miscibility gap, as the one indicated in Fig. 7.8, analogous to the one analyzed for the monotectic reaction. The Al-Zn system (for Zn between 31.6 and 78 wt.%) is an example of this type of reactions, known as *spinodals*: α solid solution (fcc) is broken down by monophasic demixing of atoms produced by diffusion, into two other solid

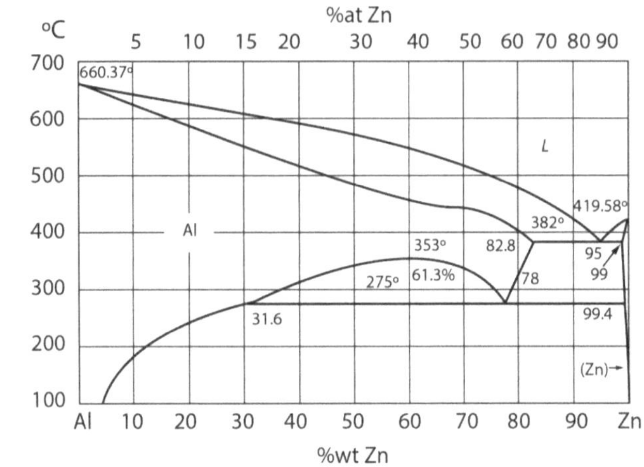

FIG. 7.8 Al-Zn phase diagram (ASM Handbook, Vol. 3, 1992).

solutions, also fcc, with composition limits defined for each temperature. Spinodal structures have a morphology similar to a weave (*tweedy*) characterized by the small distance between regions of identical composition (50–100 Å).

EXERCISE 7.5

Calculate the approximate value of the elastic limit for the Pb-10% Sb alloy (used for battery grill, plates, and accumulators, Fig. 7.9).

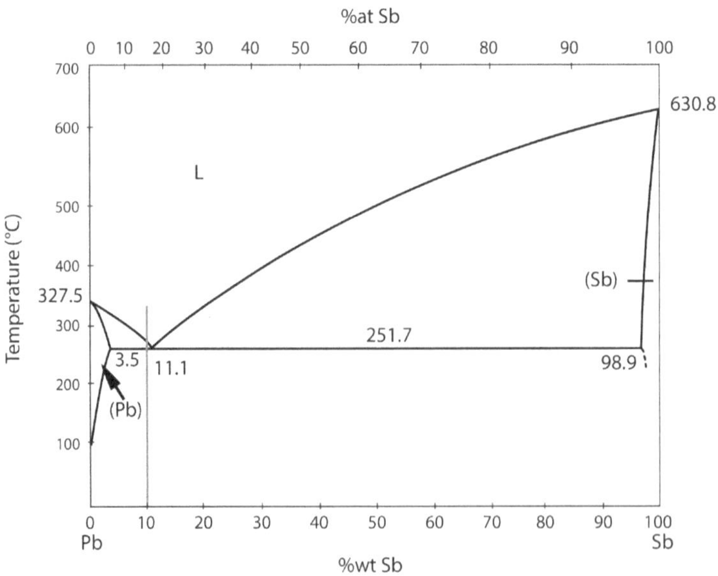

FIG. 7.9 Pb-Sb phase diagram (ASM Handbook, Vol. 3, 1992).

Solution

From the phase diagram, the amounts of matrix and disperse constituents can be obtained using the lever rule: 87% of matrix (eutectic) and 13% of disperse constituent. At the eutectic temperature, there is 3.2% β (Pb precipitated in Sb) and 3.5% α (Sb precipitated in Pb).

Since the density of Pb (11.36 g/cm^3) is almost twice that of Sb (6.62 g/cm^3), the 3.5% Sb (wt.%) precipitated in the Pb lattice equals:

$$\frac{P_{Sb}}{P_{Pb}} = \frac{3.5}{96.8} = 0.036 = \frac{\rho_{Sb} V_{Sb}}{\rho_{Pb} V_{Pb}}$$

$$\frac{V_{Sb}}{V_{Pb}} = 0.036 \cdot \frac{11.36}{6.62} = 0.062(6.2\%)$$

$$\frac{V_{Sb}}{V_{Sb} + V_{Pb}} = \frac{0.062}{1.062} = 0.058(5.8\%)$$

Considering precipitates of the nanometric type, of 10 nm (0.01 μm):

$$\sigma_y \cong \frac{G \cdot \mathbf{b}}{L} \cong \frac{(7.3 \times 10^3)(3.5 \times 10^{-4}\,\mu\text{m})}{\frac{0.01\,\mu\text{m}}{\sqrt{0.058}} - 0.01} \cong 79\,\text{MPa}$$

which indicates that precipitates of a larger size would have lower elastic limit values.

Comment: The Pb-10% Sb alloy, with a solution treatment at 235°C and subsequently aged for 1 day, has a σ_{max} of 89–90 MPa.

7.1.2 Heat Treatments and Precipitation in Age-Hardening Alloys

The nucleation and growth model described in Chapter 2 provides general guidelines to understand diffusion activated transformations. An example of this type of transformations that occur in solids is age-hardening: precipitation, either localized or continuous, allows to heat treat an alloy to either avoid the creation of the precipitates during cooling (*quenching*) or obtain a fine dispersion of precipitates (*age-hardening*):

1. *Homogenization annealing* heat treatment (Section 5.3.1) is used to reduce chemical heterogeneities during nonequilibrium solidification produced by dendritic segregation (coring). This treatment involves heating the alloy until reaching a temperature close to the solidus line, and holding for a period of time (Fig. 7.10) to obtain a uniform chemical composition (by diffusion) at all points of the solidified grain or dendrite.

   After homogenization, *quenching* implies rapidly cooling of the alloy in water avoiding the vertical maximum of the corresponding TTT curve to obtain a supersaturated solid solution (metastable), free of precipitates. This type of treatment is used, for example, in stainless steels to obtain a material

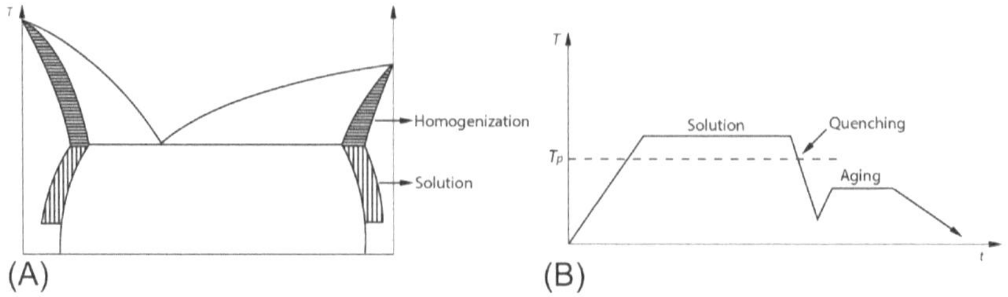

FIG. 7.10 Homogenization and solution heat treatment areas in (A) phase diagram and (B) temperatures vs. time.

"free" of carbides at room temperature and, thus, with a better corrosion behavior.

2. On the other hand, *precipitation heat treatment* (*aging, age-hardening*, or *tempering*) requires first, the solution heat treatment with subsequent quenching (shown in Fig. 7.10B). Immediately after quenching, the alloy is heated to an intermediate temperature, in order to create, by nucleation and growth, finer and more disperse precipitates; the lower the *aging* temperature, the smaller the precipitates will be. *Aging* is of the *natural* type if it is achieved by holding for long periods at room temperature.

   If aging temperature is excessively high, precipitates are usually large, as a result of *overaging*. High aging temperatures, or holding long periods of time at slightly lower temperatures, produce spheroidizing of the precipitates with polyhedral, needle, or plate shapes, in order to reduce the surface/volume ratio. This process is usually accompanied by an apparent coalescence of the precipitates: redissolution of the small precipitates and growth of the large ones (*Ostwald ripening*). In general, the rate of coarsening depends on diffusion (D), surface free energy (γ), and eutectic composition (C_E); therefore, alloys whose toughness depends on a fine precipitation dispersion must have a low value of at least one of these parameters.

   The age-hardening heat treatment is carried out to obtain a fine dispersion of precipitates; this microstructural transformation influences both, subsequent forming operations and properties of the finished parts. For example, in the Al-1% Mn alloy, the precipitation treatment decreases the required temperatures for recrystallization and inhibits the primary recrystallization grain growth. The surface appearance of sheets or parts is improved by age-hardening before an anodic oxidation treatment: commercial Al is treated at ~630°C to dissolve the precipitates, quenched, and then aged at 450°C to obtain a fine precipitation, before the anodic oxidation.

In some alloys, the formation of the stable precipitated phase is preceded by the emergence of one or more metastable and nanometric transition phases. These

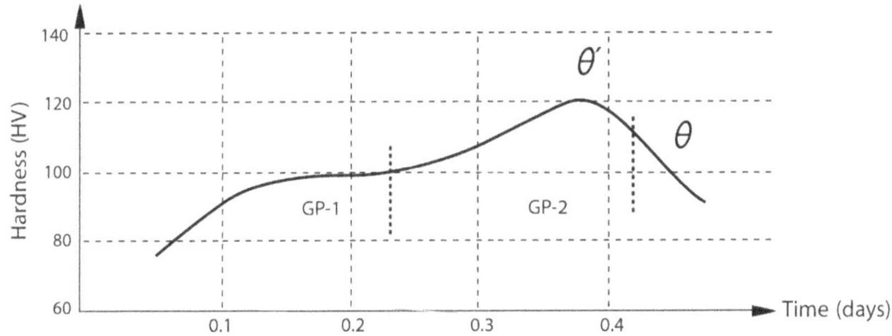

FIG. 7.11 Hardness variation for Al-4% Cu aged at 130°C.

alloys are known as *hardened* (*structural hardening*) *alloys* because precipitation considerably improves this property.

Precipitation hardening is a mechanism to increase the strength of the alloy through a fine dispersion of hard precipitated particles. Tempering generally accomplishes this; however, a noticeable increase in hardness is achieved when the space between precipitates is so small that it hinders the movement of dislocations during plastic deformation. This fine dispersion, indistinguishable by optical microscopy, is only accomplished in those alloys whose precipitation process is preceded by solution heat treatments, followed by a sequence of stages with formation of transition phases before the precipitation of equilibrium phases (according to the phase diagram). An example is duralumin (Al-4% Cu-1% Mg), named after the German city of Düren where A. Wilm discovered in 1906, its *hardening* ability by solution, quenching, and natural aging heat treatments, which are nowadays widely used in industrial applications.

Fig. 7.11 indicates the variation in hardness with time for an Al-4% Cu alloy with solution at 500°C and aging at 130°C. Maximum hardening is obtained by the formation of transition structures, before equilibrium precipitates are visible in the optical microscope. This hardness evolution is the result of formation of three transition structures[2] known as GP-1, GP-2, and θ', before the equilibrium phase Al_2Cu (θ):

- The GP-1 zones are formed by groups of atoms of Cu; their length is approximately 100 Å and their thickness is 2 or 3 interatomic distances. They exist in the (100) Al plane and are completely coherent with the matrix: atoms of Cu occupy positions of the Al lattice.
- The second transition structure, GP-2, is a precipitate shaped as a disk with a tetragonal crystalline structure and dimensions of 1500 Å in diameter and 150 Å in thickness. The diameter is coherent with the aluminum matrix,

2. The first two transition structures are known as Guinier-Preston (GP) zones, in honor of the first researchers who studied their formation using X-ray diffraction.

but the height (or thickness) of the disk is not. Its mean composition is the same as the Al_2Cu equilibrium precipitate. The maximum hardness of the alloy is achieved by the formation of these zones, which are abundant and produce large distortions.

- The structure of θ' is similar to GP-2 since it is created by the formation of dislocation loops around this precipitate, destroying the coherence with the matrix and relaxing most of the long-range stresses; thus, the moment θ' appears, and the hardness of the material drops. This phase has a tetragonal transition structure, with lattice parameters a and b having the same length as the lattice parameter of Al (4.04 Å), and parameter c with a value of 5.8 Å. The a and b axes are parallel to the <100> directions of Al.
- The Al_2Cu equilibrium phase (θ) is tetragonal and incoherent to the matrix. The a and b parameters are equal (6.06 Å) while c has a value of 4.87 Å.

Among the commercial Al alloys, structural (age) hardening can be applied to Al-Cu, Al-Zn, Al-Li, Al-Mg-Si, and Al-Zn-Mg alloys: all of them, considered *high strength light alloys*, are used for aeronautical applications.

Table 7.2 presents improvement in mechanical characteristics for Al alloys when small amounts of other elements produce structural hardening. These properties can be compared to the maximum strength of commercial Al (99.5% Al): in recrystallized state $\sigma_{max} = 100\,MPa$ and elongation of 30%, while in cold work state (even after hardening) σ_{max} can only reach 150 MPa, with an elongation of 5%.

TABLE 7.2 Mechanical Properties of Some Al Alloys With Structural Hardening

Alloy	Temperature for Quenching	Cooling Fluid	Aging Time (h)	σ_{max} (MPa)	σ_y (MPa)	A %
Al-6% Cu	535	Water	15–25 at 205°C	380	220	7
Al 1% Mg-0.5% Si	520	Water	10–12 at 170°C	300	250	8
Al 6% Zn-2% Mg	465	Water	1 day at 120°C	600	500	6

EXERCISE 7.6

Estimate the yield stress for an Al-4% Cu alloy after structural hardening, Fig. 7.12.

Data: $G_{Al} = 25.4\,\text{GPa}$, $\mathbf{b}_{Al} = 2.85\,\text{Å}$, and $\mathbf{b}_{Cu} = 2.55\,\text{Å}$

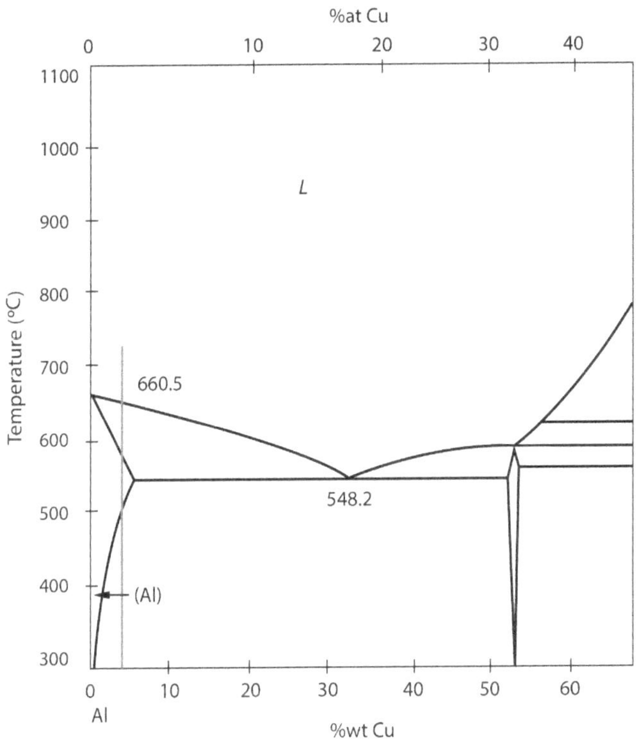

FIG. 7.12 Al-Cu phase diagram (ASM Handbook, Vol. 3, 1992).

Solution

After quenching, Cu is in a supersaturated solid solution and the elastic limit can be calculated using Eq. (7.1):

$$\sigma_y \cong \frac{25,400}{10,000} + (25,400)\left(\frac{2.85 - 2.55}{2.85}\right)(0.02) = 56.01\,\text{MPa}$$

It is a common practice to cold work the alloy before aging, so dislocation density is increased to $\rho_\perp = 1 \times 10^9\,\text{cm}_\perp/\text{cm}^3$. The elastic limit caused by dislocations is:

$$\sigma_y(\perp) \cong (25,400)(2.85 \times 10^{-8}\,\text{cm})\sqrt{1 \times 10^9 \frac{\text{cm}}{\text{cm}^3}} = 22.89\,\text{MPa}$$

After aging, all Cu forms GP nanoprecipitates (50 Å), as shown in Fig. 7.13. Using data from the phase diagram:

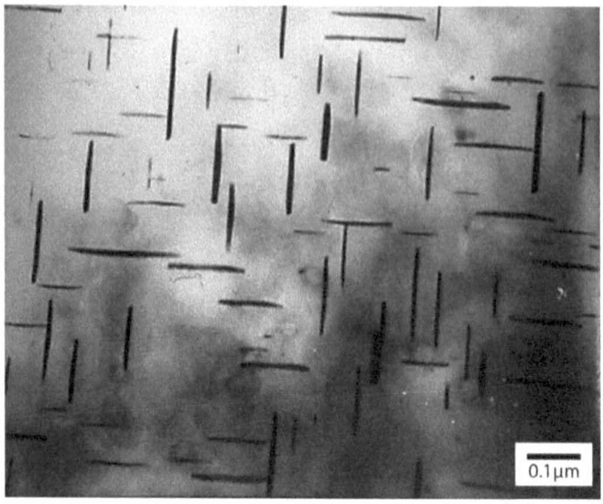

FIG. 7.13 Al_2Cu nanoprecipitates.

$$f_p = \frac{4}{53} = 0.075(7.5\%)$$

$$\sigma_y(\perp) \cong \frac{(25,400)(2.85\times10^{-8}\,\mathrm{cm})\sqrt{0.075}}{50\times10^{-8}\,\mathrm{cm}} = 396.49\,\mathrm{MPa}$$

On the other hand, reducing grain size also has a hardening effect:

$$\sigma_{Y_{gb}} = K_{gb}\cdot d^{-1/2}$$

with d in mm. Considering a grain size of ASTM 7, or $d = 30\,\mu\mathrm{m} = 0.03\,\mathrm{mm}$, K_{gb} for Al equals $1.96\,\mathrm{N/mm}^{-3/2}$, then:

$$\sigma_{Y_{gb}} = \left(1.96\frac{\mathrm{N}}{\mathrm{mm}^{-3/2}}\right)(0.03\,\mathrm{mm})^{-1/2} = 11.32\,\mathrm{MPa}$$

The elastic limit of duralumin can now be calculated again using Eq. (7.1) but considering the elastic limit of dislocations instead of the one for solid solution and the one obtained by reducing grain size:

$$\sigma_y \cong \frac{25,400}{10,000} + \sigma_y(\perp) + \sigma_{Y_{gb}} = 2.54 + 396.49 + 11.32 = 410.36\,\mathrm{MPa}$$

EXERCISE 7.7

Estimate the yield stress for the Al-4.5% Mg alloy (Fig. 5.28).

Data: the solubility of Mg in Al at room temperature is 1.8% in wt., $G_{Al} = 25.4\,\mathrm{GPa}$, $\mathbf{b}_{Al} = 2.85\,\text{Å}$, and $\mathbf{b}_{Mg} = 3.2\,\text{Å}$

Solution

If precipitates have a size of 100 nm, aging will result in:

$$f_{prec} = \frac{4.5-1.8}{35.5-1.8} = 0.08$$

$$\sigma_y = \frac{25,400}{10,000} + \left(2.54\times10^4\right)\frac{|2.85-3.2|}{2.85}(0.018) + \frac{\left(2.54\times10^4\right)\left(2.85\times10^{-8}\right)\sqrt{0.08}}{1\times10^{-5}}$$

$$= 79.16\,\text{MPa}$$

Though, a harder alloy may be achieved by quenching (with oversaturated solid solution of Mg in Al), with yield stress reaching a value of:

$$\sigma_y \cong \frac{25,400}{10,000} + \left(2.54\times10^4\right)\frac{|2.85-3.2|}{2.85}(0.045) = 142.91\,\text{MPa}$$

Many nonferrous systems (e.g., Cu-Be, Cu-Cr, and Pb-Sb-Sn) can also achieve structural hardening by quenching and aging. Among the copper-based alloys, the ones of commercial interest are Cu-1% Cr and Cu-2% Be (Table 3.1 shows the maximum solubility of Cr and Be in Cu).

Quenching a Cu-1% Cr alloy (Fig. 7.14) at 900°C and aging at room temperature (or at 500°C for 2 h), results in $\sigma_{max} = 450\,\text{MPa}$ with an elongation of 15%. On the other hand, a Cu-2% Be alloy reaches a considerable amount of structural hardening by precipitation: dissolving the precipitates at 800°C, quenching in water, and aging at 320°C; this last treatment results in $\sigma_{max} = 1225\,\text{MPa}$ with 2% elongation, which is an important increase in hardness compared to recrystallized Cu ($\sigma_{max} = 245\,\text{MPa}$ with 30% elongation). Though the alloy has a high work-hardening ability, even in its hardened state, cold work deformation can only increase the ultimate stress to 450 MPa with a 5% elongation.

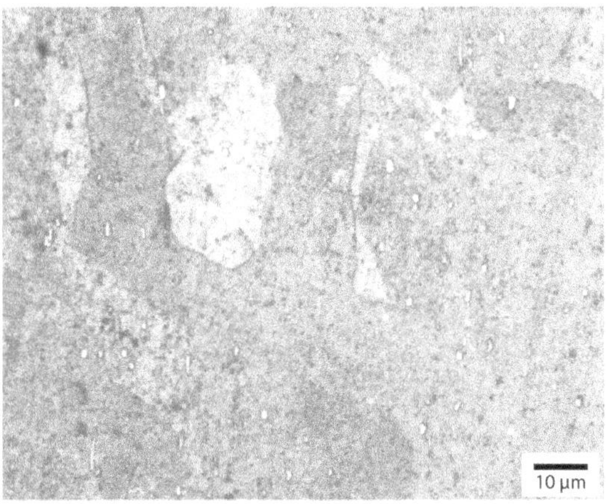

FIG. 7.14 Cu-1% Cr alloy (structural hardened state).

EXERCISE 7.8

(a) Estimate the yield stress for the Cu-2% Be alloy (Fig. 7.15) after quenching, cold work, and aging, (b) determine the volume fraction of the CuBe precipitates, and (c) their mean size.

Data: $G_{Cu} = 42.1\,\text{GPa}$, $a_{Cu} = 3.6\,\text{Å}$, $L_P = 125\,\text{Å}$, $\rho_{\perp} = 1 \times 10^{10}\,\text{cm}_{\perp}/\text{cm}^2$, $\rho_{Cu} = 63.5\,\text{g/mol}$, $a_{CuBe} = 2.71\,\text{Å}$, and $\rho_{Be} = 9\,\text{g/mol}$

Solution

(a) Yield stress of the alloy

Using in Eq. (7.1), a Burgers vector of:

$$\mathbf{b} = \frac{a\sqrt{2}}{2} = 2.55$$

$$\sigma_y \cong \frac{42,100}{10,000} + (42,100)(2.55 \times 10^{-8})\sqrt{1 \times 10^{10}} + \frac{(42,100)(2.55)}{125} = 970.4\,\text{MPa}$$

(b) Volume fraction of precipitates

The density of the CuBe intermetallic is:

$$\rho_{CuBe} = \frac{9 + 63.5}{(6.023 \times 10^{23})(2.71 \times 10^{-8})^3} \cong 6.05\frac{\text{g}}{\text{cm}^3}$$

and the fraction volume can be calculated as:

$$\frac{\rho_{CuBe}}{\rho_\alpha} = \frac{2}{10} = 0.2$$

$$\frac{\rho_{CuBe} \cdot V_{CuBe}}{\rho_\alpha \cdot V_\alpha} = 0.2$$

$$\frac{V_{CuBe}}{V_\alpha} = (0.2)\frac{9}{6.1} = 0.295(\approx 0.3)$$

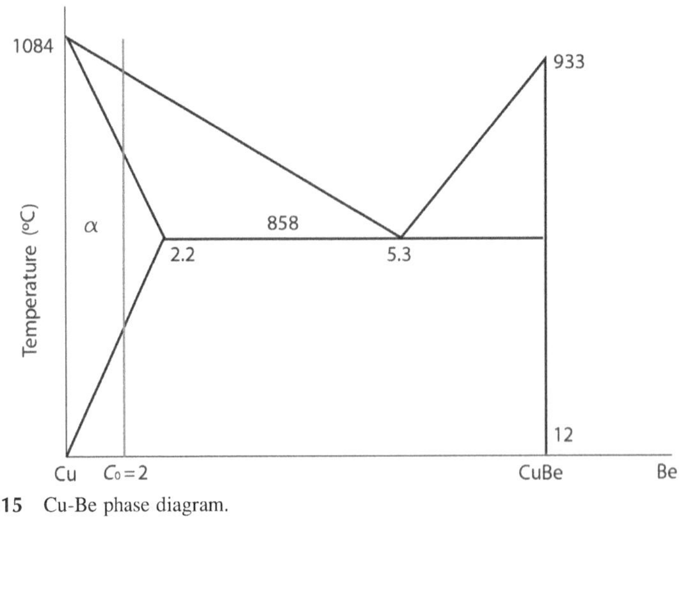

FIG. 7.15 Cu-Be phase diagram.

$$\frac{V_{CuBe}}{V_{CuBe}+V_{\alpha}}=\frac{0.3}{1.3}=0.23(23\%)$$

(c) Mean size of precipitates
The mean size of CuBe particles can be obtained by:

$$L\approx\frac{\bar{X}}{\sqrt{f_v}}-\bar{X}$$

and solving for $\bar{X}$:

$$\bar{X}=\frac{L\sqrt{f_v}}{1-\sqrt{f_v}}=\frac{(125)(0.23)}{1-0.23}=115.2\text{Å}$$

Comment: The main hardening mechanism in this alloy is precipitation. Its high strength value is a combination of precipitate size and shear stress of Cu.

Some Fe-based alloys can also be structurally hardened by tempering, supposing the material has already formed Fe-Ni martensite (body-centered cubic) by cooling in air from 850°C and aging at 480°C for 3 h. This is the case of *maraging steels (martensite age steels)*: compositions of some of these alloys are shown in Table 7.3, while their properties after tempering are indicated in Table 7.4. The main hardening constituents are Ni_3Mo, Ni_3Al, and Ni_3Ti.

There are other types of *maraging steels* with 20% Ni, 1.5% Ti, 0.25% Al, and 0.5% Nb. The precipitates responsible for their structural hardening are $Ni_3(AlTi)$ and Ni_3Nb, which are formed by cooling from 820°C and then aging

TABLE 7.3 Maraging Steels' Composition (Weight %)

Types	C	Mn	Si	S	P	Ni	Co	Mo	Al	Ti
I	<0.03	<0.10	<0.10	<0.01	<0.01	17–19	8–9	3–3.5	0.10	0.20
II	<0.03	<0.10	<0.10	<0.01	<0.01	17–19	7–8	4.6–5.1	0.15	0.40
III	<0.03	<0.10	<0.10	<0.01	<0.01	17–19	8.5–9.5	4.7–5.2	0.15	0.60

TABLE 7.4 Maraging Steels' Properties

Types	σ_y (MPa)	σ_u (MPa)	A%	Resilience at 20°C (J)
I	1373	1472	14–16	81–149
II	1668	1913	10–12	24–35
III	1962	2109	12	14–27

at 480°C for 3 h, to reach a ultimate stress of 1800 MPa, a yield stress of 1700 MPa, and 11% elongation. Higher values of σ_{max} and σ_y with similar elongation, are achieved by increasing Ni (up to 25%). However, cooling from 820°C produces austenite instead of martensite and a treatment (before aging at 480°C for 3 h) at 700°C for 4 h and then cooling to −78°C is necessary to transform austenite into martensite.

Other types of ferrous alloys that can be structurally hardened are stainless steels (*PH* or *precipitation hardened steels*), shown in Table 7.5. The elements responsible for the structural hardening are P, Mo, Cu, Nb, and/or Ti. The heat treatment includes quenching from 1200°C and aging at 700–800°C. The elastic limit reached is ~700 MPa.

7.1.3 Final and Intermediate Phases (Intermetallic Compounds)

If, after combining pure components, the free energy of the configuration of atoms results in a crystalline structure different from either one, an *intermediate phase* is formed.

Formation of precipitates with stoichiometric composition (Table 7.1) is a plausible cause of precipitation hardening or structural hardening and these constituents are known as intermetallic compounds. Their nature is partly metallic and partly similar to real chemical compounds with nonmetallic bonds (covalent and/or ionic). Their crystals (since they have more than one type of bond) are known as heterodesmic compared to pure metal or solid solutions crystals that have only one type of bond (homodesmic).

Many intermetallic compounds (intermediate phases) with weak metallic characteristics (abundance of nonmetallic bonds) can be considered as real chemical compounds, even if their formula does not entirely follow the rules of valence. Therefore, they are generally hard materials, with little plasticity and low electrical and thermal conductivities; meaning their properties lie between pure metals and chemical compounds of normal valence.

Unlike pure metals and solid solutions whose locations in a phase diagram are at both ends (*final phases*), intermetallic compounds are located at the middle of the phase diagram (*intermediate phases*), and have a stoichiometric composition: its presence in a phase diagram is determined by a vertical straight line. Some of them are formed by solid phase reactions and others are directly obtained from the melted phase and solidified at constant temperature (Chapter 3) which determines the relative maximum of the liquidus line (Fig. 3.6).

Both the intermediate and final phases are known as simple constituents of the alloys, which must not be confused with the components of an alloy (chemical elements forming it). For example, a binary alloy of two components A and B can present a microstructure formed only by grains of a single constituent (as in the case of solid solutions) or of two simple constituents (as in the case of solid solution grains with precipitates of the other phase, e.g., Figs. 7.2 and 7.3).

TABLE 7.5 Composition (Wt. %) of Stainless Steels With Structural Hardening

UNS	C	Mn	P	S	Si	Cr	Ni	Mo	Others
(S13800)	0.05	0.10	0.01	0.008	0.10	12.25/13.25	7.50–8.50	2.00/2.50	Al 0.90–1.35 N 0.010
(S15500)	0.07	1.00	0.040	0.030	1.00	14.00/15.50	3.50–5.50		Cu 2.50–4.50 Cb+Ta 0.15–0.45
(S17400)	0.07	1.00	0.040	0.030	1.00	15.50/17.50	3.00–5.00		Cu 3.00–5.00 Cb+Ta 0.15–0.45
(S17700)	0.09	1.00	0.040	0.040	1.00	16.00/18.00	6.50–7.75		Al 0.75–1.50

The properties of the alloys not only depend on the nature of the final phases, but also on the nature, amount, size, and location of the intermediate phases.

Among the most important intermetallic compounds (because of their industrial applications) are carbides, due to their mechanical and chemical contributions in alloyed steels such as microalloyed, tool-making, stainless, wear resistant, etc., as well as in electronic compounds, for their role in copper alloys such as bronzes, brasses, copper-aluminums, etc.

7.1.3.1 Carbides

The structure of intermediate phases is determined by three main factors: relative atomic size, valence, and electronegativity. Examples where atomic size determines the structure are *interstitial compounds* MX, M_2X, MX_2, and M_6X (where M can be Ti, Zr, Nb, V, Cr, etc. and X can be H, B, C, and N). In this case, M atoms form cubic or hexagonal closed-packed structures where the X atoms are small enough to fit inside the interstices.

Carbides can be classified, according to their crystalline system, into simple and complex. Examples of simple carbides are: TiC, ZrC, HfC, VC, NbC, TaC, WC, W_2C, MoC, and Mo_2C. Some of them (TiC, ZrC, HfC, VC, NbC, and TaC) have a crystalline structure similar to sodium chloride: metallic atoms located at the corners and at the center of the faces of a cube, and C located at the center of the edges and at the center of the cube. Fe is not allowed in solid solution (substituting the atoms of the carbide-forming metal) nor do Fe reacts to form ternary carbides. They are very stable particles with high melting temperatures: 3140°C (TiC), 3550°C (ZrC), 3887°C (HfC), 2830°C (VC), 3500°C (NbC), and 3875°C (TaC).

EXERCISE 7.9

Calculate the TiC density, knowing that Ti has an atomic weight of 47.9 u and lattice parameter $a = 4.3285$ Å at room temperature.

Solution

Titanium carbide has a fcc structure, with atoms of Ti in the corners and the centers of the faces of the cube (4 atoms per cell) and atoms of C in the center of the edges and center of the cell (4 atoms per cell). Thus, its density can be calculated by:

$$\rho_{\mathrm{TiC}} = \frac{n \cdot \mathrm{atomic weight}}{N_A \cdot a^3} = \frac{4(12\mathrm{u}) + 4(47.9\mathrm{u})}{6.023 \times 10^{23} \frac{\mathrm{u}}{\mathrm{g}} \left(4.3285 \times 10^{-8}\,\mathrm{cm}\right)^3} = 4.91 \frac{\mathrm{g}}{\mathrm{cm}^3}$$

On the other hand, the crystalline structures of WC, MoC, W_2C (Fig. 7.16), and Mo_2C carbides are simple hexagonal, which are very stable, with the following melting temperatures: 2867°C (WC), 2857°C (W_2C), 2692°C (MoC), and 2687°C (Mo_2C). However, compared to cubic carbides, they do react with Fe to form very complex ternary carbides such as Fe_4W_2C or Fe_3W_3C

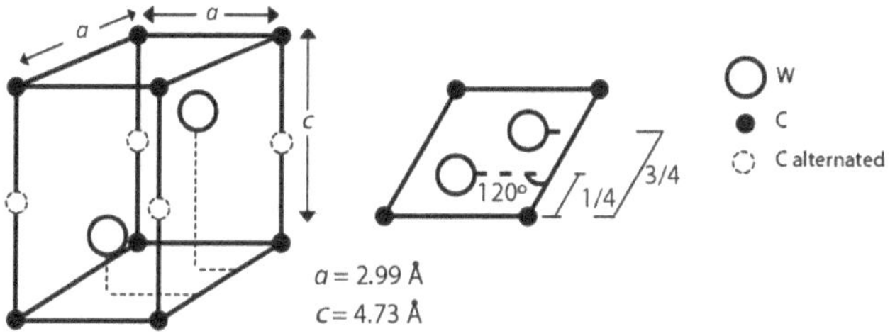

FIG. 7.16 Hexagonal unit cell of W_2C (ASM Handbook, Vol. 3, 1992).

commonly used in high speed steels for cutting tools; these carbides are known as M_6C and have a complex cubic structure with 112 atoms per unit cell (96 metallic and 16 C).

EXERCISE 7.10

Calculate the W_2C density, knowing that W has an atomic weight of 183.85 u and lattice parameters $a = 2.99$ Å and $c = 4.73$ Å at room temperature.

Solution

W_2C has a simple hexagonal structure, with atoms of C in the corners of a rhombic prism (1 atom per cell) and atoms of W in the interior of the cell at the $a \cdot \frac{1}{4}c$ and $\frac{3}{4}c$ positions, respectively (2 atoms per cell). Thus, the density is:

$$\rho_{W_2C} = \frac{n \cdot \text{atomic weight}}{N_A \cdot a^2 \frac{\sqrt{3}}{2} c} = \frac{1(12\,\text{u}) + 2(183.85\,\text{u})}{6.023 \cdot 10^{23} \frac{\text{u}}{\text{g}} (2.99 \times 10^{-8}\,\text{cm})^2 \frac{\sqrt{3}}{2} (4.73 \times 10^{-8}\,\text{cm})}$$
$$= 17.21 \frac{\text{g}}{\text{cm}^3}$$

Cr forms binary carbides (C solubility almost null) which are present in a large number of ferrous alloys, for example the complex structures $Cr_{23}C_6$ (commonly known in metallurgy as K_1), Cr_7C_3 (known as K_2), and Cr_3C_2:

- $Cr_{23}C_6$ (or K_1) has a complex face-centered cubic structure formed by 116 atoms (92 of Cr and 24 of C). Cr atoms can be substituted by other elements (e.g., Fe, W, or Mo) without losing its specific crystalline structure, though forming less stable carbides which have a tendency to decompose and form other carbides, some even more stable than $Cr_{23}C_6$.

  The melting temperature of K_1 is 1150°C, however when present as disperse constituent in steels, it decomposes before reaching that temperature and their components become a solid solution in the crystalline lattice of Fe. $Cr_{23}C_6$ practically dissolves at temperatures higher than 1100°C. On the other hand, it can be formed by precipitation during cooling, starting from Cr-rich austenite, which is the usual composition of Cr carbides at low temperatures.

- Cr_7C_3 (or K_2) crystallizes in the orthorhombic system, with 24 atoms of Cr and 16 of C per unit cell. Its melting temperature is 1890°C.
- Cr, just as Co or Mn, sometimes substitutes Fe atoms in cementite (Fe_3C) to form a ternary carbide with cementitic structure, known as M_3C or K_c carbide.

Carbides can be formed by direct solidification or by solid state precipitation during cooling, starting from complex Fe-based solid solutions. The simpler their crystalline structure, the lower the possibility to be redissolved. The delicate balance between solid-state diffusion rates and carbide stability is the main connection between properties of the steels and heat treatments.

There is a type of simple intermetallic constituent, noncarbon based, frequently found in high Cr steels, whose microstructure, when precipitation is considerable, is apparently similar to carbides and known as *sigma phases*, which, just like carbides, precipitate by nucleation and growth with C-type time-temperature-transformation curves. Their shape and distribution in the metallic matrix, even in small amounts, considerably modifies the properties of the alloy. In stainless steels, their formation is brought on by the presence of other high-Cr phases such as $M_{23}C_6$.

The Fe-Cr phase diagram (Fig. 3.20) shows the sigma phases whose Cr atomic content lies between 40% and 50%. This type of constituent also appears in other Cr-based systems, such as Cr-Mn (17%–28% at. Cr) and Cr-Co (53%–58% at. Cr). On the other hand, other binary systems without Cr that present sigma phases are: V-Mn system (17%–28% at. V), V-Fe system (48%–52% at. V), V-Ni (with at least 55% at. V), and W-Fe (50%–60% at. W).

The sigma-type constituent, with very complex crystallographic structure, plays a negative role in the alloys causing embrittlement. In stainless steels, ferritic refractories, and austenitic steels, σ usually appears after prolonged exposure (thousands of hours) at 500–900°C. However, heating at temperatures higher than 900°C can dissolve this phase and prevent embrittlement. Fig. 7.17 presents the micrographs of Cr carbides in a white casting and tool steel.

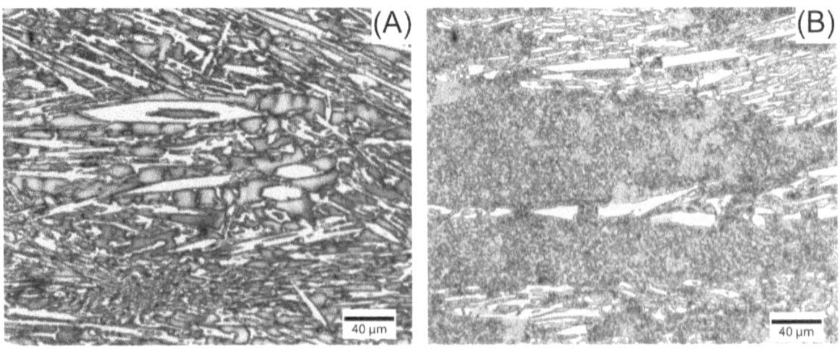

FIG. 7.17 Cr carbides in (A) a white casting and (B) a tool steel.

7.1.3.2 Electronic Compounds

Among the constituents that can be formed by Cu and Zn, as seen in Fig. 7.18, are the intermetallic compounds β, γ, δ, and ε, also known as electronic compounds (Table 7.6). Their formation is caused by an instability in the electronic cloud that becomes saturated in electrons, when dissolving a certain amount of metal B (e.g., Zn) in metal A (e.g., Cu), with B having a higher valence than A.

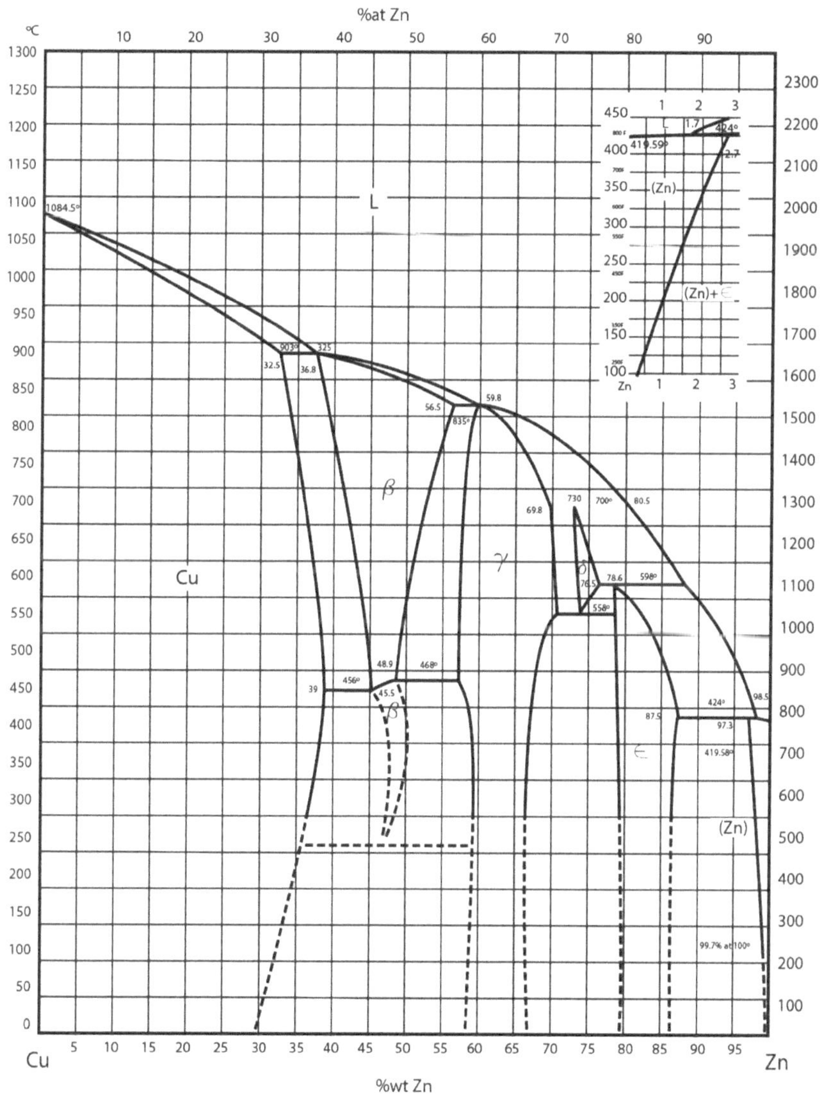

FIG. 7.18 Cu-Zn phase diagram (ASM Handbook, Vol. 3, 1992).

TABLE 7.6 Cu-Zn System Constituents

Phase	Chemical Composition	Crystalline System
α	Substitutional solid solution	fcc
β	CuZn	bcc
γ	Cu_5Zn_8	Cubic complex (52 atoms)
δ	$CuZn_3$	bcc (560° + 700°C)
ε	$Cu_{21}Zn_{79}$	hcp
η	Substitutional solid solution	hcp

The equilibrium compound between the monovalent Cu and the divalent Zn should have the composition Cu_2Zn, however the constituents β, γ, and δ (as shown in Table 7.6) and known as *abnormal valence* or *Hume-Rothery* compounds, differ as the ratio of shared electrons is: 21/14 for β, 21/13 for γ, and 21/12 for δ.

From a composition point of view, electronic compounds behave similar to solid solutions. At high temperatures, variations in composition are possible as atoms can occupy "wrong" positions or vacancies (Fig. 7.18):

- The crystalline lattice of γ phase is a complex cubic formed by 52 atoms (similar to Mn). This constituent is very hard and brittle, and because of its low toughness, Cu-Zn alloys (brasses) are not fit for industrial use if this phase is present. Thus, the industrially useful zone of the diagram, and consequently the composition of brasses, is 50%–100% Cu.
- The crystalline lattice of β is body-centered cubic. A 49% Cu alloy will be formed by β grains, and inside the grains and at their boundaries, γ will be present with shapes similar to clovers (Fig. 7.19).

  The behavior of a β-brass (50% Cu-50% Zn) at temperatures higher than 470°C is similar to a solid solution, ductile and tough; though with a crystalline lattice different from the metals that compose it. At temperatures below 468°C, the β constituent experiences an order/disorder transformation (atoms of Zn are located at the center of each cube and atoms of Cu at the corners of the cubes) creating a hard and barely deformable structure, inadequate as structural material (Fig. 7.20), but necessary in the manufacture of high-machinability $\alpha+\beta$ brasses (free-cutting brasses).

  The industrial use of β brass is limited to filler material in welding (brazing) because of its good castability, the absence of dendritic segregation during solidification (small solidification interval in the Cu-Zn diagram), and the possibility to form solid solutions with the metals to be welded using this technique.

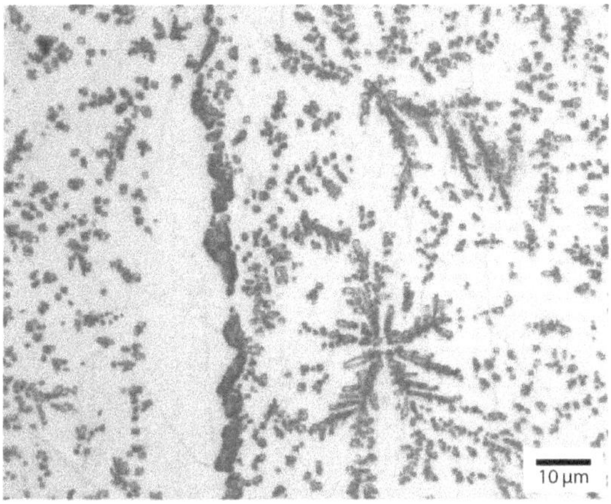

FIG. 7.19 49% Cu-51% Zn alloy.

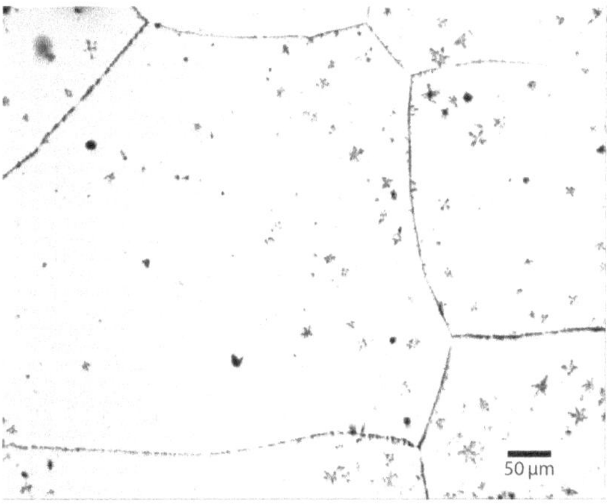

FIG. 7.20 50% Cu-50% Zn β brass.

On the other hand, electronic intermetallic materials, or Hume-Rothery intermediate compounds, are not exclusive to the Cu-Zn system. Table 7.7 presents some of the electronic compounds grouped by the R ratio between the shared electrons and the number of atoms: 21/14, 21/13, or 21/12. For the 21/14 value, the constituents adopt the bcc crystalline structure; when the ratio is 21/13 the usual lattice is "Mn-cubic type" similar to γ constituent of brasses; and for 21/12 the tendency is the hcp system.

Considering Table 7.7 for the Cu-Sn system (bronzes), there are three electronic compounds: $Cu_5Sn\,(R=21/14)$, $Cu_{31}Sn_8\,(R=21/13)$, and $Cu_3Sn\,(R=21/12)$ which correspond to the β, δ, and ε intermediate phases of the Cu-Sn diagram

TABLE 7.7 Some Electronic Compounds (Hume-Rothery) Classified by Their *R* Values

R = 21/14		*R* = 21/13		*R* = 21/12
CuBe	Ag_3Al	Cu_5Zn_8	Au_5Cd_8	$CuZn_3$
CuZn	Ag_3In	Cu_5Cd_8	Au_9In_4	$CuCd_3$
Cu_3Al	AuMg	Cu_5Hg_8	Mn_5Zn_{21}	Cu_3Sn
Cu_3Ga	AuZn	Cu_9Al_4	Fe_5Zn_{24}	Cu_3Ge
Cu_3In	AuCd	Cu_9Ga_4	Co_5Zn_{21}	Cu_3Si
Cu_5Si	FeAl	Cu_9In_4	Ni_5Be_{21}	$AgZn_3$
Cu_5Sn	CoAl	$Cu_{31}Si_8$	Ni_5Zn_{21}	$AgCd_3$
AgMg	NiAl	$Cu_{31}Sn_8$	Ni_5Cd_{21}	Ag_3Sn
AgZn	NiIn	Ag_5Zn_8	Rh_5Zn_{21}	Ag_5Al_3
AgCd	PdIn	Ag_5Cd_8	Pd_5Zn_{21}	$AuZn_3$
		Ag_5Hg_8	Pt_5Be_{21}	$AuCd_3$
		Ag_9In_4	Pt_5Zn_{21}	Au_3Sn
		Au_5Zn_8	$Na_{31}Pb_8$	Au_5Al_3

(Fig. 7.21). γ phase of the Cu-Sn system is a bcc ordered structure, different but in a certain way analogous to β phase of brasses.

When analyzing copper-aluminums (Cu-Al system, Fig. 4.5), intermediate phases β and γ_2 correspond to the electronic compounds Cu_3Al and Cu_9Al_4 with *R* values of 21/14 and 21/13, respectively.

In general, it can be concluded that electronic compounds similar to γ phase of brasses are hard and brittle; which justifies, for example, that the industrially useful zone of the binary Cu-Sn (bronzes) diagram is between 80% and 100% wt Cu. Analogously, copper-aluminums have no more than 10% wt Al.

7.2 TRANSFORMATIONS CAUSED BY ALLOTROPIC CHANGES

One of the most important properties of Sn is the degradation in cold environments, resulting in considerable changes in color and appearance, starting from a typical silver white surface, followed by superficial "pustules" and ending with complete degradation, when remaining for long periods of time at temperatures near or below zero, and also "contagion" (known as "tin pest") that occurs at low temperatures when a healthy piece is in contact with a "sick" one where "pustules" are present.

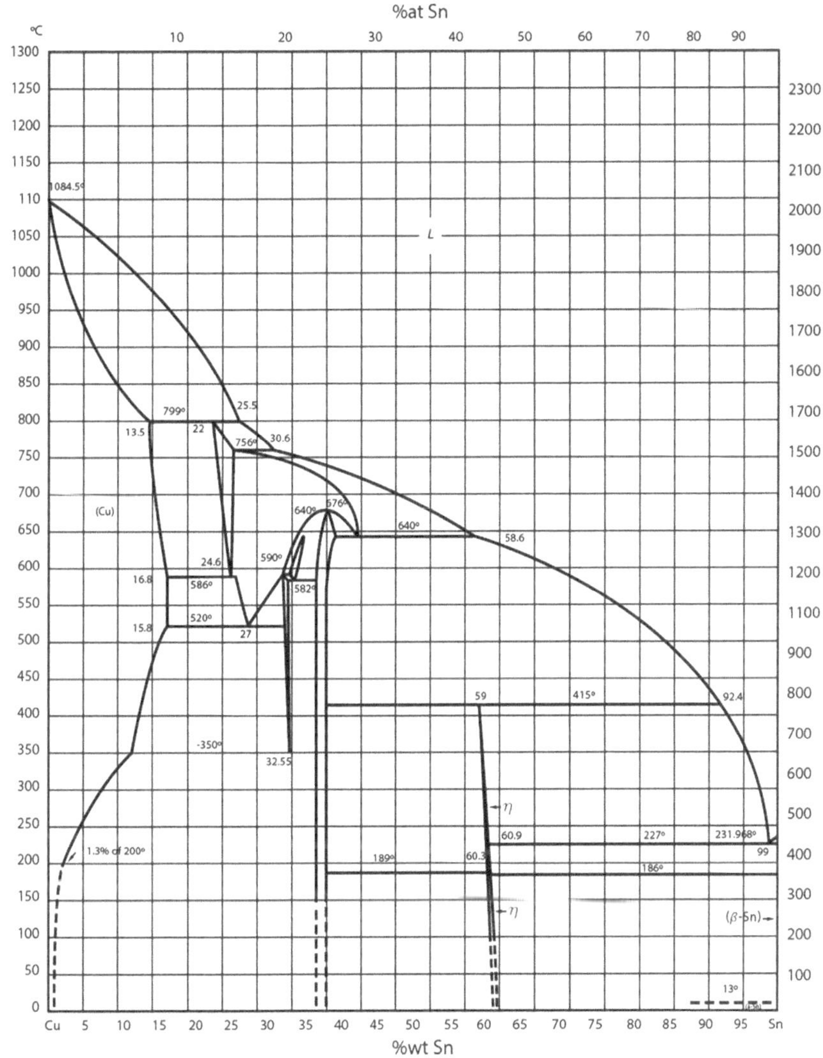

FIG. 7.21 Cu-Sn phase diagram (ASM Handbook, Vol. 3, 1992).

This phenomenon is caused by a solid-state allotropic transformation: above 20°C the stable crystalline phase of Sn is face-centered tetragonal (fct), with lattice parameters $a = 5.83$ Å, and $c = 3.18$ Å, and $\rho = 7.298\,g/cm^3$, while below 20°C, Sn crystallizes in the diamond-type cubic system resulting in a type of Sn which is gray, semiconductor and with $\rho = 5.768\,g/cm^3$. Thus, during cooling, Sn undergoes, by allotropic transformation, a relative volume change of 27%.

Below 20°C, the color change from white to gray occurs by nucleation and growth, accompanied by a considerable volume increase and a powder-like decohesion. Because of the slow rate of transformation (diffusion) at this temperature, in practice, temperatures below zero are required. Furthermore, the "contagion" phenomenon is caused by the heterogeneous nucleation of gray tin that accelerates the transformation kinetics starting with the white tetragonal phase.

EXERCISE 7.11

Calculate the volume variation occurring during the allotropic transformation of β-Sn (white, tetragonal) into α-Sn (gray, cubic) at 13°C.

Data: $a_\beta = 5.83\,\text{Å}$, $c_\beta = 3.18\,\text{Å}$, $a_\alpha = 6.47\,\text{Å}$, and $\rho_{Sn} = 118.7\,\text{g/mol}$

Solution

Calculating the densities of both phases:

$$\rho_{\beta-Sn} = \frac{(4\,\text{at})\left(118.7\frac{\text{g}}{\text{mol}}\right)}{\left(6.023\times 10^{23}\frac{\text{at}}{\text{mol}}\right)(5.83\times 10^{-8}\,\text{cm})^2(3.18\times 10^{-8}\,\text{cm})} = 7.29\frac{\text{g}}{\text{cm}^3}$$

$$\rho_{\alpha-Sn} = \frac{(8\,\text{at})\left(118.7\frac{\text{g}}{\text{mol}}\right)}{\left(6.023\times 10^{23}\frac{\text{at}}{\text{mol}}\right)(6.47\times 10^{-8}\,\text{cm})^3} = 5.82\frac{\text{g}}{\text{cm}^3}$$

Now calculating the volume variation

$$\Delta V = \frac{V_{\alpha-Sn} - V_{\beta-Sn}}{V_{\beta-Sn}} = \frac{m\left(\frac{1}{5.82} - \frac{1}{7.29}\right)}{m\left(\frac{1}{7.29}\right)} = 0.25 = 25\%$$

which is very close to the theoretical value (27%).

In many metals that present allotropic transformations, the high temperature structure of the metal can be obtained, in a metastable manner, at room temperature, through fast cooling (quenching).

Fig. 7.22 shows the solidification curve of Zr where two allotropic varieties can be identified (fcc and hcp, stable at temperatures higher and lower than 863°C, respectively). Furthermore, changes in the microstructure during cooling, result of polymorphism, are also shown. The structure obtained after solidification is known as *primary crystalline structure*, while the one obtained by cooling the primary one is known as *secondary crystalline structure*.

Since the allotropic transformation considers a change in structure, through *heat treatments (or normalizing in steels)*, grain refinement can be achieved in an as-cast part made of a polymorphic metal, by increasing the temperature of the part above the critical one until reaching the high allotropic state, followed by moderate cooling of the alloy. Furthermore, this heat treatment creates a very fine-grained microstructure whose grain size will be highly influenced by the fast heating and cooling rates (a faster cooling provides a larger undercooling) because of their effect on the nucleation and growth processes of the nuclei.

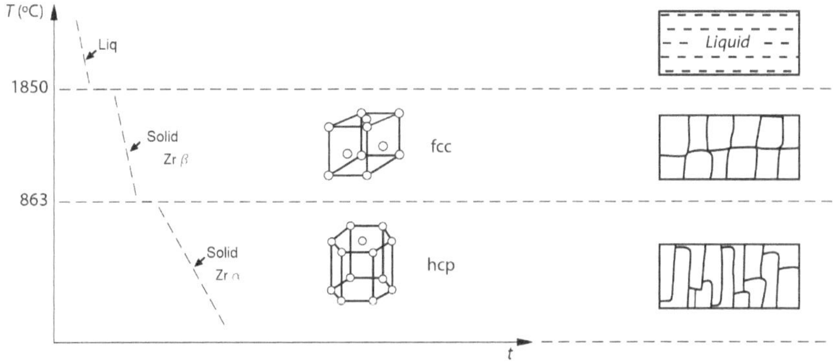

FIG. 7.22 Solid-state transformation caused by allotropic change.

Summarizing, in polymorphic metals (Fe, Ti, Zr, etc.), the as-cast microstructure can be erased by allotropic changes during cooling (Chapter 6).

Figs. 7.23 and 7.24 show the equilibrium diagrams (solidification and allotropic transformation lines) of two metals soluble in liquid phase, insoluble in solid phase and with eutectic affinity: metal A has 2 solid state allotropic varieties, the transformation line (for any of the AB system alloys) is evidently horizontal at T_{A2} (allotropic transformation of metal A). These figures also present the microstructural evolution of both alloys (I and II) until reaching room temperature.

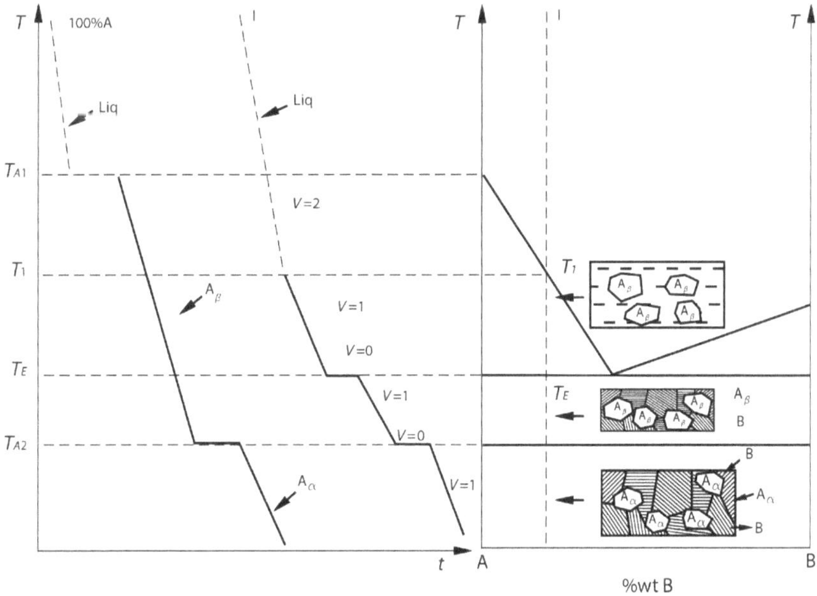

FIG. 7.23 Solidification of alloy I.

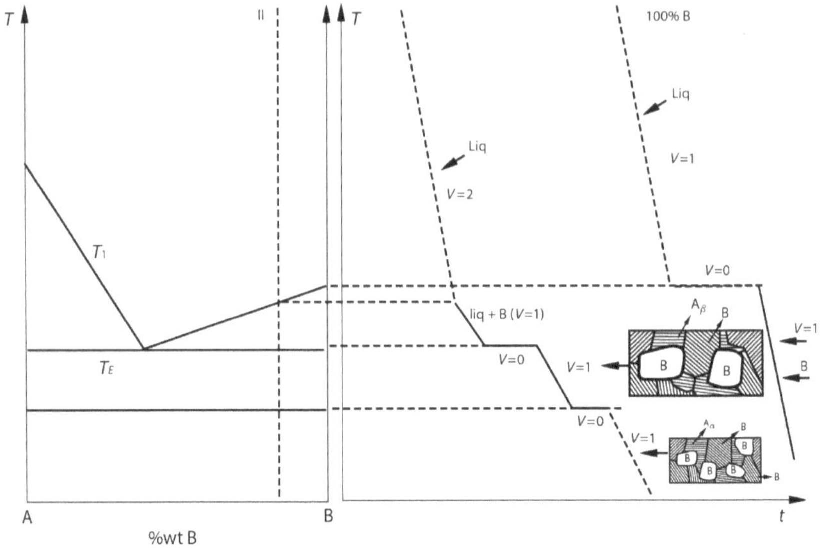

FIG. 7.24 Solidification of alloy II.

EXERCISE 7.12

Obtain the TTT curves for the transformation of austenite into ferrite, applying the generalized model for the kinetics of nucleation and growth processes (Chapter 2).

Data: $n \simeq 1 \times 10^{29}$ atoms/m^3, $\omega^* \simeq 1$, $D_0 \simeq 1 \times 10^{-4}$ m^2/s, $\mathbf{b} \simeq 10$Å, and $Q_{dif} = 282.6$ kJ/mol

Solution

Following the same procedure as Exercise 7.3:

$$
\begin{aligned}
I &= n^{\omega^*} \cdot \exp\left(\frac{-\Delta G^*}{kT}\right) \cdot \frac{D_0}{\mathbf{b}^2} \cdot \exp\left(-\frac{Q_{dif}}{RT}\right) \\
&= 1 \times 10^{44} \cdot \exp\left[\frac{-3 \times 10^{11}}{\Delta T^2 \cdot T}\right] \cdot \exp\left[\frac{-34{,}000}{T}\right]\left[\frac{\text{nuclei}}{\text{m}^3\,\text{s}}\right]
\end{aligned}
$$

$$
v = \Omega \cdot n_s \cdot \frac{D_0}{\mathbf{b}^2} \cdot \exp\left(\frac{-Q_{dif}}{RT}\right) \cdot \frac{\Delta G_v}{RT} = 3.4 \times 10^6 \cdot \frac{\Delta T}{T} \cdot \exp\left(\frac{-34{,}000}{T}\right)\left[\frac{\text{m}}{\text{s}}\right]
$$

$$
t_x = \sqrt[4]{\frac{-4 \cdot \ln(1-x)}{I \cdot v^3}}
$$

And after plotting temperature vs. time for two different shape factors, Figs. 7.25 and 7.26 are obtained.

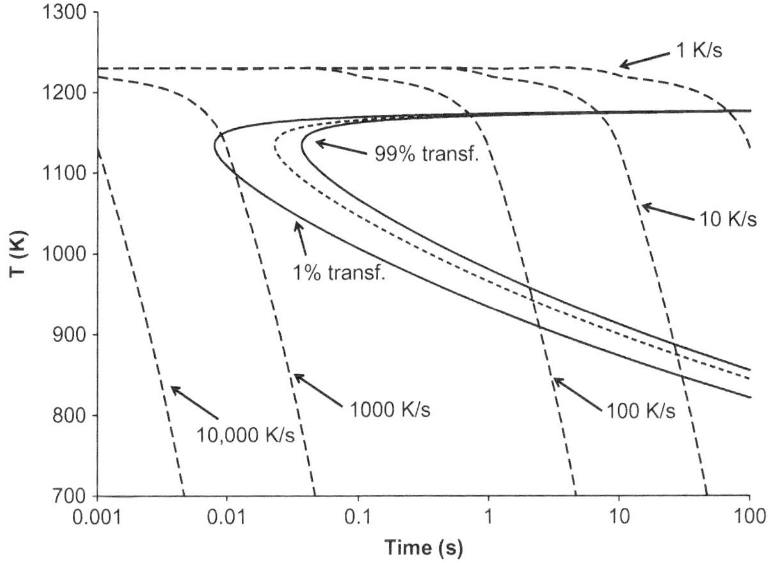

FIG. 7.25 TTT curves for a shape factor of 5 degrees.

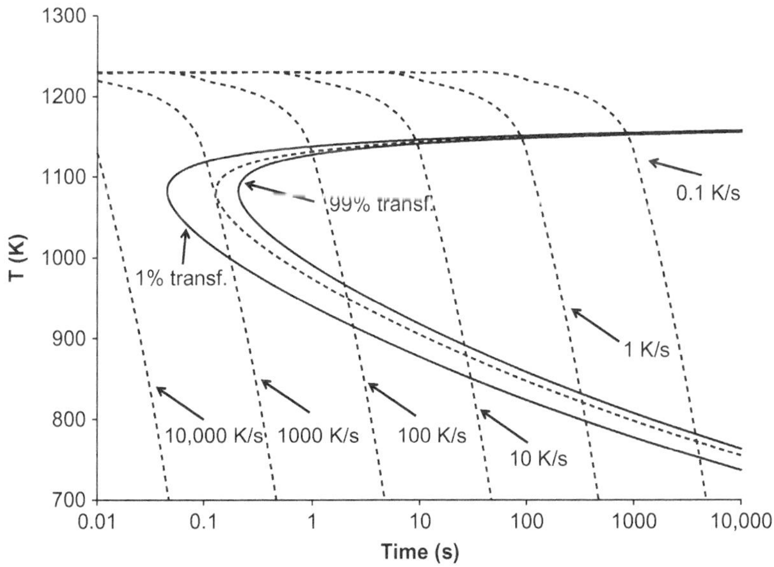

FIG. 7.26 TTT curves for a shape factor of 10 degrees.

7.3 TRANSFORMATIONS CAUSED BY ALLOTROPY AND SOLUBILITY CHANGES

The solute in a solvent metal can lose solubility as a consequence of precipitation kinetics and related processes during cooling. This mechanism is more complex when the solvent metal changes its allotropic state at a certain temperature; e.g., in the case of Fe and C (Chapter 8), austenite allotropically transforms from γ into α (ferritic structure with lesser carbon content) accompanied by the formation of the Fe_3C constituent (formed by atoms of Fe and the excess of C atoms of the initial γ), by nucleation and growth.

The equilibrium transformations by allotropic change and loss of solubility during cooling are not exclusive to the Fe-C system, as they are also present in other systems such as brasses (Cu-Zn system), bronzes (Cu-Sn system), and copper-aluminums (Cu-Al); all of them have intermediate phases (electronic compounds), as previously mentioned.

7.3.1 Cu-Zn System

Industrial brasses are classified according to their constituents in *α-brasses*, *β-brasses*, and *$\alpha+\beta$ brasses*:

α-brasses are formed by solid solution grains of Zn in Cu, with fcc crystalline structure. Because of the dissolution of Zn in Cu, the aspect of α-brasses go from the reddish color characteristic of Cu to yellow as a function of their Zn content: red brasses $<15\%$ Zn and yellow brasses $>15\%$ Zn.

Brasses with $<30\%$ wt Cu are formed by α-solid solution grains which give them the good formability properties of Cu, though in lesser degree because of the distortion the Zn atoms produce in the Cu lattice.

These brasses, deformable at both room and high temperatures, do not reach, however, the required strength level of many applications.

β-brasses (ordered β' at room temperature), difficult to deform, have better strength than α-brasses.

$\alpha+\beta$ brasses are biphasic alloys with higher strength than α-brasses and certain level of cold deformability if the β constituent proportion is not excessive.

The forming process of $\alpha+\beta$ brasses almost always occurs at high temperatures (above the transformation temperature of the pure β constituent) to avoid the ductility problems β has at room temperature.

These alloys are also known as free-cutting brasses because the hard nature of the ordered β phase turns the chips brittle and short during machining operations. This improvement in machinability depends on the amount and distribution of the β phase, as well as the amount of other elements; thus, sometimes Pb is added (1%–5%) for reasons analogous to the behavior of Cu-Pb alloys and their monotectic reactions (Chapter 4).

The most common composition for high strength biphasic $\alpha+\beta$ brasses is 60% Cu-40% Zn: in annealed state its ultimate strength is 350 MPa and its

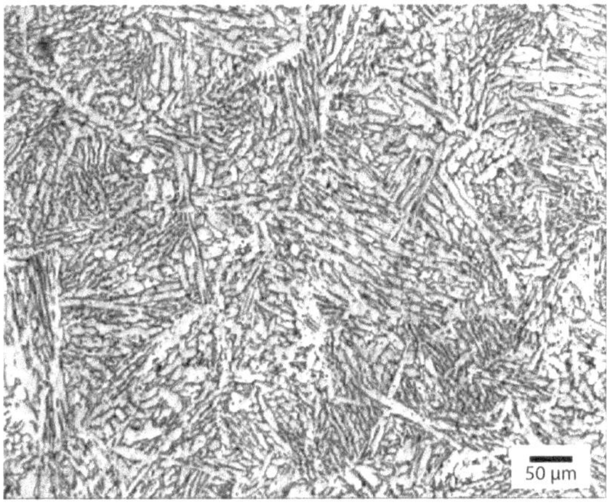

FIG. 7.27 60/40 Brass in its as-cast state.

elongation is 40%, while in cold work state (semi-hard) its ultimate strength is 500 MPa and its elongation is 10%. The phase diagram (Fig. 7.18) can be used to determine the microstructural evolution of this brass (60% Cu-40% Zn) starting from the liquid phase: the melt solidifies forming only β grains; during cooling, when reaching 800–700°C, a transformation occurs by allotropy and solubility changes of the β constituent partially transformg into α; the transformation continues during cooling until obtaining, at room temperature, a mix of α (light colored constituent) and β (dark colored constituent) phases.

Some as-cast alloys that, like 60% Cu-40% Zn, undergo allotropy and solubility variations in solid state during cooling have *acicular microstructures* (Fig. 7.27) also known as *Widmanstätten structure*[3] which is the result of minimizing the distortion free energy by maximizing diffusion kinetics, as mentioned in Section 7.1.

Heating and deforming the alloy of Fig. 7.27 at 620–680°C, induces recrystallization of the α and β phases, which result in the disappearance of the acicular structure, and originating the morphology of Fig. 7.28, which can better withstand tri-axial loads.

7.3.2 Cu-Sn System

If the Cu-Sn system is considered according to their composition these bronzes can be either monophasic (only formed by fcc solid solution of Sn in Cu) or biphasic (if the solidification originates $\alpha+\beta$ constituents):

3. This particular morphology of the solid constituent receives this name by generalization of the microstructure of Fe-Ni meteorites, and is very common in as-cast steels.

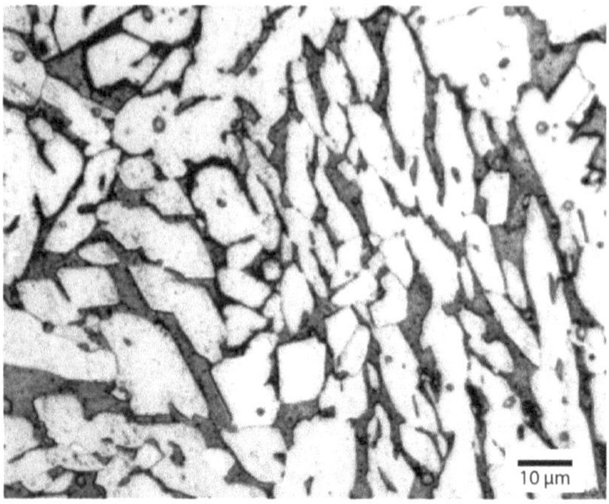

FIG. 7.28 60/40 Brass recrystallized by hot forming (Muntz alloy).

β is formed (Fig. 7.21) by a peritectic reaction at 799°C:

$$L(25.5\%\text{Sn}) + \alpha(13.5\%\text{Sn}) \rightarrow \beta(22\%\text{Sn}) \quad (7.2)$$

During equilibrium cooling, β experiences a transformation at 568°C to an $\alpha+\gamma$ eutectoid structure. Then, at 520°C, γ transforms into an $\alpha+\delta$ structure. There is a transformation at 350°C of δ into an $\alpha+\varepsilon$ aggregate, though it requires extremely long periods of time that in practice does not permit; thus, the vertical line corresponding to δ can be prolonged until room temperature.

α-bronzes with <8% Sn are commonly known as forgeables, and can be both hot and cold worked. For contents >8%, the amount of phase δ (fragile) formed during cooling because of the loss of solubility of Sn, is enough to prevent the cold plastic deformation in bronzes. Fig. 7.29 presents the micrograph of a bronze with 8% Sn whose disperse constituent is δ.

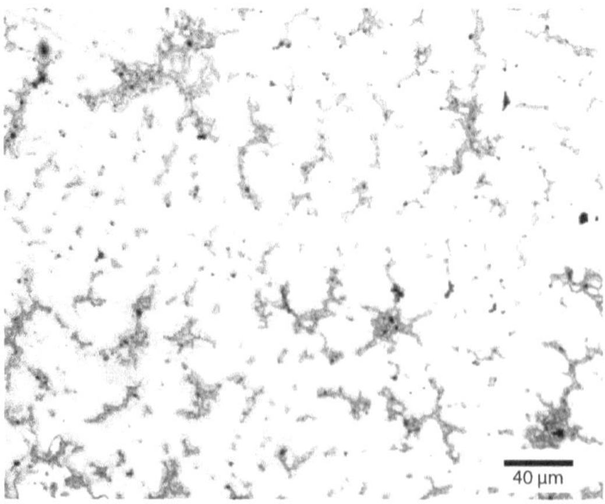

FIG. 7.29 Forgeable bronze (Cu-8% Sn).

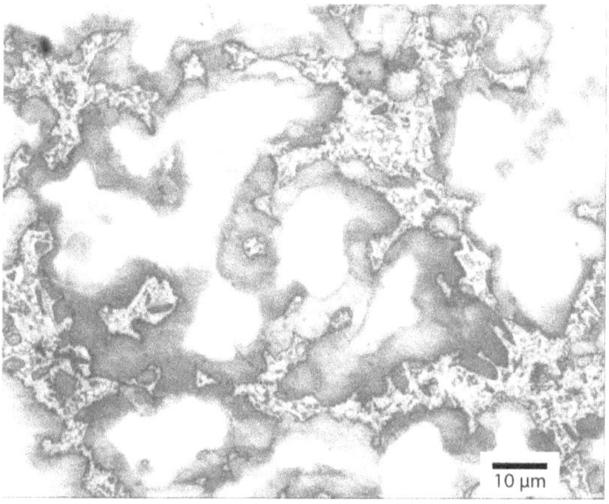

FIG. 7.30 85% Cu-15% Sn bronze.

Among the bronzes used for mold parts that do not require subsequent forming, Sn compositions can be higher: prototypes of molding bronzes can be 90% Cu-10% Sn (historically known as gun-metals) or 85% Cu-15% Sn. Both are almost exclusively used in the manufacture of bearings (by molding) because of the low elastic modulus of Cu and the good wear behavior the hard δ phase gives to the alloy (distributed in the α plastic matrix).

The differences of a 90/10 compared to a 85/15 bronze may be analyzed with the phase diagram (Fig. 7.21): in the 90/10 bronze, the δ phase formed during cooling by solubility loss is transformed into a structure similar to that of in Fig. 7.29, but with a higher proportion of δ; on the other hand, the 85/15 bronze solidifies forming an $\alpha+\beta$ microstructure, and during cooling β phase transforms into the $\alpha+\gamma$ eutectoid, subsequently γ transforms into $\alpha+\delta$ (Fig. 7.30); this structure is very resistant to wear, and its higher proportion of δ (contributing to a lower toughness) makes the 90/10 preferable to the 85/15 in the manufacture of bearings. It is important to point out that besides chemical composition, nonequilibrium cooling plays an important role in the microstructure of bronzes and thus in their properties (Chapter 3).

7.3.3 Cu-Al System

Copper-aluminums (left side of the phase diagram of Fig. 4.5), are also known as aluminum bronzes. Al, in substitutional solid solution in the Cu crystalline lattice, simultaneously provides an improvement in the mechanical strength of Cu and an increase in its resistance to corrosion to weak corrosives, specifically in acid environments, because of the formation of a thin alumina film in the surface of the alloy. As seen in the diagram, the solubility of Al in Cu at 1036°C is 7.4% and at 565°C is 9.4%.

Furthermore, Cu and Al can also form, as previously mentioned, electronic compounds, but aluminum bronzes with high Al content will have large amounts of the fragile phase γ and would be unsuitable for industrial applications, therefore Cu-Al alloys with contents higher than 12% wt Al are not used for manufacturing.

Monophasic aluminum bronzes, formed only by α solid solution grains, and consequently with compositions lower than 9% Al, are commonly used for heat exchangers, pipes, flatirons, etc., which are subjected to the corrosive environments of salt water. However, biphasic aluminum bronzes are more common because of their higher tensile stresses (analogous to $\alpha+\beta$ brasses). Fig. 7.31 shows the micrograph of an aluminum bronze with 10% Al cooled at equilibrium conditions and thus formed by phase α (light grains) and a dark one, result of the equilibrium transformation of β at rates lower than 10°C/min: β (~12% Al) is transformed by eutectoid reaction at 565°C into $\alpha(9.4\%\,\text{Al})+\gamma(15.6\%\,\text{Al})$, whose lamellar structure is shown in Fig. 7.32.

Nonequilibrium cooling processes in aluminum bronzes, just as it happens in the Fe-C system, produce nonequilibrium phases (metastable phases). For example, the alloy with 10% Al rapidly cooled from high temperatures (formed only by β) induces the formation, at the interior of these grains, an acicular metastable martensitic constituent (Fig. 7.33), that when tempered produces the α and γ equilibrium constituents with a considerably higher dispersion degree of the γ precipitate inside α compared to that obtained by equilibrium cooling, and therefore giving this alloy a significant improvement in toughness.

7.4 ORDER/DISORDER TRANSFORMATIONS

Some solid solution phases of the electronic type (β), complex interstitial compounds, and others may present order/disorder transformations: in most substitutional solid solutions, solute randomly substitutes atoms of the solvent, while

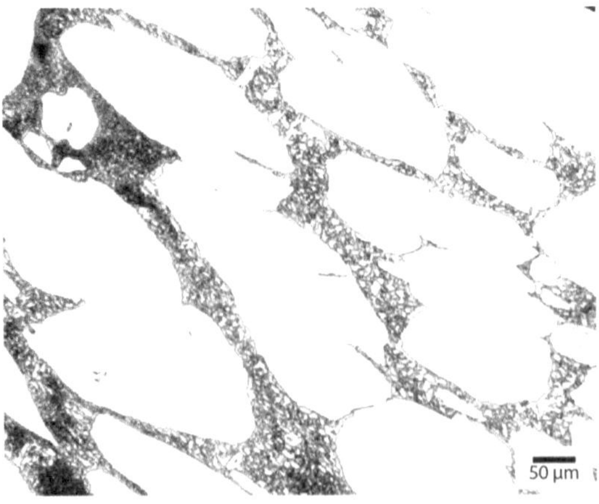

FIG. 7.31 Microstructure of a biphasic aluminum bronze alloy (Cu-10% Al), with equilibrium cooling.

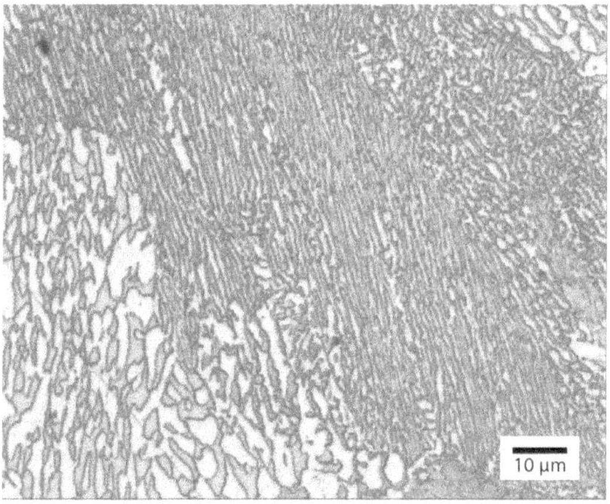

FIG. 7.32 Lamellar structure result of the β transformation into $\alpha+\gamma$ of the alloy of Fig. 7.31.

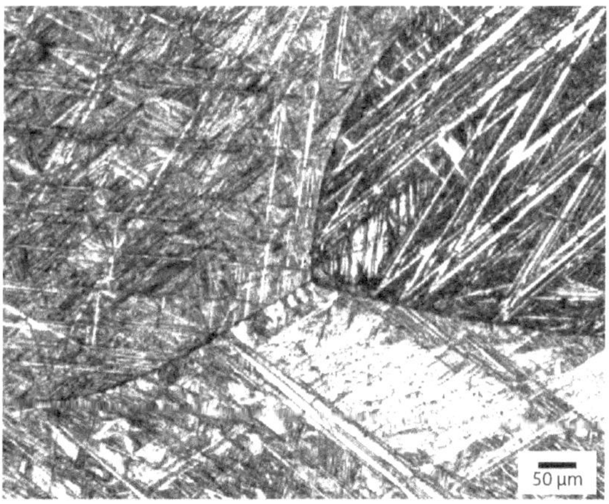

FIG. 7.33 Cu-Al martensite.

in others, the disordered atoms of the solute can, during cooling, acquire an ordered layout in the solvent lattice. This happens, for example, if the 50% Cu-50% Au solid solution is slowly cooled from 500°C: the ordered lattice (Fig. 7.34) is equivalent to the result of mixing two sublattices between them (one of Au and another of Cu); thus, the ordered structures are also known as *superlattices*. The most common ordered supperlattices are: CuZn, CuAu, Cu_3Au, Fe_3Al, and Mg_3Cd.

A superlattice is formed by two sublattices, *a* and *b*, so that all the positions of sublattice *a* are atoms of A (in Fig. 7.34, Cu) and all the positions of the sublattice *b* are B atoms (Au). In order to determine the degree of order a crystal

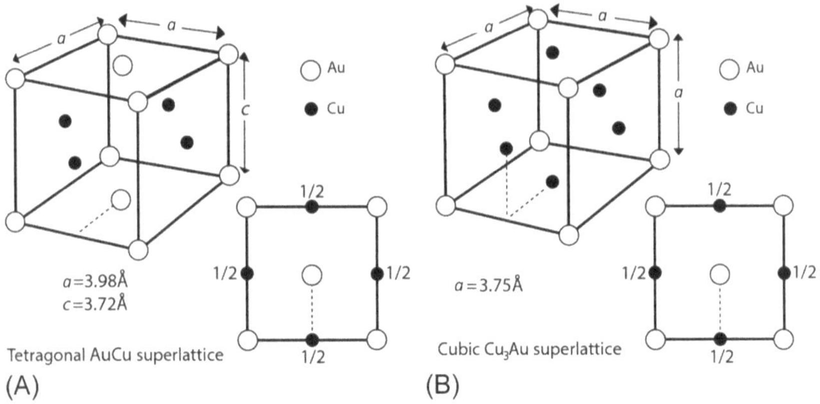

FIG. 7.34 Structures obtained from the solid solutions: (A) 50% Cu-50% Au and (B) 75% Cu-25% Au (ASM Handbook, Vol. 3, 1992).

has, when not completely ordered, the *long-range order parameter* (S) can be calculated:

$$S = \frac{F(A,a) - F(A)}{1 - F(A)} \tag{7.3}$$

where $F(A)$ is the number of atoms of A in the crystal divided by the total number of atoms and $F(A,a)$ is the number of atoms of A in the sublattice a divided by the total number of atoms in this sublattice.

When the whole crystal is perfectly ordered $S=1$, while for the nonordered crystal (with atoms positioned randomly) $S=0$. The entropy of mixing structures with long-range order is extremely small, and the higher the temperature, the smaller the degree of order will be, until reaching a critical temperature, above which there is no long-range order at all.

There is a possibility that even with $S=0$, crystals can have a certain order. Actually, a crystal can be formed by two perfectly ordered zones, each one of them a mirror image of the adjacent one (presented in two dimensions in Fig. 7.35); these two zones are known as *antiphase dominion* and the "mirror" the *antiphase boundary*. Antiphases and their boundaries can be detected using

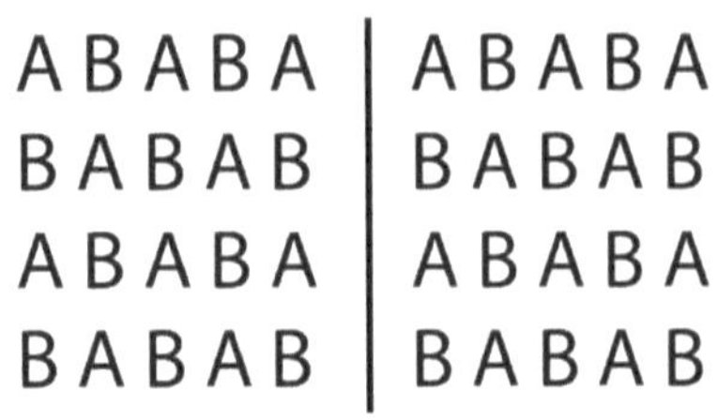

FIG. 7.35 Antiphases showing short-range order.

electronic microscopy. Even if half of the atoms do not occupy the positions they are supposed to in the sublattice, the crystal still has a high degree of order.

In an alloy, it is common that not every crystal is perfectly ordered, but has some antiphases; the properties of the material will depend mostly on the *short-range parameter* (σ):

$$\sigma = \frac{q - q_r}{q_M - q_r} \tag{7.4}$$

where q is the number of pairs of adjacent atoms (each pair is formed by an atom of A and another of B), q_M is the number of pairs of adjacent A and B atoms if all the crystals were fully ordered, and q_r is the mean number of pairs of adjacent A and B atoms if the distribution on the crystal was totally random (disordered state).

When the crystal is totally ordered, not only $S = 1$ but also $\sigma = 1$, since $q = q_M$. In all other cases, even with $S = 0$, σ would have a value lower than one (the lesser the value of σ, the lesser the level of short-range order will be) since q is lower than q_M.

The change from disordered to ordered state (by lowering the temperature below the critical one, T_C) takes place, generally, by the formation and growth of antiphases; in other words, by a mechanism similar to nucleation and growth. In some cases, however, the transformation is athermal, instantaneous, and of the martensitic type: for example the order/disorder transformation of the constituent β in brasses.

The ordered state is harder and less deformable than the disordered one. On the contrary, the electric conductivity of an ordered solid solution is higher than in the disordered state (Fig. 7.36).

Typical order/disorder cases are present in alloys of the Al-Fe system, that are paramagnetic in disordered state and ferromagnetic in their ordered one (AlFe and Fe_3Al ordered phases). Ordered structures are also present in ternary systems: a classic example is the Heussler alloy (Fig. 7.37), whose ordered state

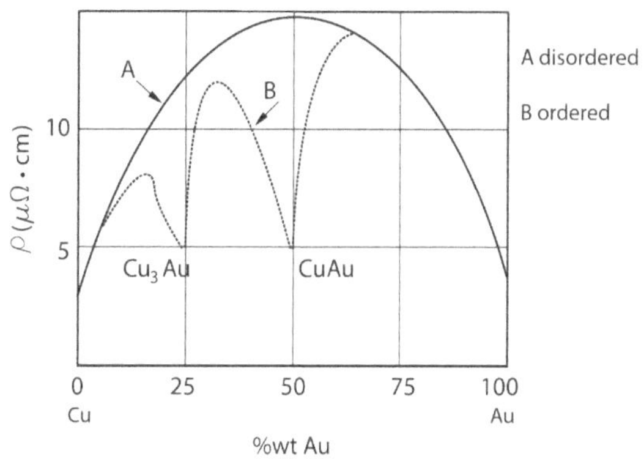

FIG. 7.36 Variation in resistivity in the Cu-Au system with order/disorder.

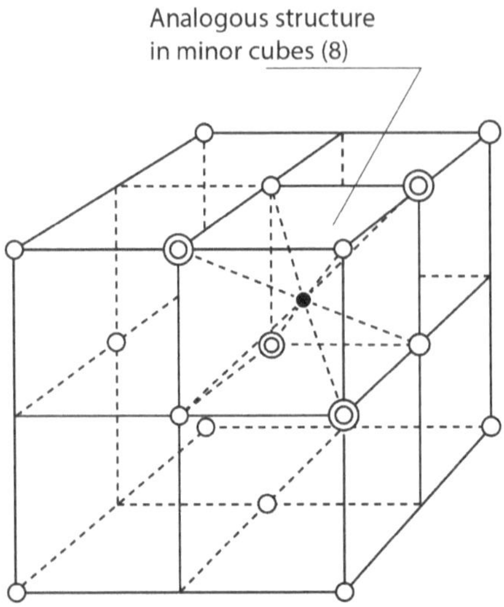

FIG. 7.37 Ordered structure of the Heussler alloy.

is ferromagnetic (it is important to point out that Cu, Mn, and Al are not ferromagnetic) and is paramagnetic in its disordered state.

Another example of an intermetallic constituent is Ni_2Al, which has an ordered state with a systematic and ordered layout of point defects in the lattice (*vacancies*). This type of constituent can be considered as an ordered ternary system, with vacancies as their third component, and known as *defective* lattices.

The phase γ' used for strain-hardening and to improve heat resistance of Ni-based superalloys is the ordered phase Ni_3Al (similar to Cu_3Au). Also, strain-hardening in maraging steels (Section 7.1.2) may be achieved by the formation of phases such as Ni_3Ti, Ni_3Mo, Ni_3Nb, etc.

EXERCISE 7.13

(a) Calculate the structural hardening of the Ni-7% Al alloy, obtained by tempering, for precipitate sizes between 50 and 100 nm and (b) approximately calculate the solution and aging temperatures of the mentioned alloy.

Data: $G_{Ni} = 79\,GPa$, $\rho_{Ni} = 8.9\,g/cm^3 = 58.7\,g/mol$, $a_{Ni_3Al} = 0.36\,nm$, T_M of Ni is 1453°C, and T_M of Ni_3Al is 1400°C, the system presents at 1385°C an invariant eutectic reaction $L(12\%Al) \leftrightarrow \alpha(9\%Al) + Ni_3Al(14\%Al)$ and the solubility of Al in Ni drops to 3.5% at room temperature.

Solution

(a) Structural hardening

Ni_3Al, in Ni-based superalloys, is known as γ', and its weight % is:

$$Ni_3Al\,wt\% = \frac{7-3.5}{14-3.5} = 0.33(33\%)$$

In order to calculate the volume fraction of this phase, its density and phase diagram must be used:

$$\rho_{Ni_3Al} = \frac{(3)\left(58.7\frac{g}{mol}\right) + 27\frac{g}{mol}}{(3.6 \times 10^{-8}\,cm)^3\left(6.023 \times 10^{23}\frac{atoms}{mol}\right)} = 7.23\frac{g}{cm^3}$$

$$\frac{m_{Ni_3Al}}{m_{Ni}} = \frac{7.23V_{\gamma'}}{8.9V_{Ni}} = \frac{7-3.5}{14-7} = \frac{1}{2}$$

$$\frac{V_{\gamma'}}{V_{Ni}} = 0.62$$

$$f_{vNi_3Al} = \frac{V_{\gamma'}}{V_{\gamma'} + V_{Ni}} = \frac{0.62}{1+0.62} = 0.38(38\%)$$

The volume fraction can be used to calculate the distance between precipitates (L):

$$L \approx \frac{\bar{X}}{\sqrt{f_{vNi_3Al}}} - \bar{X}$$

For $\bar{X} = 50\,nm \rightarrow L = 31\,nm$ and for $\bar{X} = 100\,nm \rightarrow L = 62\,nm$. Thus the yield stress for aging is:

$$\sigma_{y_{aging}} = \frac{G\mathbf{b}}{L} = \frac{G}{L}\left(\frac{\sqrt{2}}{2}a_{Ni_3Al}\right)\frac{(79{,}000\,MPa)}{L}\left(\frac{\sqrt{2}}{2}0.36\right)$$

So for $\bar{X} = 50\,nm \rightarrow \sigma_y = 648.7\,MPa$ and for $\bar{X} = 100\,nm \rightarrow \sigma_y = 324.3\,MPa$

(b) Solution and aging temperatures

The solution temperature can be obtained using the phase diagram:

$$\frac{1385 - T_p}{9-7} = \frac{1385-20}{9\ \ 3.5}$$

$$T_p = 888.6°C$$

or, in other words higher than 900°C. For the aging temperature, it is preferable that Al is dissolved in Ni, only to a degree, as there are vacancies in the Ni matrix created by quenching, thus:

$$T_{aging} = \frac{T_M}{3} = \frac{1453+273}{3} = 575.3\,K = 302.3°C$$

which is corroborated by industrial practice, which indicates a temperature higher than 300°C.

Comment: The heat-resistant Nimonic alloys (Ni-Cr-based and with additions of Al and Ti) obtain their high strength from a fine dispersion of the ordered fcc $Ni_3(Al,Ti)$ or γ', precipitated in the fcc Ni-rich matrix. The interfaces Ni/γ' are fully coherent and the interfacial free energy is very low (10–30 mJ/m^2), enabling the alloys to maintain a fine-grained microstructure at high temperatures and improve their creep resistance.

REFERENCE

ASM Handbook, 1992. 10th ed., vol. 3, Alloy Phase Diagrams. ASM International, Metals Park, OH.

BIBLIOGRAPHY

AISI, 1974. Steel Products Manual. American Iron and Steel Institute (AISI), Washington, DC.

Calvo Rodes, R., 1948. Metales y Aleaciones. INTA, Madrid.

García Poggio, J., 1972. Aleaciones Industriales de Aluminio. INTA, Madrid.

Goldschmidt, H.I., 1967. Interstitial Alloys. Butterworths, London, pp. 214–231.

Hansen, M., Elliot, R., 1965. Constitution of Binary Alloys, second ed. McGraw-Hill, New York.

Hume-Rothery, W., 1969. The Structure of Metals and Alloys, fifth ed. The Institute of Metals, London.

Massalski, T., 2001. Binary Alloy Phase Diagrams, second ed. American Society for Metals, Metals Park, OH.

Newkirk, 1973. Structures resulting from aging and precipitation. Met. Hand. 8, 177.

Porter, D., Easterling, K., 2009. Phase Transformations in Metals and Alloys, third ed. CRC Press, Boca Raton, FL.

Chapter 8

Solid-State Transformations in the Fe-C System

8.1 EQUILIBRIUM TRANSFORMATIONS IN THE METASTABLE Fe-C SYSTEM

Steels and white castings are examples of materials that experience constitutional transformations due to solubility variation and allotropic solid-state transformation. Iron has three allotropic varieties: δ, γ, and α. Pure Fe solidifies at 1538°C in the form of δ (bcc unit cell). At 1394°C, δ transforms into γ (fcc) which has a lattice parameter that decreases with temperature (3.548 Å at 913°C). Another allotropic state is reached at 912°C: γ turns into α which is bcc, just as δ though with a smaller lattice parameter (2.8664 Å).

The $\gamma \rightarrow \alpha$ transformation results in a volume increase as the γ lattice (four atoms) breaks into α (two atoms); and the volume of two α unit cells is higher than the original γ cell. This volume increase (Exercise 1.2) is close to 1%. It is also important to point out that solidification and allotropic transformations of pure Fe are carried out at constant temperature, according to thermodynamics.

8.1.1 Constituents of the Metastable Fe-C System

The Fe-C metastable system constituents are austenite (γ), ferrite (α), and cementite (Fe_3C). The first two are solid solutions of octahedral insertions of C in γ and α, respectively (Figs. 3.11 and 3.14). On the other hand, cementite is an intermetallic compound with an orthorhombic structure as shown in Figs. 8.1 and 8.2.

If the system is formed by Fe-Fe_3C it is considered metastable and used in the analysis of steels and white castings, while the Fe-C one is reserved for gray castings which are formed by austenite, ferrite, and graphite (and sometimes cementite if the solid-state cooling is metastable).

The difference between stable and metastable behaviors depends particularly on cooling rate and alloying elements. In the case of steels, the metastable behavior is the most common one.

Solidification and Solid-State Transformations of Metals and Alloys. http://dx.doi.org/10.1016/B978-0-12-812607-3.00008-5

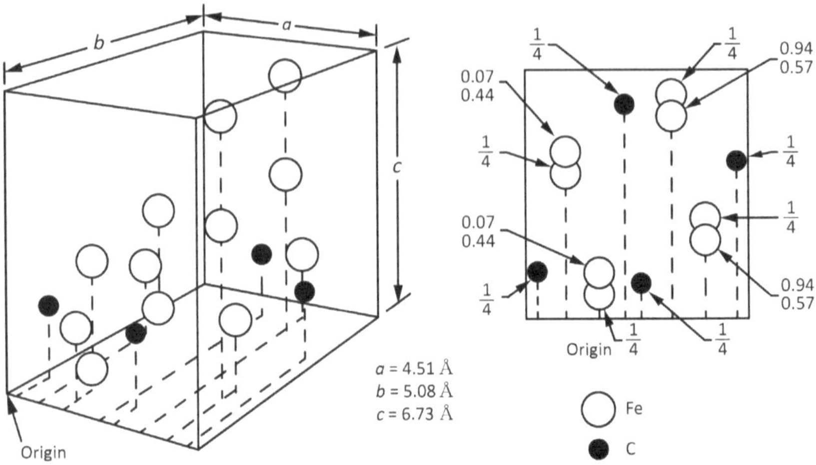

FIG. 8.1 Orthorhombic structure of cementite (ASM International, 1992).

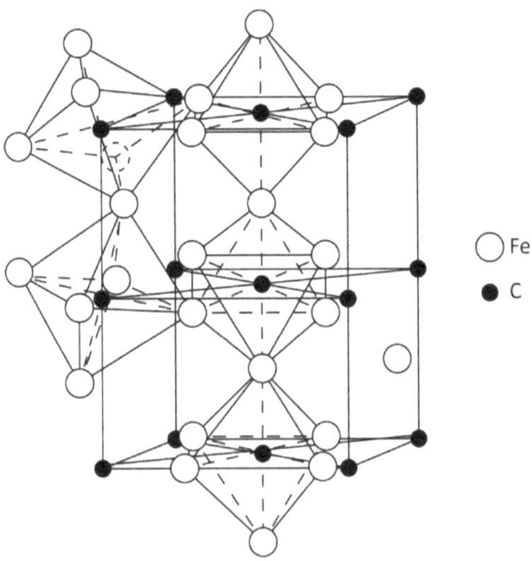

FIG. 8.2 Cementite. Detail of the relative position of Fe and C atoms (Hendricks, 1930).

The most important phases of the Fe-C diagram are:

- *Austenite* (γ): in this structure, C atoms are randomly positioned in the middle of the edges or at the center of the cube (interstices). This means that γ could add C atoms until all interstitial positions are occupied, although saturation is reached with only four C atoms per cell, and considering that there are also four Fe atoms in a unit cell, a 50% atomic C amount would be reached (23% wt C).

Nevertheless, the maximum amount of C in austenite is much lower than 50% due to the diameter ratio between C and Fe (Exercise 3.3). Fig. 3.11 indicates that this ratio should be 0.414 in order for the C atoms to occupy interstitial positions without distorting the γ lattice and as the Goldschmidt relation between diameters of these two atoms is 0.63, insertion of one C atom in austenite results in considerable lattice distortion proportional to the amount of dissolved C. Specifically, the austenite lattice is a function of C content:

$$a\left[\text{Å}\right] = 0.448(\%\text{C}) + 3.548 \tag{8.1}$$

If the concentration of C is increased (and thus the distortion of the lattice), the austenite is incapable of withstanding further changes, though increasing the temperature (causing lattice expansion) will partially compensate the differences in size and therefore, an increasing solubility with temperature may be observed. Saturation of C in austenite is reached at 1148°C with a 2.11% wt C content.

Though some authors indicate 1.7% C as the austenite saturation limit at 1130°C, this disparity in results is caused by the extrapolation measurements performed over saturated austenite at temperatures below 1148°C. Binary Fe-C alloys with C content below the maximum saturation limit at this temperature, are known as steels, which do not present austenite at room temperature by slow cooling unless using alloying elements (12% Mn or 8% Ni).

Austenite is easy to identify as it does not show magnetic properties. It is also soft, ductile, and fracture resistant. Its mechanical properties vary with chemical composition, yet approximate mean values are: hardness of 300 HV, tensile stress of 880–1100 MPa, and elongation of 30%–60%.

- *Ferrite* (α): is an octahedral insertion solid solution of C in α or δ. Carbon atoms may position themselves randomly at the middle of the edges or at equivalent positions at the center of the faces, with each C atom in contact with two Fe atoms. The atomic radius ratio between C and Fe required to form a theoretical octahedral solid solution insertion is 0.154 (Fig. 3.14), while the real diameter ratio is 0.63. This means that austenite admits a C amount much higher than ferrite: α allows a maximum of 0.0218% wt C at 727°C and almost null at lower temperatures.

  Ferrite is magnetic below 770°C and nonmagnetic above this temperature, and is the softest constituent of steels, with the following mechanical characteristics: hardness of 90 HV, tensile stress of 300 MPa, and elongation of 40%. It is softer than austenite as a result of its lower C content and more plastic even though it crystallizes in a system (bcc) that does not present compact planes.

- *Cementite* (Fe_3C): thermodynamically unstable, can be decomposed into Fe_3 and graphite at certain conditions (temperature and time), meaning a number of hours so high, that, in practice, it never occurs in Fe-C binary steels at room temperature. However, holding for long periods of time (thousands of hours at $T > 450°C$), low carbon or low-alloyed steels (e.g., $< 0.15\%C, 0.5\%Mn$) can partially *graphitize* by decomposing cementite into ferrite and graphite.

  In orthorhombic cells, Fe atoms can sometimes be substituted by other atoms of Cr, Mo, Mn, etc. Since cementite is an intermetallic compound with covalent-ionic bond, it is brittle and thus, the hardest constituent of steels (68 HRC). It is ferromagnetic below 210°C, and with an almost impossible to determine melting point (1227°C) since it decomposes at high temperatures before reaching fusion. Its density is 7694 kg/m^3.

EXERCISE 8.1

Calculate the density of cementite and compare it with the theoretical value 7694 kg/cm^3.

Data: $a = 5.08$ Å, $b = 4.51$ Å, and $c = 6.73$ Å.

Solution

Cementite (Fe_3C) cristallizes in the orthorhombic system and each cell has 4 atoms of C and 12 atoms of Fe, thus:

$$\rho_{Fe_3C} = \frac{(12 \text{ at Fe})\left(55.847 \frac{g}{mol_{Fe}}\right) + (4 \text{ at C})\left(12 \frac{g}{mol_C}\right)}{\left(6.023 \times 10^{23} \frac{at}{mol}\right)(5.08 \times 10^{-8}\text{cm})(4.51 \times 10^{-8}\text{cm})(6.73 \times 10^{-8}\text{cm})}$$
$$= 7.733 \frac{g}{cm^3}$$

which has a difference of 0.5% with respect to the theoretical value.

8.1.2 Austenite-Ferrite Transformation Kinetics

Room temperature microstructures of steels are the result of austenite transformations: during slow cooling, the allotropic $\gamma \rightarrow \alpha$ transformation is related to the diffusion of Fe and C. The movement of Fe atoms results in nucleation of new bcc cells. Once ferrite crystallites have formed, their growth occurs simultaneously with the diffusion of C atoms towards the surrounding austenite.

When analyzing austenitic transformations by slow cooling, besides the *gamma-forming* (or gamma-phase stabilizer) nature of C, the following must be considered:

- The transformation of δ into γ takes place above 1394°C; the higher the C content (within a limit) in γ, the higher the transformation temperature will be.
- The austenite into ferrite transformation occurs below 912°C; the higher the C content of austenite, the lower the transformation temperature will be.
- For C contents higher than 0.77%, stabilization of γ is such that the allotropic transformation into α does not occur by slow cooling until temperature reaches 727°C.

Examples of three different transformations of austenite can be analyzed:

1. *Steel with 0.3% C* (hypoeutectoid): at temperatures above 912°C, the alloy is fully formed by γ. Decreasing the temperature below 912°C, austenite does not transform, as a consequence of the γ-forming nature of C, and it is until reaching temperature A_3 that the $\gamma \rightarrow \alpha$ transformation begins. This starting point of the transformation occurs with a volume increase, so A_3 can be experimentally obtained through dilatometry tests.

   The $\gamma \rightarrow \alpha$ transformation does not fully take place at A_3: according to the phase rule, at this temperature, the nucleation of α starts at the austenitic grain boundaries, with migration of C (by diffusion) towards the interior of the grains. Thus, the center of the grain, having a higher C content, would require an even lower temperature so that transformation into α can occur.

   When lowering the temperature, the proportion of α in the grain boundaries increases, while the amount of untransformed γ, enriched with C, decreases.

   At 727°C, a third of the amount of γ will remain untransformed compared to the original amount. At this moment, γ, now with 0.77% C and known as *eutectoid binary austenite*, will transform while still dilating (process that started at A_3) until reaching 727°C. The result, at room temperature, is the microstructure shown in Fig. 8.3, formed by α and pearlite (eutectoid structure of $\alpha + Fe_3C$).
2. *Steel with 1% C* (hypereutectoid): when cooling from high temperatures, the contraction of the crystalline lattice will promote an increase in the crystal distortion caused by the dissolved C, and, at a certain temperature, this lattice will eventually become unstable. The atoms of C expelled from the γ lattice will form, by affinity with Fe, cementite. Temperature A_{cm} corresponds, in each steel, to the Fe_3C precipitation (which occurs by C solubility loss, without the allotropic $\gamma \rightarrow \alpha$ transformation as a consequence of the gamma-forming nature of C).

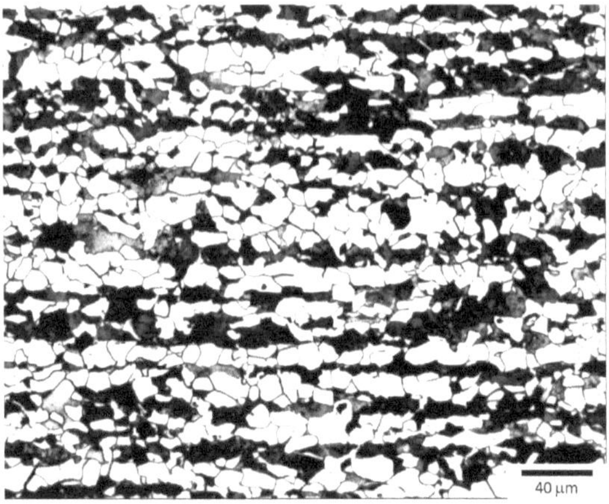

FIG. 8.3 Binary steel (0.3% C) equilibrium microstructure.

When the 1% C austenite reaches A_{cm}, C atoms move towards the austenitic grain boundaries and form cementite by reaction with Fe atoms. Therefore, austenitic grains remain with low C content, and will be stable. A decrease in temperature is necessary to produce a new migration of C towards the grain boundaries (which prevent the increasing distortion caused by C when the austenite parameter decreases with temperature). This process continues during cooling until temperature reaches 727°C: the Fe_3C in the grain boundaries has 4% wt C, while the remaining austenite has an average of 0.77% C. The result, at room temperature, is the micrograph of Fig. 8.4, formed by Fe_3C and pearlite.

FIG. 8.4 Binary steels (1% C) equilibrium microstructure.

In other words, the two cooling processes analyzed (austenite with 0.3% and with 1% C) are summarized in:

- Austenite with less than 0.77% C starts to transform into α at A_3. If the C content is higher than 0.02% the $\gamma \rightarrow \alpha$ transformation does not end until reaching 727°C.
- Austenite with more than 0.77% C will begin forming Fe_3C precipitates at A_{cm}, and ferrite will only start to appear at temperatures below 727°C.
- In both cases, when reaching 727°C, the C content in austenite is 0.77% C.

3. *Steel with 0.77% C* (eutectoid): γ with this composition does not experience any transformation during slow cooling until reaching 727°C. At this temperature, C in supersaturated solid solution is unstable, diffuses and, after an "incubation" period, starts to form Fe_3C nuclei, mainly at the austenitic grain boundaries. Zones close to cementite crystallites, depleted of C, allotropically transform into α.

   The cementite crystallites continue to grow at expense of the C absorbed from the adjacent austenite. The final result is a sandwich microstructure formed by lamellae of cementite and ferrite, which was defined by Sorby as *pearlite* (Fig. 4.31).

The solid-state $\gamma(0.77\%\text{C}) \rightarrow \alpha + Fe_3C$ transformation is known as eutectoid reaction (by analogy to eutectic reactions). The constant temperature at which this reaction occurs is known as A_e (or A_{123}). The eutectoid temperature can be calculated, according to Andrews (1965), as a function of the composition of the steel:

$$A_e(°\text{C}) = 727 - 10.7(\%\text{Mn}) - 16.9(\%\text{Ni}) + 29.1(\%\text{Si}) + 16.9(\%\text{Cr}) + 290(\%\text{As}) + 6.38(\%\text{W}) \quad (8.2)$$

Furthermore, the A_{3C} (Section 8.1.3) can likewise be calculated with a formula, also obtained by Andrews (1965), considering the alpha- and gamma-stabilizing natures of the alloying elements of the steel (Section 8.1.7):

$$A_{3C} = 912 - 203\sqrt{\%\text{C}} - 30(\%\text{Mn}) - 15.2(\%\text{Ni}) - 11(\%\text{Cr}) - 20(\%\text{Cu}) + 44.7(\%\text{Si}) + 31.5(\%\text{Mo}) + 13.1(\%\text{W}) + 104(\%\text{V}) + 120(\%\text{As}) + 400(\%\text{Ti}) + 400(\%\text{Al}) + 700(\%\text{P}) \quad (8.3)$$

EXERCISE 8.2

Calculate the C content in a steel, using the dilatometry curve of Fig. 8.5, with the following alloying elements: 0.54% Mn, 0.33% Si, 0.78% Cr, 2.40% Ni, and 0.30% Mo.

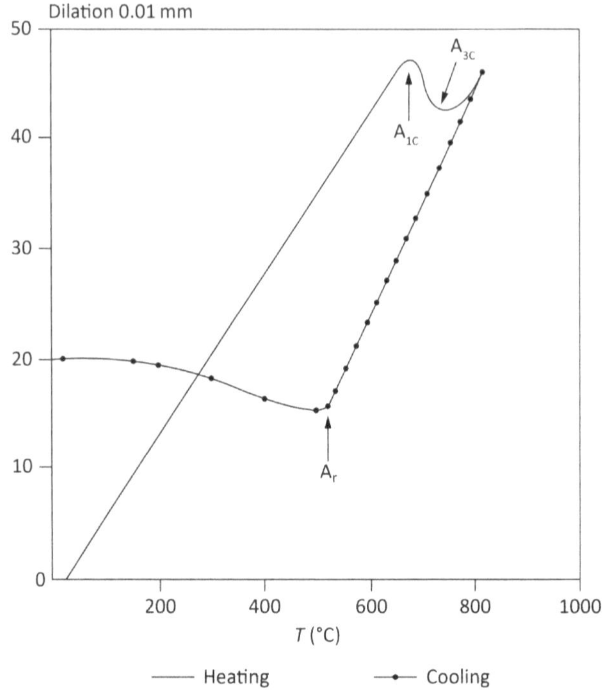

FIG. 8.5 Dilatometry curve of a steel.

Solution

From Fig. 8.5, the steel suffers contraction (A_{1C}) at 680°C and dilation (A_{3C}) at 740°C, during heating. Using Eq. (8.3) with the alloying amounts, and solving for C content:

$$A_{3C} = 912 - 203\sqrt{\%\mathrm{C}} - 30(0.54) - 15.2(2.40) - 11(0.78) - 20(0) + 44.7(0.33)$$
$$+ 31.5(0.30) + 13.1(0) + 104(0) + 120(0) + 400(0) + 400(0) + 700(0)$$

$$A_{3C} = 912 - 203\sqrt{\%\mathrm{C}} - 16.2 - 36.48 - 8.58 + 14.75 + 9.45 = 740$$

$$134.94 = 203\sqrt{\%\mathrm{C}}$$

$$\%\mathrm{C} = 0.44$$

8.1.3 Fe-Fe_3C Metastable Diagram

Fig. 8.6 shows the solid state transformation lines of the metastable Fe-Fe_3C diagram, including the solvus limit of C in ferrite for temperatures below 727°C, with the precipitation of *tertiary or vermicular cementite*.

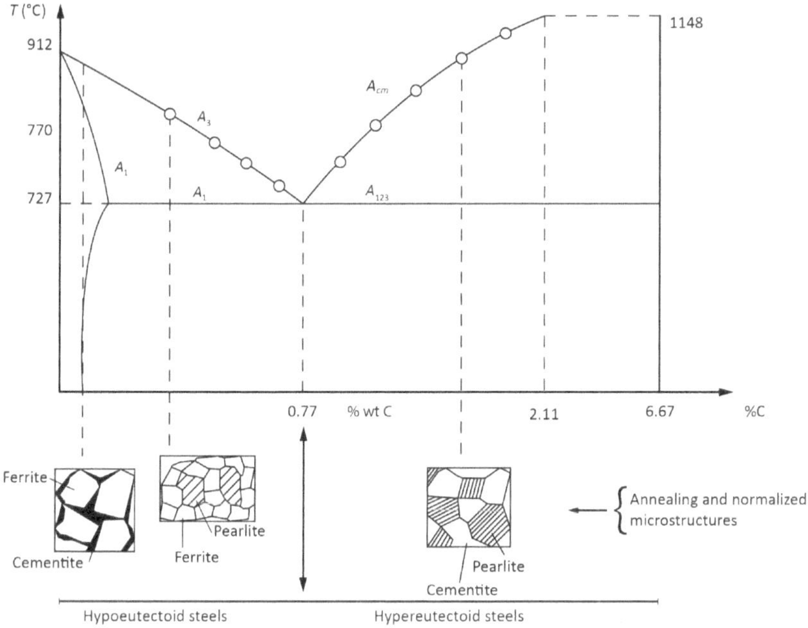

FIG. 8.6 Lower part of the Fe-Fe_3C metastable diagram.

Steels are classified according to their C content into: *hypoeutectoids* (lower than 0.77% C) or *hypereutectoids* (between 0.77% and 2.11% C); while Fe-C alloys with more than 2.11% C are known as *white cast irons*.

Fig. 8.6 indicates the geometric location of the A_3 and A_{cm} "*critical temperatures*"; below these, alloys undergo austenitic transformation either into ferrite (hypoeutectoid steels) or into cementite (hypereutectoid steels). Critical temperatures sometimes include the subindex r (from the French *refroidissement*) when related to the cooling process or the subindex c (from the French *chauffage*) when related to heating. The difference between the A_{3r} and A_{3c} critical points for the same hypoeutectoid steel (or between the $A_{cm\,r}$ and $A_{cm\,c}$ in the case of a hypereutectoid steel) is known as *thermal hysteresis*, and can be obtained by dilatometry tests.

The A_1 transformation line defines the critical temperatures below which, all austenite has transformed into ferrite, and is divided into: a sloped line (below 0.0218% C) and a horizontal one (between 0.0218% and 0.77% C).

In the case of hypereutectoid steels, the allotropic transformation points of austenite are known as A_{123}, to indicate the intersection between the $\gamma \rightarrow \alpha$ and the magnetic transformation lines. Therefore, the A_3 and A_2 critical points (A_2 indicating the magnetic transformation of ferrite) match with the A_1 point; occurring at constant temperature (727°C).

The diagram of Fig. 8.7 indicates that a 4.3% C alloy will present an eutectic of austenite (2.11% C) and cementite. This eutectic is known as *ledeburite* and

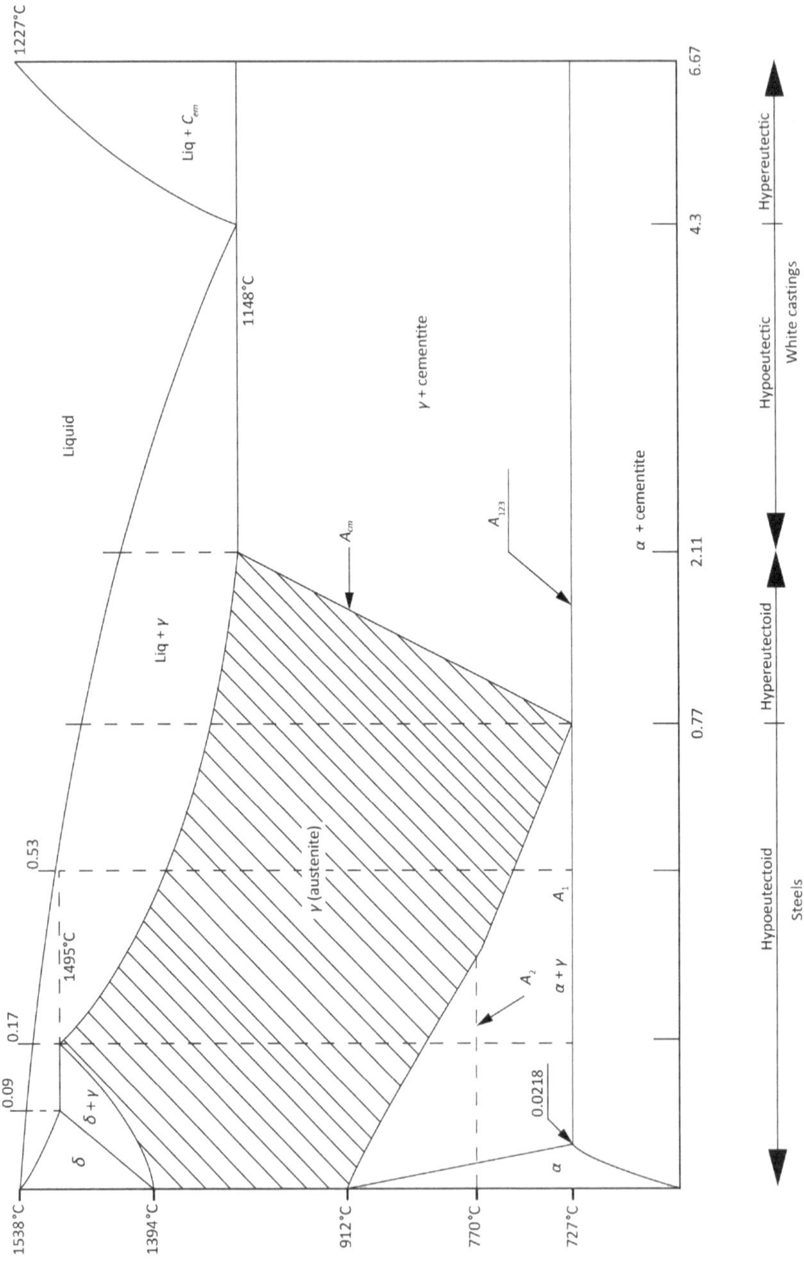

FIG. 8.7 Fe-Fe_3C metastable diagram (distorted to improve clarity).

melts at 1148°C. Furthermore, steels with 0.17% C have a peritectic reaction at 1495°C:

$$\delta(0.09\%\mathrm{C}) + \mathrm{L}(0.53\%\mathrm{C}) \rightarrow \gamma(0.17\%\mathrm{C}) \quad (8.4)$$

At the same zone, the transformation limits of δ (with initial composition below 0.17% C) into γ (below 1495°C) are shown.

EXERCISE 8.3

Calculate the solidification temperatures (starting from liquid) and the proportions, until room temperature, of all the phases forming steels with 0.1%, 0.2%, and 0.4% C.

Solution

(a) From liquid phase

Using Fig. 8.8:

- Steel with 0.1% C

The initial solidification temperature (T_L), can be obtained by similar triangles:

$$\frac{1538-1495}{0.5} = \frac{1538-T_L}{0.1}$$

$$T_L = 1529°\mathrm{C}$$

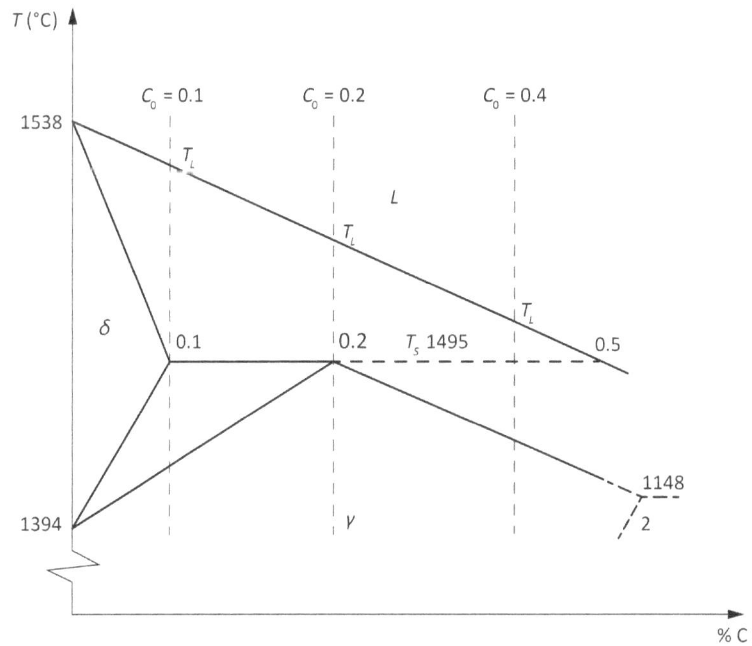

FIG. 8.8 Peritectic zone of the Fe-Fe_3C phase diagram.

The final solidification temperature (T_S) is the peritectic one 1495°C. The microstructure is formed by 100% δ.

- Steel with 0.2% C

  Following the same procedure:

$$\frac{1538-1495}{0.5}=\frac{1538-T_L}{0.2}$$

$$T_L=1521°C$$

  T_S is also 1495°C. At this temperature, the peritectic reaction occurs:

$$L(0.5\%\,C)+\delta(0.1\%\,C)\leftrightarrow\gamma(0.2\%\,C)$$

  and applying the lever rule, also at 1495°C, the weight fractions of liquid and δ are:

$$f_L=\frac{0.2-0.1}{0.5-0.1}=0.25\,(25\%)$$

$$f_\delta=\frac{0.5-0.2}{0.5-0.1}=0.75\,(75\%)$$

- Steel with 0.4% C

  Following the same procedure as for the 0.1% C steel:

$$\frac{1538-1495}{0.5}=\frac{1538-T_L}{0.4}$$

$$T_L=1504°C$$

And in this case, the material will be formed by liquid and γ, with the following fractions (obtained by the lever rule):

$$f_L=\frac{0.4-0.2}{0.5-0.2}=0.67\,(67\%)$$

$$f_\gamma=\frac{0.5-0.4}{0.5-0.2}=0.33\,(33\%)$$

Now, the solidification temperature (below which the microstructure will be fully austenitic) is calculated by similar triangles:

$$\frac{1495-1148}{2-0.2}=\frac{1495-T_S}{0.4-0.2}$$

$$T_S=1456°C$$

(b) Eutectoid temperatures

In order to continue the analysis of cooling until room temperature, Fig. 8.9 is used:

- Steel with 0.1% C

  By similar triangles, A_3 (beginning of α formation from γ), can be calculated:

$$\frac{912-727}{0.77}=\frac{912-A_3}{0.1}$$

$$A_3=888°C$$

At A_1 (eutectoid), the equilibrium phases are $\alpha(0.0218\%\,C)$ and $\gamma(0.77\%\,C)$ in the following proportions (by applying the lever rule):

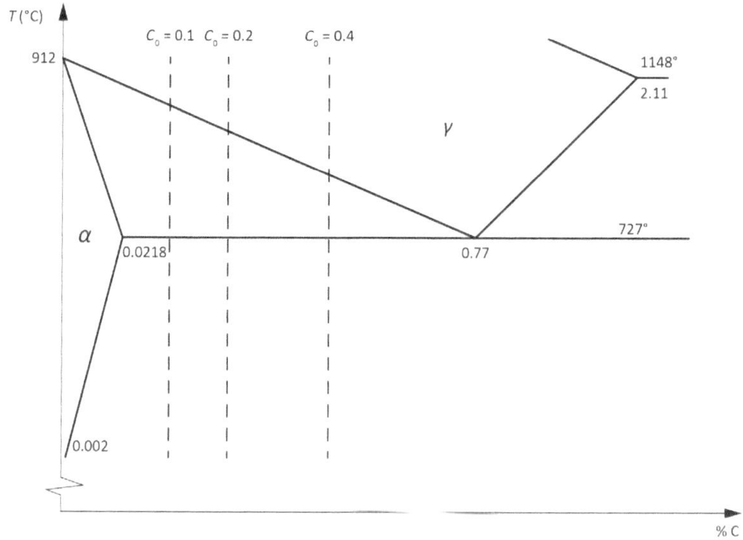

FIG. 8.9 Fe-Fe_3C phase diagram (eutectoid reaction zone).

$$f_\alpha = \frac{0.77 - C_0}{0.77 - 0.0218} = \frac{0.77 - 0.1}{0.77 - 0.0218} = 0.90\,(90\%)$$

$$f_\gamma = (1 - f_\alpha) = 0.10(10\%)$$

with ferrite being the proeutectoid constituent. At 727°C the eutectoid reaction:

$$\gamma(0.77\%\,C) \leftrightarrow \alpha(0.0218\%\,C) + Fe_3C(6.67\%\,C)$$

takes place forming a sandwich structure known as pearlite. By slowly cooling until room temperature, cementite (tertiary) grows in boundaries and triple points of ferritic grains, consequence of the solubility loss of C in α from A_1, in the following amounts:

$$\text{Total eutectoid } Fe_3C = \frac{0.1 - 0.0218}{6.67 - 0.0218} = 0.012(1.2\%)$$

And at room temperature:

$$\text{Total } Fe_3C\,(\text{room temperature}) = \frac{0.1 - 0.002}{6.67 - 0.002} = 0.015(1.5\%)$$

The amount of tertiary cementite formed is 0.3%, and results from the solubility loss of C in the bcc lattice of ferrite as temperature drops (from 0.0218% at 727°C to 0.002% at room temperature).

- Steel with 0.2% C

  In this case A_3 is:

$$\frac{912 - 727}{0.77} = \frac{912 - A_3}{0.2}$$

$$A_3 = 864°C$$

At A_1, the ferrite and austenite proportions are:

$$f_\alpha = \frac{0.77 - 0.2}{0.77 - 0.0218} = 0.76\,(76\%)$$
$$f_\gamma = (1 - f_\alpha) = 0.24\,(24\%)$$

while at 727°C, the amount of cementite is:

$$\text{Total eutectoid Fe}_3\text{C} = \frac{0.2 - 0.0218}{6.67 - 0.0218} = 0.027\,(2.7\%)$$

And at room temperature:

$$\text{Total Fe}_3\text{C (room temperature)} = \frac{0.2 - 0.002}{6.67 - 0.002} = 0.03\,(3\%)$$

Once again, the amount of tertiary cementite is 0.3%.

- Steel with 0.4% C

In this case A_3 is:

$$\frac{912 - 727}{0.77} = \frac{912 - A_3}{0.4}$$

$$A_3 = 808°\text{C}$$

At A_1, the ferrite and austenite proportions are:

$$f_\alpha = \frac{0.77 - 0.4}{0.77 - 0.0218} = 0.49\,(49\%)$$
$$f_\gamma = (1 - f_\alpha) = 0.51\,(51\%)$$

while at 727°C, the amount of cementite is:

$$\text{Total eutectoid Fe}_3\text{C} = \frac{0.4 - 0.0218}{6.67 - 0.0218} = 0.057\,(5.7\%)$$

And at room temperature:

$$\text{Total Fe}_3\text{C (room temperature)} = \frac{0.4 - 0.002}{6.67 - 0.002} = 0.06\,(6\%)$$

In this case, the amount of tertiary cementite is also 0.3%.

8.1.4 Equilibrium Cooling Microstructures of Steels and White Cast Irons

8.1.4.1 Hypereutectoid Steels

The microstructure of these steels (Fig. 8.4) is formed by pearlite colonies surrounded by a more or less continuous cementitic matrix. Free cementite, precipitated at the temperature interval between A_{cm} and 727°C, is known as *proeutectoid* or *secondary*, and its amount can be calculated using the lever rule in the Fe-Fe_3C diagram.

8.1.4.2 Hypoeutectoid Steels With Less Than 0.0218% C

The $\gamma \rightarrow \alpha$ transformation starts at A_3; and the higher the amount of C in the steel, the lower will be A_3.

The enrichment of C in the austenite during cooling is less than 0.77%; and therefore, the allotropic transformation ends at A_1 (above 727°C) producing a fully ferritic structure.

Cooling below 727°C dissolves C in the ferrite exceeding solvus and reacting with Fe, precipitating in the shape of *vermicular* or *tertiary* cementite. This precipitation usually occurs in the shape of small worms (Fig. 8.10) at the ferritic grain boundaries, or at triple point joints between grains (or at the interior of the grains when cooling rate is higher).

8.1.4.3 Hypoeutectoid Steels With More Than 0.0218% C

These steels, just as the one in Section 8.1.2 (0.3% C steel), present a microstructure formed by allotriomorphic ferrite grains, or *proeutectoid* ferrite, surrounded by pearlite colonies. The ferrite and pearlite proportions can be deduced by the lever rule applied on the Fe-Fe_3C phase diagram.

In forged or hot-rolled plain C steels (0.55%–0.77% C), proeutectoid ferrite appears as the matrix constituent (in the form of a net that surrounds pearlite). However, for steels with less than 0.55% C, bands appear: ferrite crystals intermixed with pearlite colonies or a ferritic-pearlitic banded microstructure.

Furthermore, when the austenitic grain size is very large (e.g., steels in the as-cast state) ferrite usually appears as *Widmanstätten* plates or needles (Fig. 8.11).

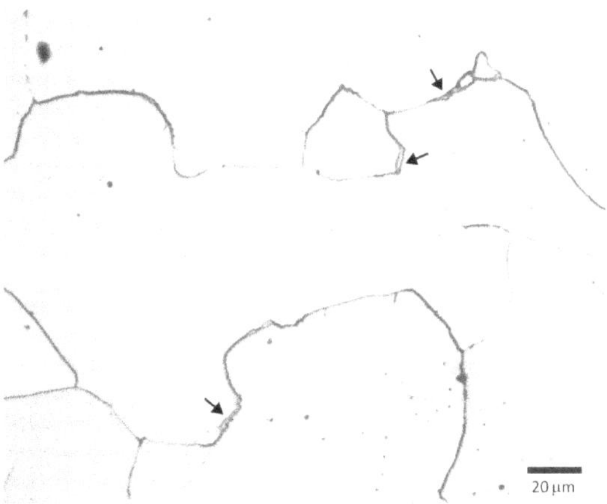

FIG. 8.10 Ferrite and tertiary cementite *(arrows)*.

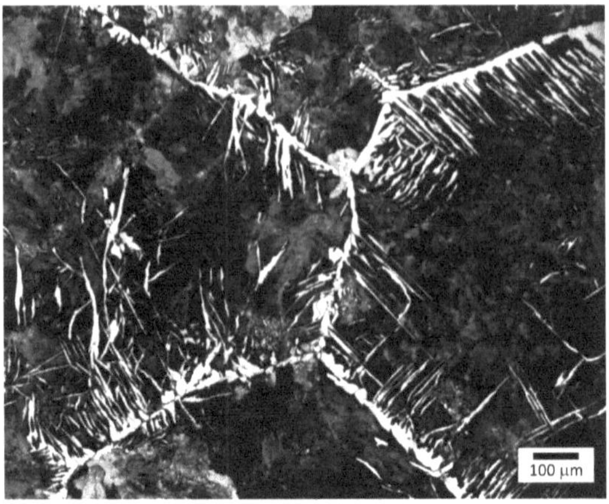

FIG. 8.11 Widmanstätten structure in steels.

8.1.4.4 Hypoeutectic White Castings

Fig. 8.12 presents the solidification and cooling curve of one of these alloys (3% C), considering the phase rule. Solidification starts at a temperature considerably lower compared to steels (<2.11% C), forming austenite as primary constituent in the temperature interval between T_1 and 1148°C, resulting at the end, in 2.11% C austenite embedded in the eutectic matrix known as *ledeburite*: formed by 52 wt% of austenite (2.11% C) and 48 wt% of Fe_3C.

If temperature decreases, with the material already in solid state, the solubility of C in γ also decreases (both at the primary austenite and the one forming the eutectic); and the excess of C, expelled from the γ lattice, starts precipitating as Fe_3C, either at the austenitic grain boundaries or the interior of the austenitic grains.

Applying the lever rule, C content of the austenite for each temperature can be determined, as well as the weight proportions of both austenite (eutectic plus primary γ) and Fe_3C (eutectic cementite plus the one formed by solubility loss of C in austenite, known as *secondary* or *proeutectoid cementite*).

For temperatures higher or very close to 727°C, the carbon content in the austenite will be close to 0.77%. Fig. 8.7 indicates that at this temperature, with equilibrium cooling, the invariant eutectoid transformation of the 0.77% C austenite into pearlite aggregate (fine lamellae of ferrite and cementite) occurs, both at the interior of primary austenitic grains and at the interior of eutectic austenite. Below 727°C, ferrite (for this 3% C alloy, α will be found in the pearlite) will also lose carbon because of a lack of solubility and will originate a new amount of cementite known as *tertiary*.

The microstructure of the 3% C alloy (shown in Fig. 8.13) is formed by an eutectic matrix of cementite (white constituent) and dark areas corresponding to eutectic austenite already transformed into pearlite. The dark constituent (both primary and eutectic) is pearlite produced by the eutectoid austenitic transformation during cooling.

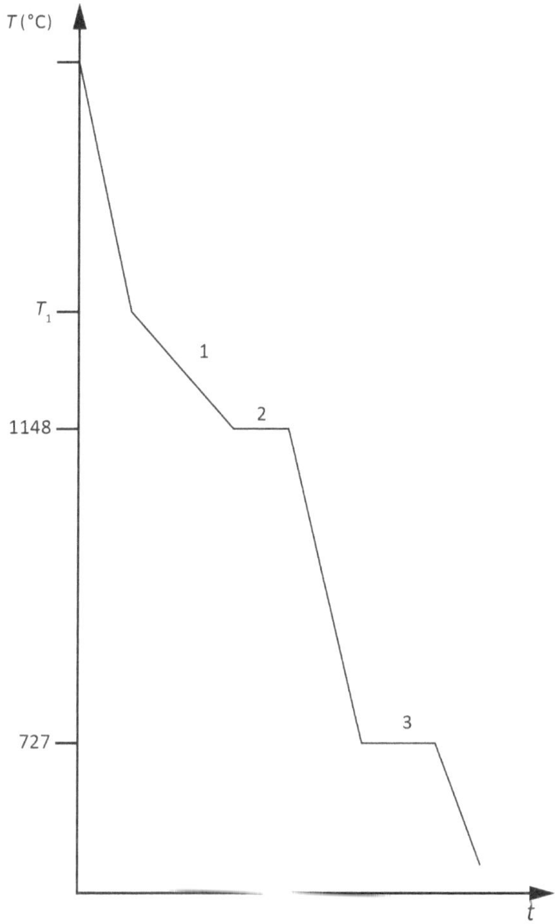

FIG. 8.12 Solidification and cooling curve of an hypoeutectic white casting: (1) austenite + liquid, (2) formation of ledeburite, and (3) formation of pearlite.

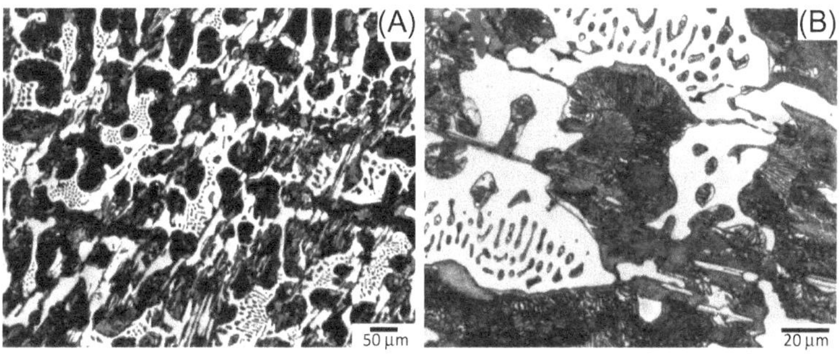

FIG. 8.13 (A) Hypoeutectic white casting (3% C) and (B) seen at higher magnification.

8.1.4.5 Hypereutectic White Castings

For *hypereutectic white castings* (>4.3% C), both solidification and cooling curves (Fig. 8.14, extremely similar to Fig. 8.12) and microstructures (Fig. 8.15) are analogous to hypoeutectic white castings with the difference of presenting as primary or disperse constituent, Fe_3C needles (*primary cementite*) that formed before solidification of the eutectic.

White castings, analyzing Fig. 8.7, are the alloys with the lowest melting point of the Fe-Fe_3C metastable system and have a small solidification interval. Their properties include good castability and weak shrinkage, which are both favorable for molded parts, justifying their name of "*ferrous castings*."

Their microstructures are related to the properties of these alloys: matrix constituent is ledeburite (and inside it, cementite). The model to explain their behavior could be a cementite sponge with the empty spaces filled by pearlite. As a

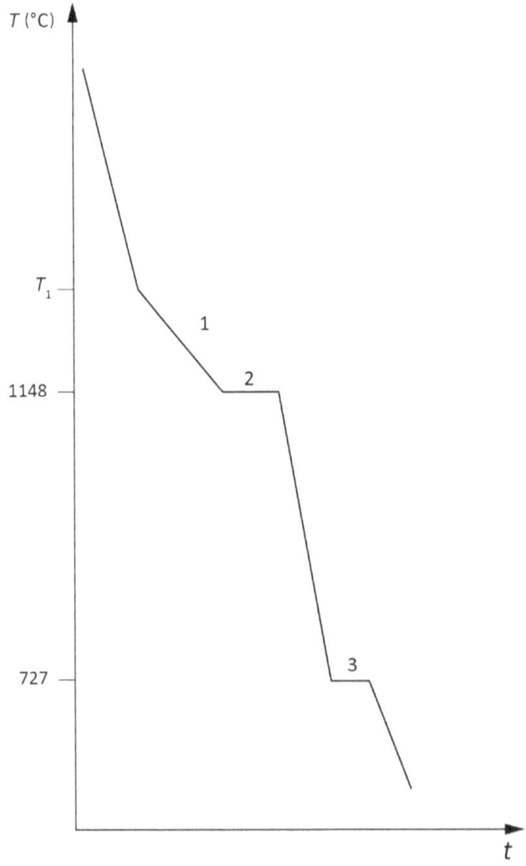

FIG. 8.14 Solidification and cooling curve of a hypereutectic white casting: (1) primary cementite + liquid, (2) formation of ledeburite, and (3) formation of pearlite.

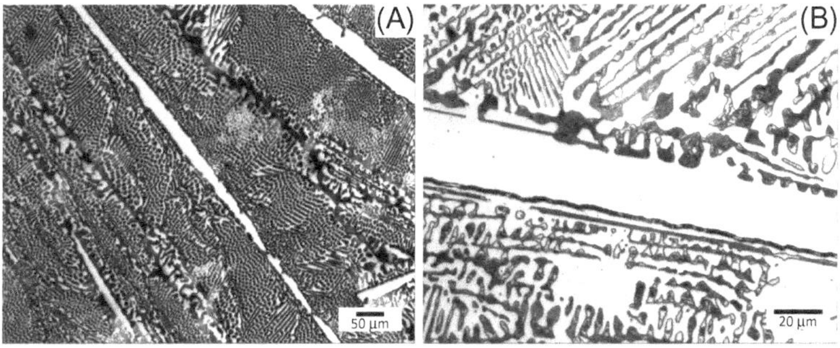

FIG. 8.15 (A) Hypereutectic white casting and (B) seen at higher magnification.

consequence (due to the elevated hardness and brittleness of cementite) their mechanical properties include low toughness, almost null elongation in tension and high wear resistance. Their wear behavior determines their use in industrial applications and also explains the poor machinability of white castings.

8.1.5 Ultimate Stress Estimation in Binary Ferritic-Pearlitic Steels

While eutectic binary aggregates always appear as a matrix constituent, eutectoids form disperse constituents with convex shapes surrounded by other phases.

Since mechanical properties of multiphasic alloys mainly depend on the matrix constituent, *annealed* hypereutectoid steels (slowly cooled from the full austenitic state) will not, in general, be deformable below 727°C and will not present plastic deformation before fracture. This is due to the nature of cementite (hard and fragile) that usually surrounds pearlite, and must absorb stresses. The ductility of the steel can be improved using an adequate heat treatment (Section 8.3.2) to break the continuity and globulize the cementite.

On the contrary, *annealed* hypoeutectoid binary steels (*normalized* if cooling is made in air from the austenization temperature) have larger elongations since the matrix constituent is soft and deformable. The cold-work ability of these steels is inversely proportional to their C content, since the higher the % C, the lower the ferrite proportion will be. Ferritic-pearlitic steels with less than 0.3% C can be cold worked by rolling, extrusion, drawing, etc. with relative easiness.

A *first approximation* of the tensile strength of an annealed or normalized hypoeutectoid steel considers the tensile strength of pearlite (800 MPa) and ferrite (300 MPa), as well as the ferrite and pearlite proportions, and applying the lever rule on the $Fe\text{-}Fe_3C$ diagram, the fraction of pearlite (1.30 C_1) and ferrite (1.00–1.30 C_1) in the steel can be obtained as a function of the percentage of carbon (C_1):

$$\sigma_{max}\,(\text{in MPa}) = 300 + 650\ C_1 \qquad (8.5)$$

A better approximation to the tensile stress was obtained by Pickering (1971):

$$\sigma_{max} = f_\alpha^{\frac{1}{3}}\left[246.43 + 1142.81\sqrt{N} + 18.17d^{-\frac{1}{2}}\right] + \left(1 - f_\alpha^{\frac{1}{3}}\right)\left[719.26 + 3.54S_0^{-\frac{1}{2}}\right] + 97.03\,\mathrm{Si} \tag{8.6}$$

where f_α is the ferrite volumetric fraction, d is the mean grain size of ferrite (mm), and S_0 is the interlamellar spacing of pearlite (mm). In this formula, the tensile stress does not vary linearly with pearlite content. For carbon contents lower than 0.2%, the amount of pearlite barely influences σ_{max} (the grain size of the proeutectoid ferrite is more determinant) and, instead, for higher carbon contents the pearlite is the factor that determines tensile stress.

EXERCISE 8.4

Estimate the tensile stress of four ferritic-pearlitic steels with the following carbon contents: 0.1%, 0.2%, 0.4%, and 0.6% C, using (a) Eq. (8.5), (b) Pickering equation, and (c) graphically represent the tensile stress as a function of pearlite fraction for these four steels.

Data: $N = 0.01\%$, $d = 7\,\mathrm{ASTM}\,(28\,\mu\mathrm{m})$, $S_0 = 0.2\,\mu\mathrm{m}$

Solution

(a) Using Eq. (8.5)

$$\sigma_{max}(0.1\%\,\mathrm{C}) = 300 + 650(0.1) = 365\,\mathrm{MPa}$$

$$\sigma_{max}(0.2\%\,\mathrm{C}) = 300 + 650(0.2) = 430\,\mathrm{MPa}$$

$$\sigma_{max}(0.4\%\,\mathrm{C}) = 300 + 650(0.4) = 560\,\mathrm{MPa}$$

$$\sigma_{max}(0.6\%\,\mathrm{C}) = 300 + 650(0.6) = 690\,\mathrm{MPa}$$

(b) Using Pickering equation (8.6)

The ferrite fraction must be calculated as:

$$f_\alpha(0.1\%\,\mathrm{C}) = \frac{0.77 - 0.1}{0.77 - 0.0218} = 0.895$$

$$\sigma_{max}(0.1\%\,\mathrm{C}) = (0.895)^{\frac{1}{3}}\left[246.43 + 1142.81\sqrt{0.0001} + 18.17(0.028)^{-\frac{1}{2}}\right] + \left[1 - (0.895)^{\frac{1}{3}}\right]\left[719.26 + 3.54(0.0002)^{-\frac{1}{2}}\right] + 97.03(0)$$

$$= (0.895)^{\frac{1}{3}}(366.44) + \left[1 - (0.895)^{\frac{1}{3}}\right](969.58) = 388.34\,\mathrm{MPa}$$

$$f_\alpha(0.2\%\,\mathrm{C}) = \frac{0.77 - 0.2}{0.77 - 0.0218} = 0.762$$

$$\sigma_{max}(0.2\%\,\mathrm{C}) = (0.762)^{\frac{1}{3}}(366.44) + \left[1 - 0.762^{\frac{1}{3}}\right](969.58) = 418.83\,\mathrm{MPa}$$

$$f_\alpha(0.4\%\,\mathrm{C}) = \frac{0.77 - 0.4}{0.77 - 0.0218} = 0.495$$

$$\sigma_{max}(0.4\%\,C) = (0.495)^{\frac{1}{3}}(366.44) + \left[1 - (0.495)^{\frac{1}{3}}\right](969.58) = 492.47\,MPa$$

$$f_\alpha(0.6\%\,C) = \frac{0.77 - 0.6}{0.77 - 0.0218} = 0.227$$

$$\sigma_{max}(0.6\%C) = (0.227)^{\frac{1}{3}}(366.44) + \left[1 - (0.227)^{\frac{1}{3}}\right](969.58) = 601.65\,MPa$$

(c) Graphical representation of tensile stress as a function of pearlite fraction

Plotting data from the Pickering estimation, Fig. 8.16 is obtained. Analyzing the change in the slope, it can be observed that pearlite fractions <20% have a small influence on tensile stress modifications, while only higher percentages can considerably increase this property. Furthermore, as ferrite is the matrix constituent in hypoeutectoid steels, the smaller the grain size of this phase, the higher the tensile stress will be.

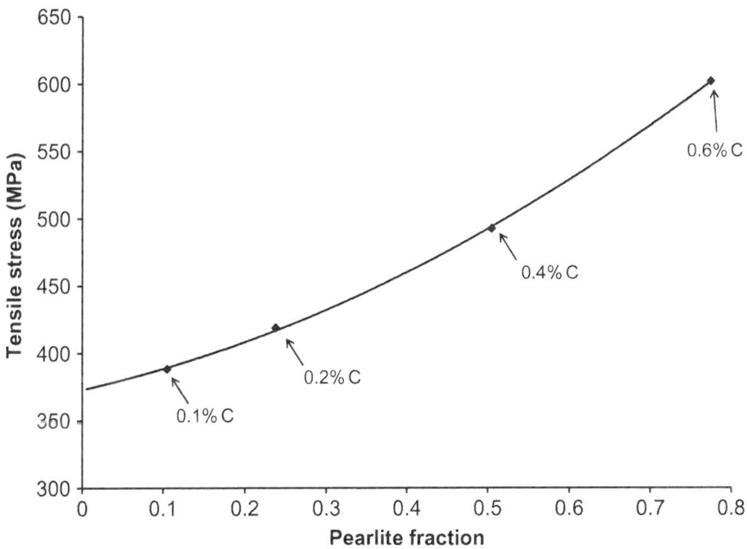

FIG. 8.16 Tensile stress as a function of pearlite fraction in ferritic-pearlitic steels.

8.1.6 Cooling at a Rate Faster Than Equilibrium

If two samples of the same steel are compared, both heated to the same austenization temperature, one of them cooled at equilibrium conditions and the other one cooled slightly faster, the amount of pearlite at room temperature is higher in the second sample.

In general, increasing the cooling rate, since it creates proeutectoid α and pearlite as constituents, will imply a decrease in A_{3r} caused by the formation of proeutectoid α (nucleation and growth) during cooling.

Analogous to the solidification of a pure metal, increasing the cooling rate of austenite with certain C content (Fig. 8.17), decreases the temperature when ferrite starts to form (T_1 for V_1 rate, T_2 for V_2 rate, etc.). The critical size of the ferrite nucleus also decreases with cooling rate, thus the higher the cooling rate, the finer the proeutectoid α grains will be.

The transformation lines of the Fe-Fe_3C diagram are modified to those indicated in Fig. 8.18 for higher cooling rates: the eutectoid point for the V_1 cooling rate corresponds to the temperature and composition of E_1; analogously, for V_2 the eutectoid point is E_2, for V_3 is E_3, etc. It is important to point out that once austenite has transformed, the higher the cooling rates, the higher the pearlite/ferrite ratio will be.

As a consequence, the tensile stress σ_{max} of an annealed steel ($20\%\alpha_{\text{proeutectoid}} + 80\%$pearlite) increases with cooling rate: high rates will decrease the pearlite lamellar spacing S_0 and the C content in the pearlite, though the resistance of this phase is decreased by the lower C content, the small lamellar spacing causes tensile stress of ~ 800MPa.

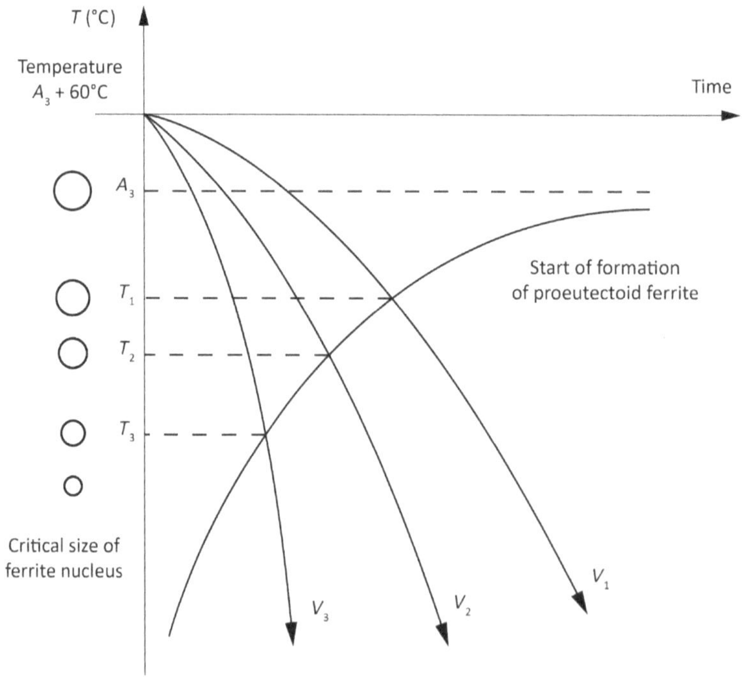

FIG. 8.17 Transformation of austenite by continuous cooling for a steel with <0.77% C.

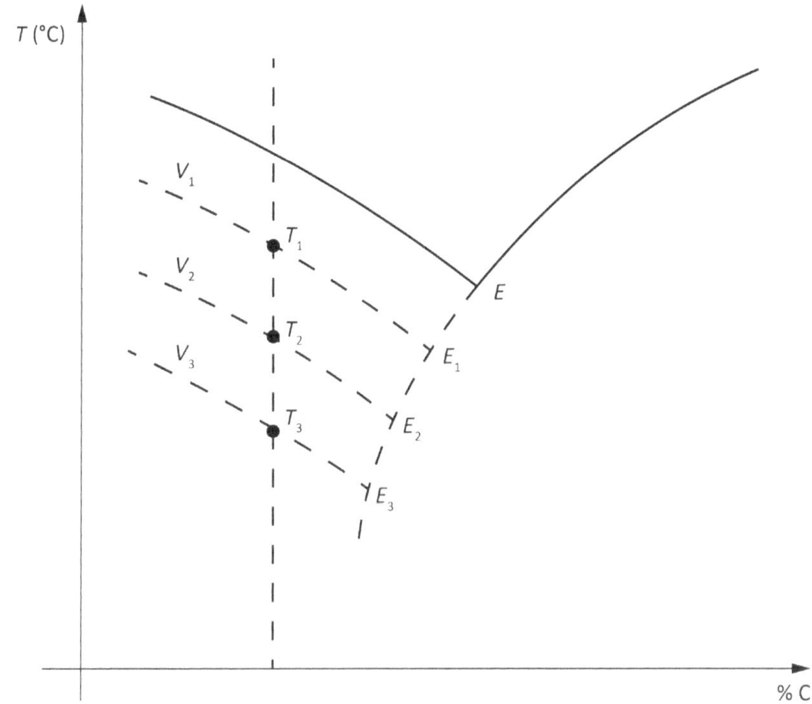

FIG. 8.18 Displacement of both A_3 and E (eutectoid) when increasing the cooling speed ($V_1 < V_2 < V_3$).

If a ferritic-pearlitic structure is obtained during cooling, thin parts will have a more regular phase distribution along their transverse section compared to those with large diameter or width: lower tensile stress at the center than at the surface.

Ferrite crystals will only acquire polyhedral shapes if the latent heat and excess of C removal rates are sufficient; such is the case of fine-grained slowly cooled austenite. When cooling rate increases, and thus transformation rate, the amount of latent heat that must be transferred per time unit through the ferrite-austenite interfaces, also increases. The growth of tips and edges in the ferrite crystals allow the heat to be released in multiple directions since the growth of a plate element, parallel to itself, favors heat release and diffusion (e.g., the *Widmanstätten* microstructure).

For large austenitic grain sizes, fast cooling rates promote the *Widmanstätten* structure (Fig. 8.11). The growth of a ferrite crystal depends on the rate of latent heat released during the $\gamma \rightarrow \alpha$ transformation and the rate of excess C transferred from the already formed ferrite.

On the other hand, since C diffusion rate is higher in directions parallel to the (1 1 1) planes of austenite, ferrite needles preferentially grow with their (1 1 0) planes matching up the (1 1 1) austenite planes (Table 7.1).

Since cooling rates (in air) of plain-carbon steels are usually higher than equilibrium ones, it is common to find acicular ferrite in their as-cast, rolling or forging states, as well as in welding seams ("beads").

Because of their low toughness, *Widmanstätten* structures are undesirable from the industrial point of view and *normalizing* heat treatments are necessary to erase the acicular morphology.

8.1.7 Influence of Alloying Elements in the Metastable Fe-Fe_3C Diagram

8.1.7.1 Gamma-Stabilizing Elements

The elements that stabilize phase γ of Fe, such as C, are known as gamma-stabilizing elements:

- In the presence of gamma-stabilizing elements in solid solution, the $\delta \rightarrow \gamma$ transformation takes place above 1394°C: the higher the content of gamma-stabilizing elements in solid solution, the higher A_4 will be.
- The $\gamma \rightarrow \alpha$ transformation starts below 912°C: the higher the content of gamma-stabilizing elements in the austenite, the lower A_3 will be.

Furthermore, considering the Fe-Mn phase diagram of Fig. 8.19, for contents higher than 40%, Mn (a strong gamma-stabilizing element) causes the disappearance of the A_3 transformation ($\gamma \rightarrow \alpha$). The solidification of these steels takes place directly in γ phase without the appearance of α. Other gamma-stabilizing elements such as Ni, Co, N, Zn, Cu, and Au behave similar to Mn.

Considering Fig. 8.20 for the same C content, the higher the Mn content solubilized in austenite, the lower the A_3 point will be, result of the gamma-stabilizing influence of Mn.

Furthermore, the maximum saturation limit of C in austenite remains unaltered irrespective of the content of Mn, Ni, or Co, because of the small distortion created in the γ cell when substituting atoms of Fe with Mn, Ni, or Co. The same behavior can be translated to the precipitation of cementite below the A_{cm} solvus line.

Both, eutectoid temperatures and C content of the Fe-C-Mn eutectoid austenite vary according to the Bain curves of Fig. 8.21, and since the eutectoid point has lower C content and lower temperature when the percentage of Mn (or Ni) increases, the following is obtained: if Mn or Ni contents are high, the eutectoid transformation occurs below room temperature and results in austenitic steels.

For gamma-stabilizing amounts unable to move the eutectoid transformation below room temperature, the microstructure of a *normalized hypoeutectoid steel* (cooled in air from the austenitic state) will still be formed by ferrite and pearlite. The drop in A_3 induces a grain refinement of the proeutectoid ferrite and, for equal carbon contents, the pearlite ratio is proportional to the Mn content.

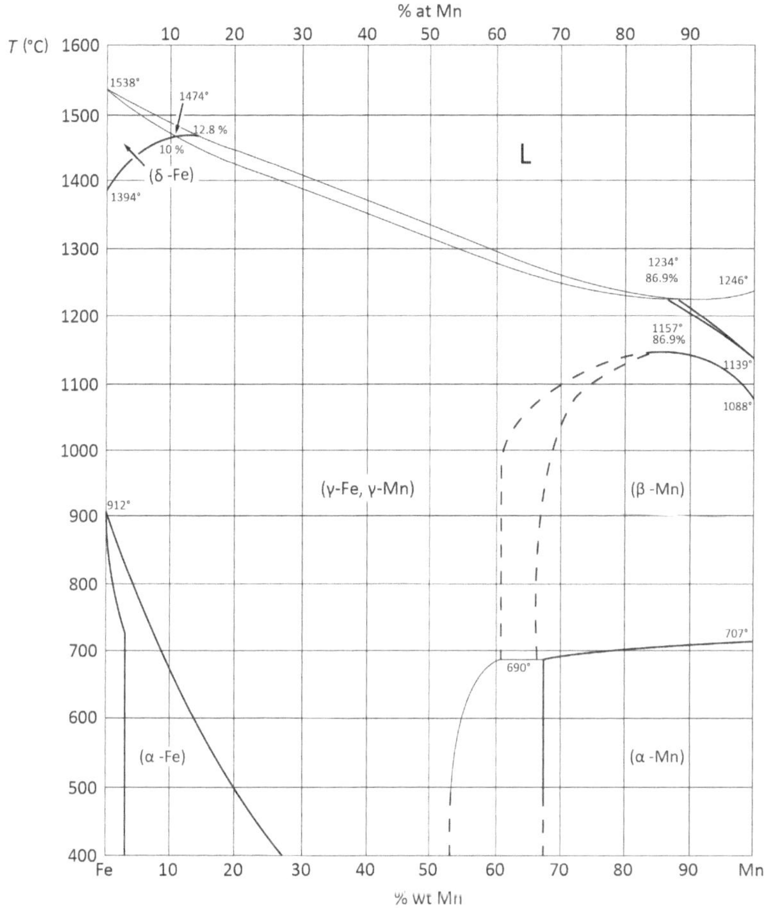

FIG. 8.19 Fe-Mn phase diagram (ASM International, 1992).

If tensile stress is analyzed, the higher the percentage of Mn, the higher the tensile stress will be. For an hypoeutectoid steel with ferritic-pearlitic structure, the tensile stress can be estimated as the sum of two products: fraction of ferrite multiplied by the tensile stress of the alloyed ferrite plus the fraction of pearlite multiplied by its tensile stress. Since increasing the Mn content also increases the pearlite fraction, the tensile stress of the steel will be higher for steels alloyed with Mn, than for a plain-carbon ones. On the other hand, the presence of Mn in solid solution in the ferrite somewhat increases the tensile stress.

The modification of A_3 has other industrial consequences: for the same C amount, increasing Mn content reduces A_3 and allows forming processes of the steel (e.g., rolling, forging, drawing, etc.) in its austenitic state at

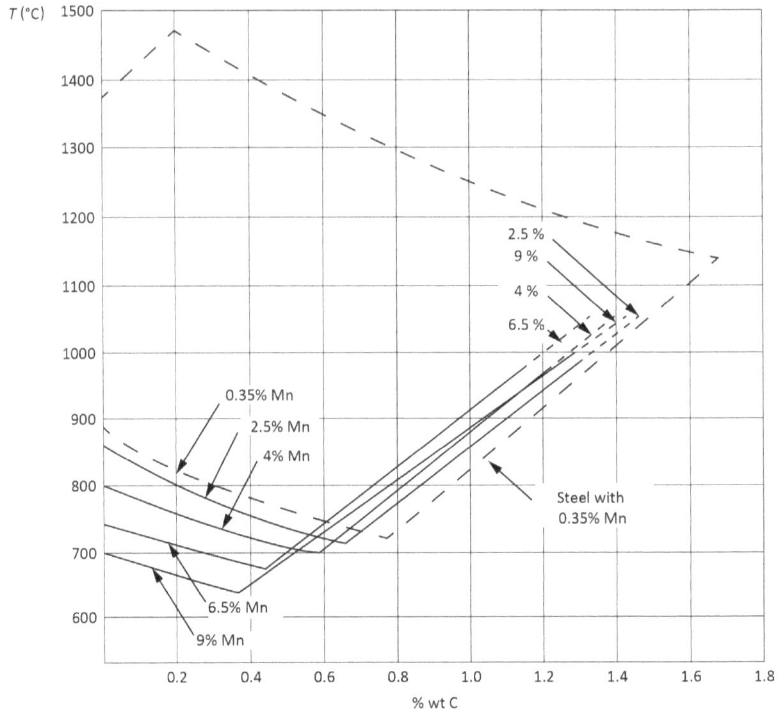

FIG. 8.20 Austenite section for the simplified Fe-C-Mn phase diagram (Bain and Paxton, 1966).

temperatures lower than those for a plain-carbon steel without Mn; furthermore, steels with Mn can be quenched at lower temperatures. Otherwise, if the same austenization temperature is used (for forming or quenching) for two steels with the same % C, one with Mn and one without, the steel with Mn has a higher tendency to suffer *overheating* (Section 8.3.3).

8.1.7.2 Alpha-Stabilizing Elements

Some alloying elements produce an opposite effect to that of gamma-stabilizing elements: Si, P, Al, Be, Sn, Sb, As, Ti, Nb, V, Ta, Mo, W, and Cr are stabilizers of Fe in its body-centered cubic form and are known as *alpha-stabilizing elements*.

Fig. 8.22 shows the Fe-Si phase diagram (low amount of Si section). The influence of Si on the amount of C that saturates austenite at high temperatures is shown in Fig. 8.23. For the same C content, the higher the Si content:

- the higher the A_3 critical line,
- the lower the A_4 critical line, and
- the higher the temperature of the eutectoid point

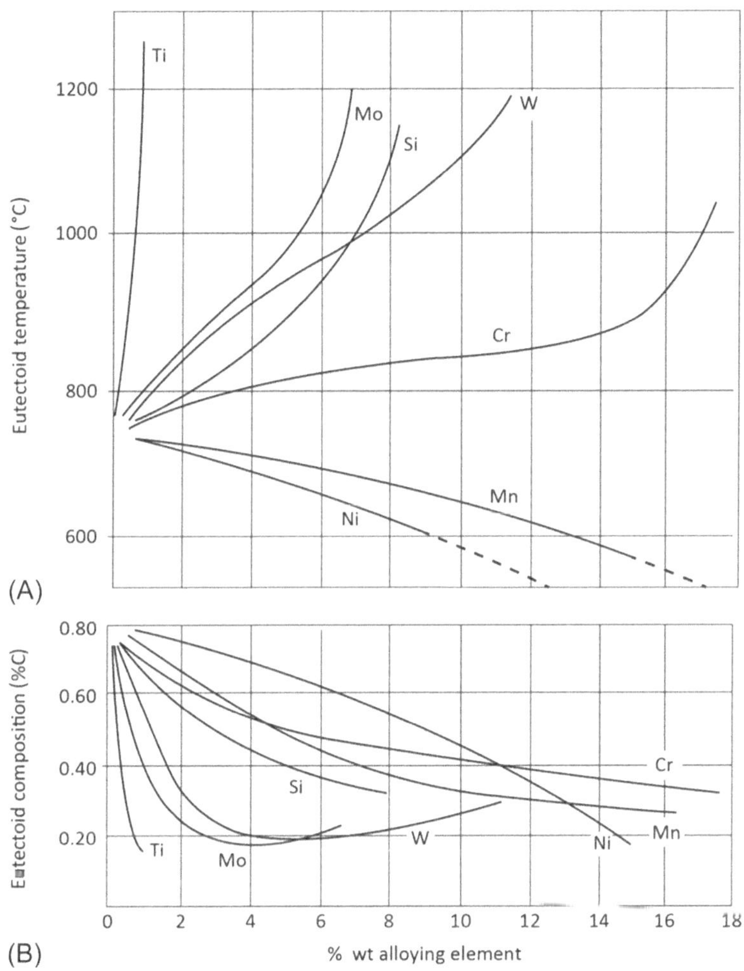

FIG. 8.21 Influence of elements on eutectoid temperature (A) and composition (B) (Bain and Paxton, 1966).

will be (lower C contents). Binary hypoeutectoids steels without Si (e.g., 0.6% C) will become eutectoid when the Si content is 2%, and hypereutectoid for even larger Si contents.

The higher the amount of Si, the lower the amount of C that saturates the eutectic austenite will be (2.11% C at 1148°C in plain-carbon steels), due to the distortion produced by Si in the γ lattice (in solid solution). In fact, Si, as other alpha-stabilizing elements, does not crystallize on the face-centered cubic system and, as such, can only form partial (not total) solid solutions in austenite. The temperature necessary to achieve maximum solubility in the austenite is also increased, as seen in Figs. 8.23–8.25 for alpha-stabilizing elements Si, Mo, and Ti.

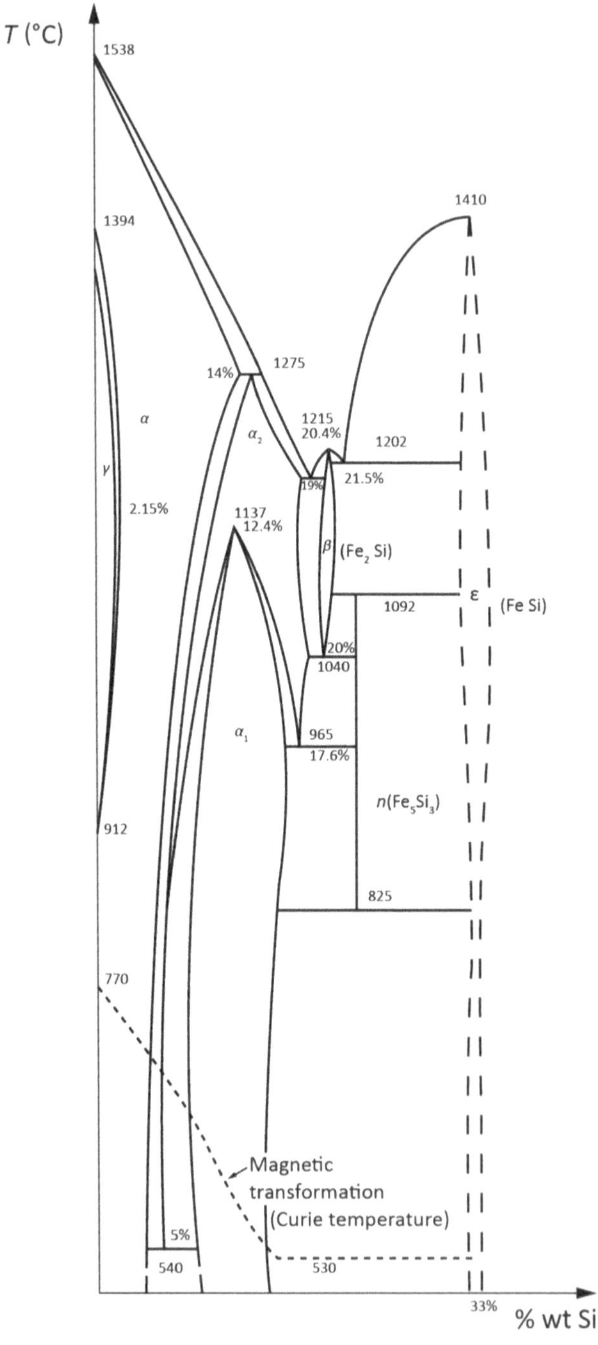

FIG. 8.22 Fe-Si phase diagram showing the Curie transformation temperature (ASM International, 1992).

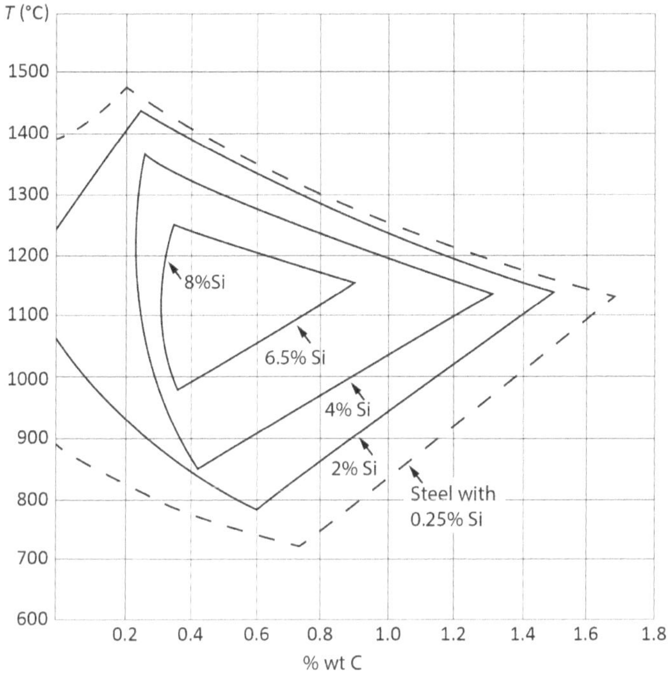

FIG. 8.23 Austenite section for the simplified Fe-C-Si phase diagram (Bain and Paxton, 1966).

Cr differs from other alpha-stabilizing elements in the following ways:

- In the $\delta \rightarrow \gamma$ transformation, Cr is alpha-stabilizing, as seen in the Fe-Cr diagram (Fig. 3.20).
- In the $\gamma \rightarrow \alpha$ transformation, Cr is an alpha-stabilizing element only for contents higher than 7%, while for less than 5% it is gamma-stabilizing.

For low Cr contents, Eq. (8.3) considers the gamma-stabilizing nature of this element, and it is evident that this expression is not valid for highly alloyed steels. High Cr ferritic stainless steels ($17\%\text{Cr}, \text{C} \leq 0.003\%$) and refractory steels ($25\% - 30\%\text{Cr}, \text{C} \leq 0.35\%$), solidify as δ-Fe, which remains as bcc until room temperature: the $\delta \rightarrow \gamma \rightarrow \alpha$ transformations are suppressed and only the transformations by carbon solubility loss to form carbides, by nucleation and growth, occur.

Among the alpha-stabilizing elements, there are those with *carbide-stabilizing*[1] nature such as Ti, Zr, Nb, V, Ta, W, Mo, and Cr (in descending order according to their carbide stability), while other alloying elements that do not promote the

1. Carbide-stabilizing elements are not exclusively alpha-stabilizing. Some gamma-stabilizing ones, such as Mn, have slight carbide-forming tendencies (carbide stability).

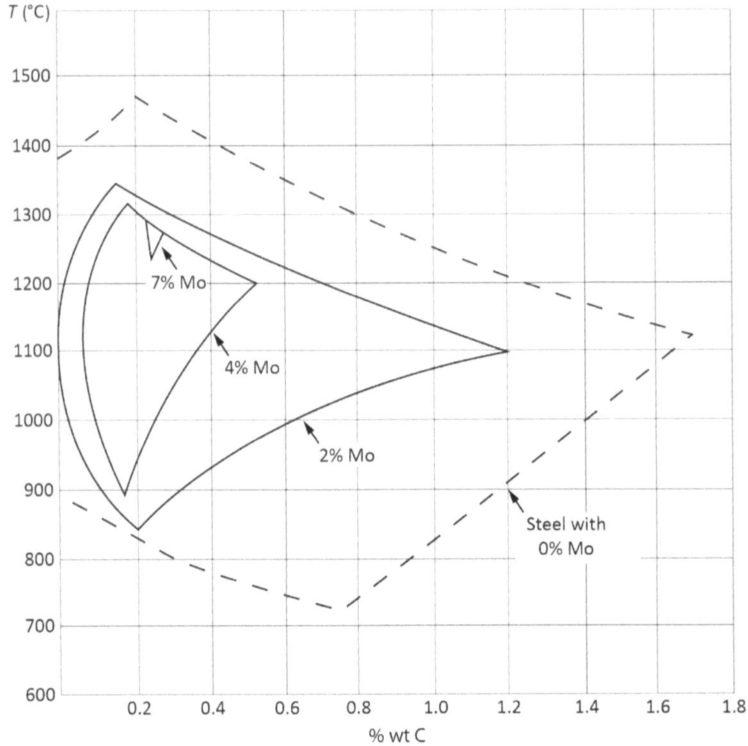

FIG. 8.24 Austenite section for the simplified Fe-C-Mo phase diagram (Bain and Paxton, 1966).

formation of carbides, because of their null or negative affinity with C, are of the *graphite-stabilizing* nature, such as Si, P, Al, Ni, and Cu (in descending order). These last ones tend to favor the isolated formation of γ (or α) and graphite, inhibiting, by a dilution and affinity effect, the reaction between atoms of Fe and C to form cementite. Thus, ferrous castings with an adequate Si content (higher than 1.5%) solidify without forming ledeburite, resulting in an eutectic aggregate of austenite and graphite (*gray castings*).

Once a gray casting has solidified, if sufficient amounts of graphite-stabilizing elements are present, the Fe-C "stable" transformations are promoted resulting in ferrite and graphite as constituents, instead of ledeburite. The proportions of graphite-stabilizing elements always involve a compromise between:

- Stable solidification (gray casting) and the metastable one (white casting), that sometimes can be solved with the simultaneous presence of graphite and ledeburite (*mottled cast irons*).
- During cooling, the austenite of a gray casting transforms into ferrite and graphite, or into cementite and pearlite (gray castings with pearlitic matrix).

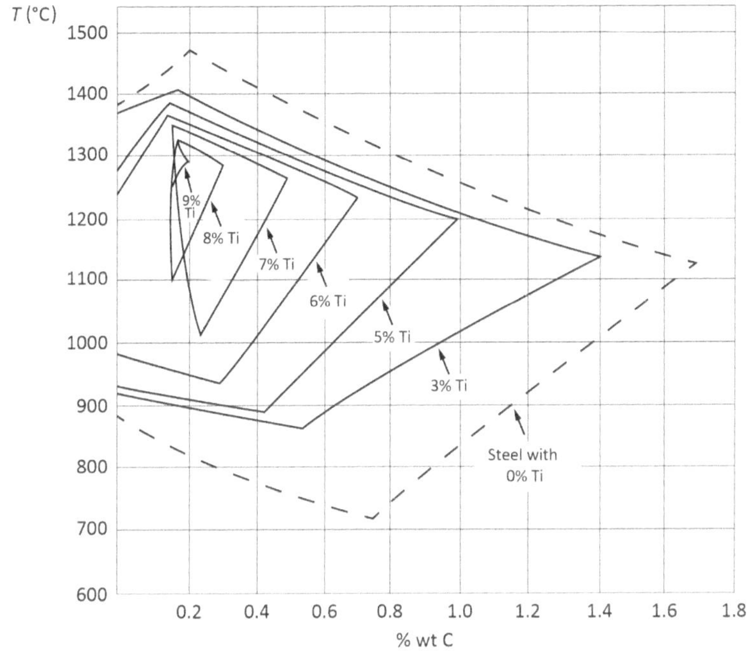

FIG. 8.25 Austenite section for the simplified Fe-C-Ti phase diagram (Bain and Paxton, 1966).

Both gamma- and alpha-stabilizing elements have a hardening effect on ferrite when they form solid solutions. This hardening is considerably higher in interstitial solid solutions (C and N) than in substitutional ones (P, Sn, Si, Cu, Mn, and Mo, in decreasing order). On the other hand, Ni and Al barely harden ferrite, while Cr, possibly because of the removal of interstitial solutes, apparently softens it.

Increase in hardness is generally accompanied with a loss in toughness. N and P considerably increase the ductile-brittle transition temperature (DBTT) and, as a consequence, significantly increase the fragility of ferrite. For 1% N in solid solution, the DBTT would be increased by 700°C, 1% P by 400°C, 1% Sn by 150°C, and 1% Si only by 44°C. On the other hand, Cu, Mn, Mo, and Cr, dissolved in the ferrite, barely modify toughness.

8.1.7.3 Precipitate-Forming Elements (Microalloying)

The production of high-strength low-alloy (HSLA) structural steels, with low cost and improved properties, is based on the use, in small amounts, of alloying elements such as V, Ti, Mo, and Cr, capable of forming precipitates by reacting with C and N present in the steel, and taking advantage of normalizing, quenching, and tempering treatments, as well as transformations during deformation processes such as hot forging or hot rolling (or drawing).

Thermomechanically controlled rolling processes combined with continuous annealing lines are used to produce steel sheets with very fine ferrite grain sizes ($d_{\alpha} \approx 5\,\mu m$) and improved fracture toughness and weldability, taking advantage of the hardening effect of precipitates, both by inhibiting grain growth and acting as obstacles to dislocation movement.

The combination of controlled rolling, controlled cooling, and direct quenching during thermomechanical processes may result in grain sizes even lower than $\sim 5\,\mu m$ and extremely good properties, with these ultrafine grained materials classified as dual-phase (DP) or HSLA. Yet, rolling conditions must be very strict to take full advantage of the thermomechanical phenomena, and, as Fig. 8.26 indicates, different stages of deformation depending on transformation temperatures of the steel must be controlled in order to achieve the desired microstructure.

Controlled deformation processes usually consist of a first stage where coarse austenite is reduced in size through repeated deformation and recrystallization. Ti plays a key role as it allows refinement of austenite before rolling. A second stage results in elongated and nonrecrystallized austenite bands due to deformation, which allows ferrite to nucleate at both deformation bands and γ grain boundaries. The ferrite grains produced by this mechanism are small and Nb can favor accumulation of deformation inside the austenite when the process takes place below the nonrecrystallization temperature (T_{nr}). The precise determination of this last temperature is a very important factor in these industrial processes.

A last stage consists of deformation in the ferrite-austenite DP region to produce a soft structure. Precipitation of V in the form of V_4C_3 in the ferrite grains appears during this stage.

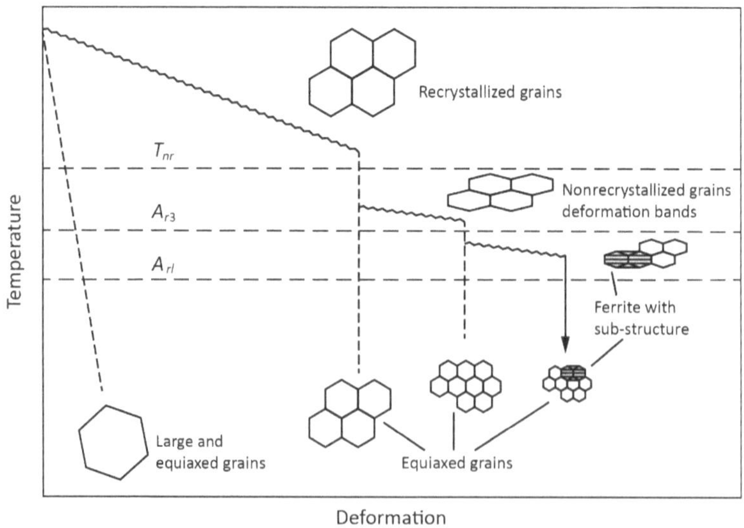

FIG. 8.26 Schematic diagram of the controlled rolling process (Quintana and Gonzalez, 2016).

Furthermore, control of time periods by delaying the start of the second stage after the first one has ended, as well as rapidly cooling after the last stage, results in fine ferritic and martensitic (or bainitic) grain sizes in the form of bands (Fig. 8.27), improving strength and toughness of the final product.

Though the effect of precipitates is a smaller grain size, the reduction in ductility as well as in hardenability (formation of Lüders bands) is an important disadvantage. Thus, industrial processes and commercial products cannot only focus on reducing grain size and increasing the amount of precipitates, but obtaining a material with adequate ductility and strain-hardening coefficient (n in the $\sigma = C\varepsilon^n$ equation) as steel sheets must be subsequently stamped, folded, drawn, etc.

The role of precipitates depends on the elements that form carbides, nitrides, or carbonitrides: Ti nitrides are responsible for maintaining austenitic grain size during homogenization before rolling. Both Ti nitrides and Nb carbonitrides are obstacles to grain growth of recrystallized austenite during rolling passes.

The large amount of Nb carbonitrides precipitation, before roughening and finishing mills, delays static recrystallization of the austenite. Furthermore, Nb and Ti nanoprecipitates increase the elastic limit of the steel without a negative effect on stiffness.

Fig. 8.28 shows a Ti carbonitride with rectangular shape which served as substrate for a Nb precipitate, as Ti(C, N) has a higher precipitation temperature (~ 1300°C) and is located at a grain boundary. Also, Fig. 8.29 shows a string of this type of precipitates located at the pearlitic band as this was a zone of elongated austenitic crystals.

Precipitation of V carbides (rectangular shapes) at the ferrite grain boundary is shown in Fig. 8.30, while Ti carbonitride may act as a substrate for both Nb and V carbides (Fig. 8.31). It is evident that precipitates have different shapes and sizes (even in the form of nanoprecipitates only observable in transmission

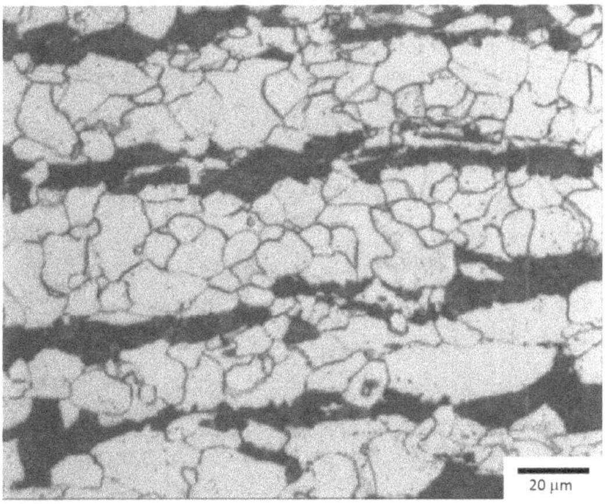

FIG. 8.27 Microstructure of the steel in its hot rolled raw state (Quintana and Gonzalez, 2016).

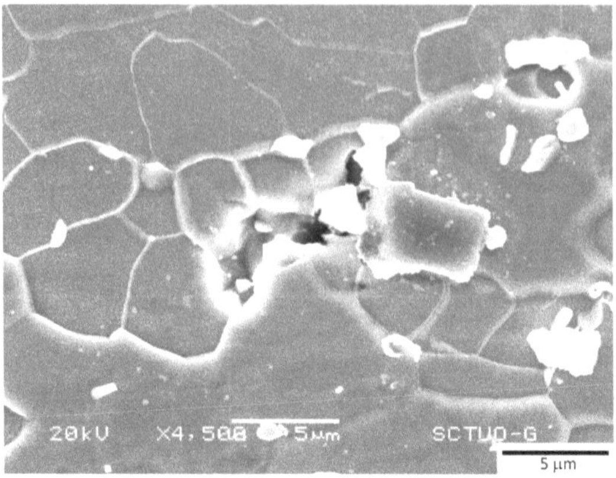

FIG. 8.28 SEM micrograph showing a titanium carbonitride and niobium carbide (Quintana and Gonzalez, 2016).

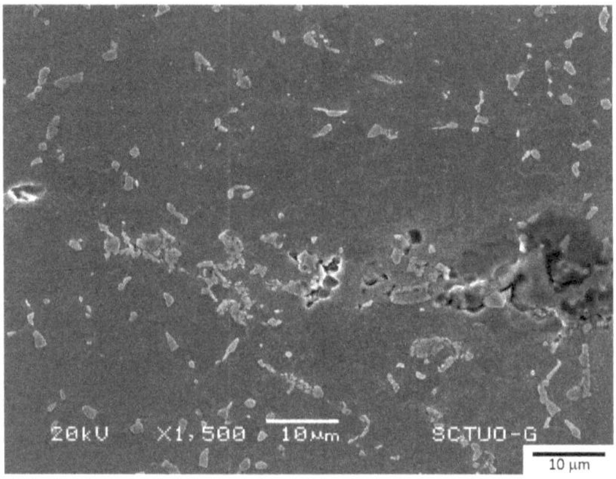

FIG. 8.29 SEM micrograph showing a string of titanium carbonitrides and niobium carbides (Quintana and Gonzalez, 2016).

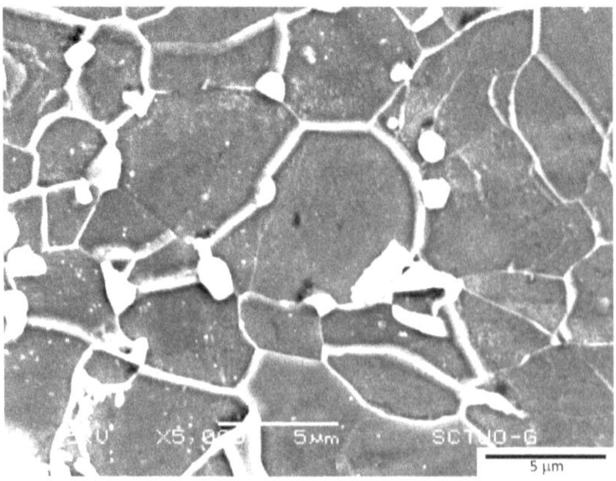

FIG. 8.30 SEM micrograph showing a vanadium carbide (Quintana and Gonzalez, 2016).

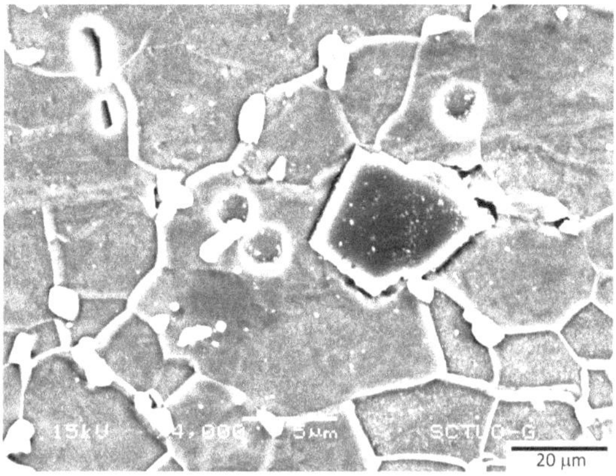

FIG. 8.31 SEM micrograph showing a titanium carbonitride, a niobium carbide, and vanadium carbide (Quintana and Gonzalez, 2016).

microscopy) and that quality control must be very strict in order to increase strength without compromising ductility, toughness, and/or hardenability of the steel.

8.2 NONEQUILIBRIUM TRANSFORMATIONS OF AUSTENITE DURING COOLING

8.2.1 Pearlitic Transformation

Cooling rate determines microstructures and properties that result from the transformation of austenite. When a 0.77% C austenite is cooled (isothermal transformation) to a temperature slightly lower than A_e, γ is transformed into pearlite (by nucleation and growth, same as solidification, allotropic precipitation and recrystallization, among others).

The lower the transformation temperature, the smaller the cementite and ferrite nuclei will be and thus, the smaller the separation between the lamellae of pearlite and the higher its hardness will also be.

When temperature is very low, C diffuses with difficulty in the interior of the bulk austenitic mass, and the nucleation and growth rates of cementite decrease: the lower the temperature, the lower the nucleation and growth rates of cementite will be.

Transformation rate (or percentage of austenite isothermically transformed as a function of time) is the product of nucleation and growth rates. Thus, the curve corresponding to 1% austenite transformed into pearlite will adopt the typical C-shape (if there are no alloying elements); and the same occurs for the 100% transformed austenite curve (Fig. 8.32). In general, the vertical maximum of this curve (100%) will not match with the vertical maximum of the 1% curve, for the same isothermal transformation temperature.

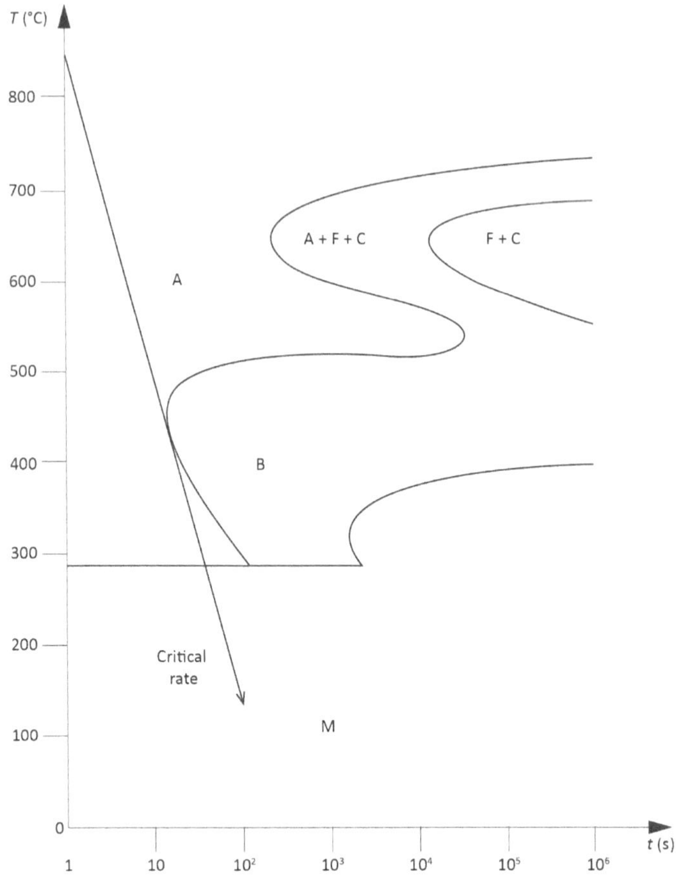

FIG. 8.32 TTT curve of a 0.57% C, 0.7% Mn, 0.2% Si, 0.7% Cr, 1.7% Ni, 0.3% Mo, and 0.1% V steel for 880°C austenization temperature.

Pearlite will adopt different morphologies known as coarse pearlite, fine pearlite (or sorbite), or troostite, depending on the isothermal transformation temperature:

- *Coarse pearlite* (Fig. 8.33): formed between 650–727°C, with interlamellar spacing S_0 of 0.25–0.5 μm and tensile strength σ_{max} of 800 MPa.
- *Fine pearlite*: formed in the 600–650°C range, with interlamellar spacing of 0.1–0.2 μm and tensile strength in the 900–1400 MPa interval. It has a strength similar to tempered martensite, however, its fatigue limit is lower because of the lower refinement in cementite dispersion.
- *Troostite* (Fig. 8.34): formed at lower isothermal transformation temperatures (500–600°C) where austenite transforms into a very fine pearlite, practically indistinguishable in the optical microscope with interlamellar

FIG. 8.33 Coarse lamellar pearlite.

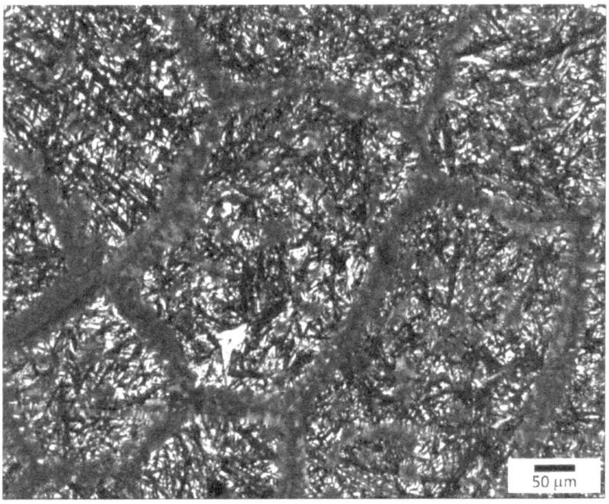

FIG. 8.34 Troostite (at the grain boundaries) over martensite.

spacing lower than 0.1 μm. This pearlitic aggregate acquires nodular and radial morphologies, and has a tendency to form at the austenitic grain boundaries. Its mechanical properties lie between those of fine pearlite and bainite, with a strength of 1400–1750 MPa, hardness of 400–500 HV, and elongation of 5%–10%.

If the composition of the initial austenite is not 0.77% C, its decomposition by lower cooling rates will primarily cause the formation, by nucleation and growth, of proeutectoid structures: ferrite or cementite depending on whether the steel is hypo or hypereutectoid.

In isothermal transformations, proeutectoid steels will also have a C-type transformation curve. If the steel is hypoeutectoid, it will form proeutectoid ferrite between A_3 and A_e. Only when the temperature is lower than A_e, the pearlitic reaction occurs: after an "incubation" period, austenite starts to decompose, first forming proeutectoid ferrite, and after a certain time (another incubation period for the pearlitic reaction), the transformation of austenite into pearlite takes place. For transformation temperatures lower than the one corresponding to the pearlitic nose (vertical maximum), austenite usually transforms directly, without previous formation of proeutectoid ferrite.

An analogous analysis can be made for hypereutectoid steels, with the difference being the formation of proeutectoid cementite preceding the pearlitic transformation.

EXERCISE 8.5

Wires of circular transverse section are produced by drawing them through a die or plate (generally with a conical shape), with diameters of the undrawn wire being 5–20 mm.

The steels used are mild to eutectoid, or even slightly hypereutectoid, while the alloying limits are 1.65% Mn, 0.60% Si, and 0.40% Cu. Mild steels are drawn starting from an as-rolled state or after subcritical annealing intermediate processes. Medium- and high-carbon steels, especially those for high-resistance applications (tires, piano wires, etc.) are drawn starting from a pearlitic structure as fine, homogeneous, and perfect as possible, by isothermal transformation at ~ 550°C or by controlled cooling from the finishing rolling temperature ("Stelmor" process or similar). The final interlamellar spacing of the pearlite is lower than 0.1 μm.

(a) Obtain an expression for plastic deformation as a function of wire diameter, (b) calculate the interlamellar spacing of the initial pearlite for a $\sigma_{max} = 1400\,\text{MPa}$, and (c) calculate the σ_{max} of the drawn steel after a diameter reduction ratio $d_0/d = 10$.

Solution

(a) Expression for plastic deformation

Plastic deformation implies constant volume, thus:

$$A_0 l_0 = A_f l_f$$

$$\frac{l_f}{l_0} = \frac{A_0}{A_f}$$

$$\varepsilon = \ln\frac{l_f}{l_0} = \ln\frac{A_0}{A_f} = \ln\frac{\frac{\pi}{4}d_0^2}{\frac{\pi}{4}d_f^2} = \ln\left(\frac{d_0}{d_f}\right)^2 = 2\ln\frac{d_0}{d_f}$$

(b) Interlamellar spacing

Considering a simplification of Eq. (8.6):

$$\sigma_{max} = 719.26 + 3.54 S_0^{-\frac{1}{2}}$$

$$1400 = 719.26 + 3.54 S_0^{-\frac{1}{2}}$$

$$S_0 = (1/(1400 - 719.26))^2 = 2.16 \times 10^{-6}\,\text{mm} = 0.0022\,\mu\text{m}$$

See Fig. 4.31.

(c) Maximum tensile stress

Considering experimental data from steels (chemical composition, strength, pearlite and/or ferrite fractions, ferritic grain size, interlamellar spacing, and size of pearlite colonies, etc.), Pickering formula (8.6) can be modified as follows:

- For ferritic steels:

$$\sigma_{max}\,(\text{ferritic steels}) = 470 + 150\varepsilon = 470 + 150(2\,\ln 10) = 1160.78\ \text{MPa}$$

- For pearlitic steels:

$$\sigma_{max}\,(\text{pearlitic steels}) = 70 + 1330\varepsilon = 70 + 1330\exp(2\,\ln 10/4) = 4275.83\,\text{MPa}$$

Comment: Highly drawn pearlite is the structural material with the highest tensile strength. The maximum stress values are around $\sim G/5$ (for steels $G \approx 64\,\text{GPa}$), which theoretically are those values withstood by a solid by decohesion or cleavage of dense atomic planes.

8.2.2 Bainitic Transformation

The transformation that takes place at temperatures lower than those of the pearlitic zone is known as *bainitic* reaction: the low diffusivity of C in the austenite at those temperatures will prevent the C atoms from migrating by diffusion and concentrating at certain locations to create cementite nuclei. However, γ at that temperature is very far from the equilibrium condition (temperatures higher than A_e) and thus the $\gamma \rightarrow \alpha$ transformation is promoted by the large undercooling. The interval between temperatures higher than A_3 and the isothermal transformation ones is enough to activate the formation of *ferrite nuclei*, by the simple allotropic transformation of austenite. The initial *leading* constituent of the bainitic transformation is ferrite (unlike pearlitic transformations in which the leading constituent is cementite).

Considering temperatures lower than the pearlitic transformation, such that $T_1 > T_2 > T_3 > T_4$:

- At T_1, the formation of ferrite and the simultaneous rejection of C towards the untransformed γ are accompanied by a fast growth of ferrite that progressively surrounds austenitic zones enriched in C (gamma-stabilizing tendency). These austenitic zones are later transformed into α and granules of cementite preferably formed in the γ-α interface or in surrounding zones (although some may precipitate inside those zones). The microstructure has a morphology known as *granulite* and appears more easily in continuous cooling conditions than isothermal ones, and is common in low-carbon steels. This hard constituent is unfavorable to machining operations.
- At T_2, in general, C expelled from the ferrite diffuses and enriches the untransformed γ; but the growth of α is slower at this temperature and does

not surround the austenitic zones. Since C stabilizes γ, this reaction stops with time, and α needles (laths, plates, etc.) appear over a background of yet-untransformed austenite. This structure is usually known as *Davenport's X constituent*, and is common in alloyed steels. It is rather similar, though with finer α needles (laths, plates), to the Widmanstätten structure. The formation of the X constituent is promoted with large austenitic grain sizes.

- At T_3, α nuclei are also formed, but expelled C diffuses with more difficulty and does not enrich, in general, all the austenitic zones. The formed α grows as needles (laths, plates) rejecting all the excess C from its sides and resulting in the formation of cementite in the C-enriched adjacent γ. The latter, when losing C, is transformed into α, and the previously formed needles grow laterally until reaching relatively important sizes. Although only visible under electronic microscopes, Fe_3C crystals tend to be rather large and almost parallel to the axis of the ferrite needles. The result is the formation of a structure with a morphology similar to feathers (Fig. 8.35), known as *upper bainite*.
- At T_4, diffusion of C is even harder and atoms are unable to move the small distances required for the upper bainite. Carbon diffusion is so slow that α plates thicken rapidly, by repeated precipitation of cementite within them. The carbide dispersion becomes finer, similar to tempered martensite.

  At this temperature, small α lamellae are formed, oversaturated in C, with growth in their own plane, with C precipitation in its interior, over the (1 1 0) planes of α. The constituent formed is known as *lower bainite* (Fig. 8.36).

  There are epitaxial ratios (Kurdjumov expressions) between the austenitic phase and ferrite formed in lower bainite:

$$(1\,1\,1)\gamma/(1\,1\,0)\alpha \text{ and } (1\,1\,0)\gamma/(1\,1\,1)\alpha \tag{8.7}$$

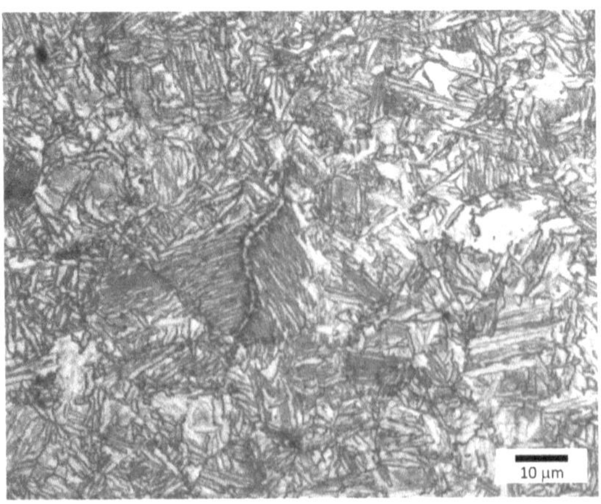

FIG. 8.35 Upper bainite.

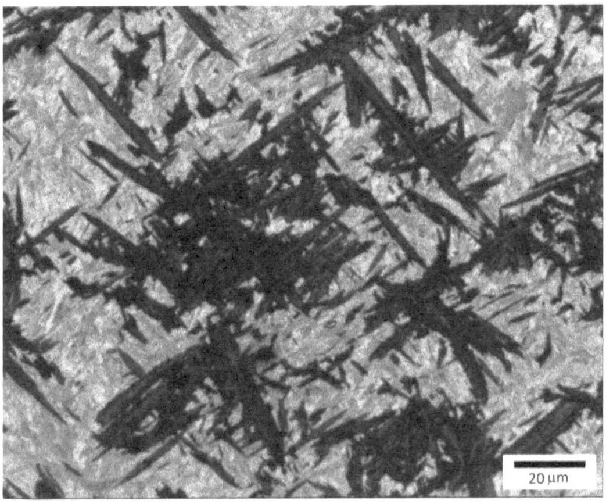

FIG. 8.36 Lower bainite.

The temperature range at which α is the main constituent of the γ transformation, is known as bainitic zone. The 1% and 100% curves of bainite are also of the "C" type (Fig. 8.32).

Furthermore, the hardness of bainite depends on C content of the steel and undercooling: lower bainite is harder than the upper one (bainite hardness is typically 40–60 HRC). Sometimes, particularly in high-carbon steels (0.6 %–0.7% C), lower bainite is more resilient than tempered martensite (for the same hardness value).

The formation process of lower bainite is similar to the transformation of γ into martensite (Section 8.2.3), in which the Kurdjumov relations are also valid, with the difference that, just as it happens in martensite, C is retained in a "distorted α" lattice (*tetragonal*) formed by shearing of γ phase; in the lower bainite there is also shearing, but with simultaneous release of C. A confirmation of shearing taking place in the lower bainite, is the topography that can be observed in a polished surface, which is not present in proeutectoid or pearlitic α.

Like pearlite, bainite is a mixture of ferrite and cementite, but it is microstructurally very different from pearlite and can be characterized by its own C-curve on the time-temperature-transformation (TTT) diagram. In plain-carbon steels, this curve overlaps with the pearlite one: at temperatures close to 500°C, both pearlite and bainite form, one at the expense of the other. However, in some alloyed steels, the two curves are clearly separated (Fig. 8.32) by a *bainitic bay*. The microstructure of bainite mainly depends on the temperature at which it is formed. Steven and Haynes (1956) calculated a B_s temperature at which the bainitic transformation takes place:

$$B_s(°C) = 830 - 270(\%C) - 90(\%Mn) - 37(\%Ni) - 70(\%Cr) - 83(\%Mo) \quad (8.8)$$

valid with a ±25°C approximation, and a 90% confidence level, for steels whose compositions are within the following limits: 0.1%–0.55%C, 0.2%–1.7%Mn, $\leq$5%Ni, $\leq$3.5%Cr, and $\leq$1%Mo.

8.2.3 Martensitic Transformation

Austenite transforms, by nucleation and growth (diffusion controlled), in pearlitic or bainitic structures, depending on the temperature. However, fast cooling (quenching) prevents the formation of pearlite or bainite and will take the austenite to temperatures lower than that of the bainitic reaction, and a transformation without diffusion (at a rate approximately equal to that of sound) occurs resulting in a constituent, with identical C content as austenite, known as *martensite*. Crystallographically, martensite has a tetragonal structure (or bct, body-centered tetragonal) with atoms of C in their interstitial positions (Fig. 8.37).

Fig. 8.38A and B presents the variation of the martensitic crystalline parameters, *c* and *a*, as a function of C percentage, as well as the variation of the *a* parameter of austenite. After a comparison, the following may be concluded:

- Variation of the *c* parameter of martensite (function of the C percentage) presents a slope three times higher than the *a* parameter of austenite (also as a function of the C percentage).

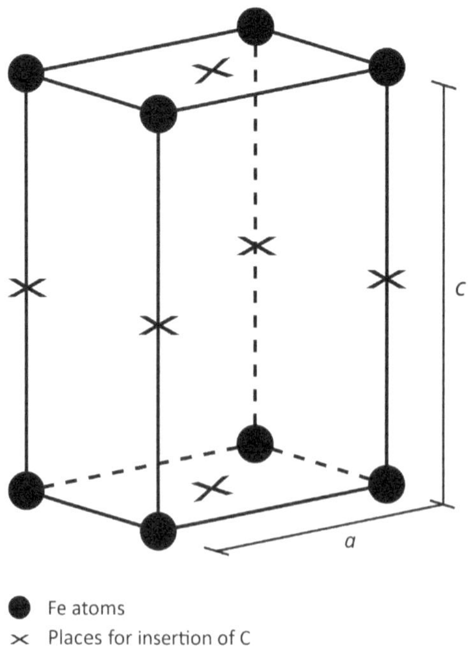

FIG. 8.37 Martensite crystalline structure.

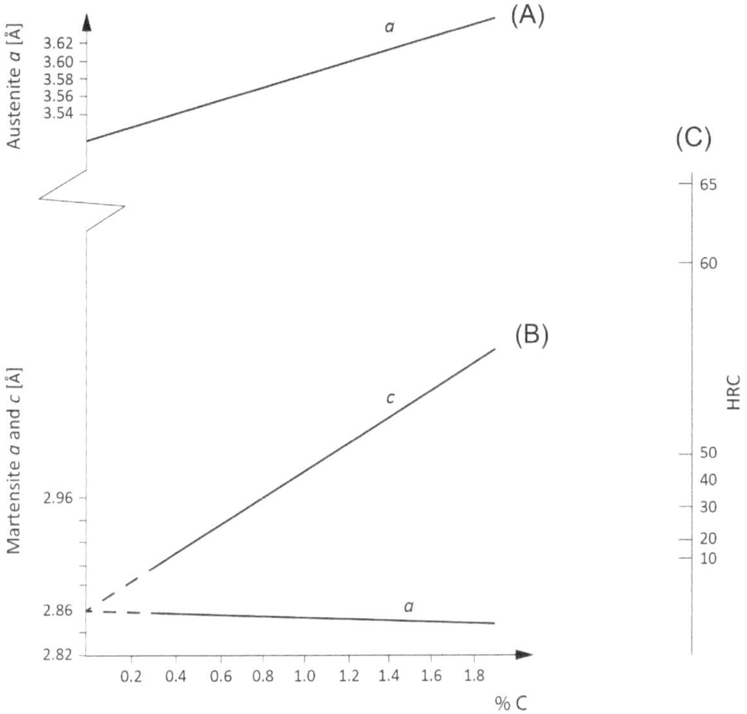

FIG. 8.38 Austenite parameter (A), martensite parameters (B), and martensite hardness (C) as a function of C%.

- Though c increases with C percentage, a constantly decreases in a linear manner:

$$c = 2.861 + 0.116(\%C) \tag{8.9}$$

$$a = 2.861 - 0.013(\%C) \tag{8.10}$$

$$\frac{c}{a} = 1 + 0.045(\%C) \tag{8.11}$$

The extrapolated curves for 0% C, intersect at the same value of 2.86 Å, which is precisely the lattice parameter for pure α-Fe. Thus, martensite has a ferrite lattice but deformed, mainly in the c axis direction, which explains the *metastable* nature of martensite. Furthermore, by increasing the temperature (tempering), elimination of interstitial C atoms can be achieved; these atoms will react with Fe, resulting in decomposition of martensite into the equilibrium constituents, ferrite and cementite.

On the other hand and since the elevated martensite hardness value (Fig. 8.38C) is caused by C (and not by the alloying elements), the mechanisms that promote its mechanical behavior, including the presence of defects in bct crystals (vacancies and dislocations) are researched and debated.

The transformation of austenite into martensite is always accompanied by a *volume increase*. For example, in a binary steel with 0.6% C, the parameter of austenite is 3.588 Å and the ones for martensite are $a = 2.86$ Å and $c = 2.94$ Å; austenite has four atoms of Fe per unit cell, while martensite has only two; N austenite cells transform into $2N$ martensite cells, and after calculating the corresponding volumes for these cells, in a value of $N \cdot 3.588^3 = 46.191N$ which is lower than $2N \cdot 2.86^2 \cdot 2.94 = 48.096N$. The relative volume change is $\sim 4\%$, and is almost independent of C amount.

EXERCISE 8.6

Calculate the volume variation occurring in the transformation of austenite into martensite for the following steels: 0.2%, 0.4%, and 0.6% C.

Solution

(a) Steel with 0.2% C

The lattice parameter of austenite as a function of the carbon amount is:

$$a_\gamma = 3.548 + 0.0448(\%C) = 3.548 + 0.0448(0.2) = 3.557\,\text{Å}$$

The lattice parameters of martensite (Eqs. 8.9 and 8.10) result in:

$$a_M = 2.861 - 0.013(\%C) = 2.861 - 0.013(0.2) = 2.858\,\text{Å}$$

$$c_M = 2.861 + 0.116(\%C) = 2.861 + 0.116(0.2) = 2.884\,\text{Å}$$

And the volume variation can be calculated through:

$$\Delta V_{\gamma \to M} = \frac{2V_M - V_\gamma}{V_\gamma} = \frac{2\left[a_M^2 c_M\right] - a_\gamma^3}{a_\gamma^3} = \frac{2\left[(2.858)^2(2.884)\right] - (3.557)^3}{(3.557)^3}$$

$$= 0.047(4.7\%)$$

(b) Steel with 0.4% C

Following the same procedure as in (a):

$$a_\gamma = 3.548 + 0.0448(0.4) = 3.566\,\text{Å}$$

$$a_M = 2.861 - 0.013(0.4) = 2.856\,\text{Å}$$

$$c_M = 2.861 + 0.116(0.4) = 2.907\,\text{Å}$$

$$\Delta V_{\gamma \to M} = \frac{2\left[(2.856)^2(2.907)\right] - (3.566)^3}{(3.566)^3} = 0.046(4.6\%)$$

(c) Steel with 0.6% C

And once again, the same operations as in (a) and (b):

$$a_\gamma = 3.548 + 0.0448(0.6) = 3.575\,\text{Å}$$

$$a_M = 2.861 - 0.013(0.6) = 2.853\,\text{Å}$$

$$c_M = 2.861 + 0.116(0.6) = 2.931\,\text{Å}$$

$$\Delta V_{\gamma \to M} = \frac{2\left[(2.853)^2(2.931)\right] - (3.575)^3}{(3.575)^3} = 0.044(4.4\%)$$

Additionally, the martensitic transformation takes place *without diffusion* (*without nucleation and growth*):

- Rate of martensite formation by rapidly cooling the austenite is very high (the transformation is almost instantaneous).
- Martensite (when observed through the microscope) has an acicular geometry with a topography produced by *shearing mechanisms* (Fig. 8.39) in certain crystallographic planes of the austenitic lattice,[2] which is called *invariant plane strain transformation*.
- Martensite transforms by shearing over the {1 1 1} planes of austenite in the ⟨2 1 1⟩ direction followed by a second shearing in the ⟨1 0 1⟩ direction of austenite (or the ⟨1 1 1⟩ direction of martensite) and, finally, some slight modifications in the atomic positions (shuffling) to exactly obtain the dimensions of the martensite lattice.

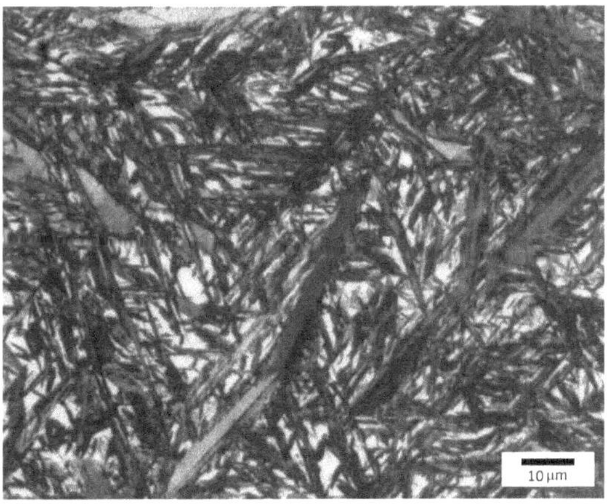

FIG. 8.39 Optical micrograph of martensite.

2. Kurdjumov and Sachs (1930) established the orientation relationship between martensite and austenite in a steel with $<1.4\%$ C and explained the change from one lattice to another by a simple shearing mechanism, demonstrating that the denser {1 1 0} planes of tetragonal martensite are almost parallel to close-packed {1 1 1} planes and the close-packed ⟨1 1 0⟩ directions of austenite are practically parallel to the ⟨1 1 1⟩ ones of martensite. Considering the symmetry of the lattice, this implies the existence of 24 different orientations of martensite with regard to austenite.

- Martensite has the same C content as austenite, and the location of C atoms in the martensite lattice confirms that individual atomic movements are smaller than one interatomic spacing.

Fig. 8.40 shows the mechanism proposed by Bain and Paxton (1966) to explain the martensitic transformation: two fcc cells will produce one elongated or bct (body-centered tetragonal) unit cell. In these conditions, the c/a ratio of the bct cell equals $\sqrt{2}$ (higher than the c/a ratio of martensite, as seen in Eqs. (8.9)–(8.11)), so the transformation to a true martensitic cell occurs by:

(a) contracting the cell in the z axis (where c is measured) or
(b) expanding the cell along the x and y axes,

which clearly explains that a certain amount of martensite can be obtained by simply deforming the austenite.

However, though the Bain mechanism (1966) was historically the first explanation proposed for the formation of martensite, it is not entirely true since it fails to explain the crystallographic relationship between the high-temperature phase

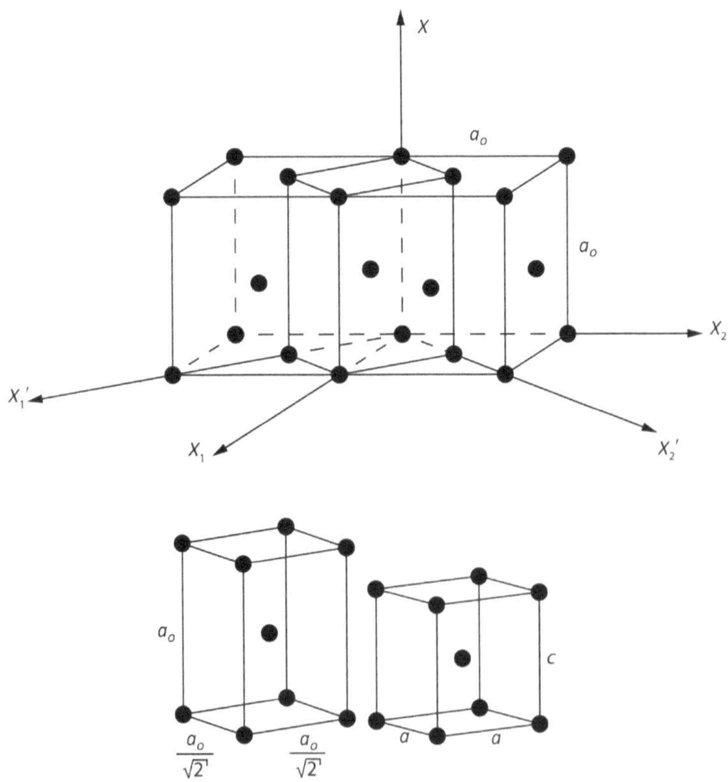

FIG. 8.40 Formation of martensite: Bain mechanism.

and martensite (Fig. 8.40): the real parallelism between the $\langle 1\,1\,0\rangle$ direction of austenite and the $\langle 1\,1\,1\rangle$ direction of martensite, and the $\{1\,1\,0\}$ planes of martensite being almost parallel to the $\{1\,1\,1\}$ planes of austenite (Kurdjumov relationship), and the invariant plane close to the $\{225\}$ one (*habit plane*).

Nevertheless, apart from the crystallographic mechanisms transforming austenite into martensite, a first approximation of the thermodynamic possibility of this reaction, at a temperature T, is determined by a balance between:

- *Decrease of energy* caused by the chemical free energy release associated with the $\gamma \rightarrow \alpha$ transformation.
- *Increase in mechanical energy* stored by distortion of the α-shaped grains that will adopt a tetragonal structure with volume increase and large strains in the surrounding austenite.

Furthermore, if alloying elements are considered, the higher the C and alloying elements' contents (dissolved in the austenite), the larger the required undercooling to produce martensite will be. To form *1% of martensite*, the temperature M_S (*martensite start*) depends, almost exclusively, on chemical composition (Fig. 8.32) and it does not depend, for example, on cooling rate (given that the rate is higher than a certain critical value to prevent the formation of pearlite or bainite).

Nehrenberg (1946), Hollomon and Jaffe (1945), and Steven and Haynes (1956) proposed expressions to determine M_s as a function of chemical composition:

- Nerenberg formula:

$$M_s(°\text{C}) = 500 - 350(\%\text{C}) - 40(\%\text{Mn}) - 22(\%\text{Cr}) - 17(\%\text{Ni}) - 11(\%\text{Si}) - 11(\%\text{Mo}) \quad (8.12)$$

- Hollomon and Jaffe formula:

$$M_s(°\text{C}) = 500 - 350(\%\text{C}) - 40(\%\text{Mn}) - 35(\%\text{V}) - 20(\%\text{Cr}) - 17(\%\text{Ni}) - 10(\%\text{Cu}) - 10(\%\text{Mo}) - 5(\%\text{W}) + 15(\%\text{Co}) + 30(\%\text{Al}) \quad (8.13)$$

- Though the one by Steven and Haynes, corrected by Irvine et al. (1969), is the most precise:

$$M_s(°\text{C}) = 561 - 474(\%\text{C}) - 33(\%\text{Mn}) - 17(\%\text{Ni}) - 17(\%\text{Cr}) - 21(\%\text{Mo}) - 11(\%\text{W}) - 11(\%\text{Si}) \quad (8.14)$$

which determines M_s with a $\pm 2°\text{C}$ margin of error when the steel composition is: 0.1 %–0.55 % C, $< 5\%$Ni, 0.1 %–0.35 % Si, $< 3.5\%$Cr, 0.3 %–1.7 % Mn, and $< 1\%$Mo.

Determining M_s is very important in industrial operations, as it gives information on the behavior of the steel towards deformation and cracking during quenching (the lower the M_s, the higher the number of cracks will be, which is particularly important when quenching steels with $M_s < 200°C$).

An expression to calculate the volume fraction of retained (or untransformed) austenite, V_a, was formulated by Koistinen and Marburger (1959):

$$V_a = \exp|-0.011(M_s - T_m)| \tag{8.15}$$

where T_m is the average temperature of the quenching medium (water, oil, air, etc.).

EXERCISE 8.7

Determine which of these two steels require more precise industrial temperature conditions during quenching:

Material	%C	%Mn	%Ni	%Cr	%Mo
Steel I	0.30	0.60		1.00	0.20
Steel II	0.35	0.80	2.20	0.90	0.30

Solution

(a) Steel I

Estimating the austenization temperature according to Andrews (Eq. 8.3):

$$A_{3C} = 912 - 203\sqrt{\%C} - 30(\%Mn) - 15.2(\%Ni) - 11(\%Cr) - 20(\%Cu) + 44.7(\%Si) + 31.5(\%Mo) + 13.1(\%W) + 104(\%V) + 120(\%As) + 400(\%Ti) + 400(\%Al) + 700(\%P)$$

$$= 912 - 203\sqrt{0.30} - 30(0.60) - 11(1) + 31.5(0.20) = 778.11°C$$

According to the Steven and Irvine equation (8.14):

$$M_s(°C) = 561 - 474(\%C) - 33(\%Mn) - 17(\%Ni) - 17(\%Cr) - 21(\%Mo) - 11(\%W) - 11(\%Si)$$

$$= 561 - 474(0.30) - 33(0.60) - 17(1) - 21(0.20) = 377.8°C$$

And according to the Koistinen and Marburger equation (8.15), the amount of retained austenite is:

$$V_a = \exp|-0.011(M_s - T_m)| = \exp|-0.011(377.8 - 25)| = 0.021 = 2.1\%$$

(b) Steel II

Following the same procedure as in (a):

$$A_{3C} = 912 - 203\sqrt{0.35} - 30(0.80) - 15.2(2.20) - 11(0.90) + 31.5(0.30) = 734°C$$

$$M_s(°C) = 561 - 474(0.35) - 33(0.80) - 17(2.2) - 17(1) - 21(0.30) = 308°C$$

$$V_a = \exp|-0.011(M_s - T_m)| = \exp|-0.011(308 - 25)| = 0.044 = 4.4\%$$

After analyzing the results, Steel II has the lowest austenization temperature (A_{3C}) which can increase the risk of overheating during austenization, and decrease the mechanical properties due to austenitic grain growth, larger grain sizes, thus requiring a strict control of austenization temperature. Also, this steel has the lowest martensitic start temperature (M_s), requiring the selection of adequate cooling agents to prevent the formation of cracks. The amount of retained austenite in this steel is higher than in Steel I.

8.2.4 TTT (Time-Temperature-Transformation) Curves

Pearlitic, bainitic, and martensitic zones are strongly altered by intrinsic factors such as carbon content and alloying elements. In general, elements that form solid solutions with austenite, either of the substitutional (Mn, Ni, Cr, etc.) or interstitial (B, N, etc.) types, delay the isothermal transformations (pearlitic and bainitic), as these elements have a barrier or obstruction effect on the diffusion of carbon; and thus *cementite* (in the pearlitic zone) or *ferrite* (in the bainitic one) will require longer times for nucleation and growth.

Gamma-forming elements (particularly Mn and Ni) reduce both, the austenitic transformation temperatures (A_3 and A_e) and the pearlitic transformation temperatures. Furthermore, Mn and Ni, delay the pearlitic and bainitic "noses" of the TTT curve.

On the other hand, carbide-forming elements (Cr, Mo, V, Nb, Ti, etc.) have a stronger effect on delaying the pearlitic transformation than the bainitic one since they are alpha-forming elements and tend to increase A_3 and A_e, which in turn increase the diffusion in the reactions accelerating them; however, this effect is counteracted by their affinity for carbon and the formation of carbides. Nucleation of cementite is delayed and the "incubation" period is increased. Furthermore, the pearlitic curve is displaced upwards and to the right, narrowing the transformation zone.

The influence of carbon is, in theory, similar to any other element: for *hypoeutectic steels*, the higher the carbon content, the larger will be the time required for the pearlitic and bainitic transformations.

In the case of 0.0005 %–0.003 % B in interstitial solid solution in austenite, the proeutectoid transformation and the pearlitic zones are delayed in a considerable manner, favoring the bainitic transformation.

For low-alloyed carbon steels, the minimum average period, in seconds, for the beginning of the pearlitic (m) and bainitic (n) transformations are, respectively:

$$m=(0.254\%\,\mathrm{C})(1+4.1\%\,\mathrm{Mn})(1+2.83\%\,\mathrm{P})(1-0.62\%\,\mathrm{S})(1+0.64\%\,\mathrm{Si})$$
$$(1+2.33\%\,\mathrm{Cr})(1+0.52\%\,\mathrm{Ni})(1+3.14\%\,\mathrm{Mo})(1+0.27\%\,\mathrm{Cu}) \tag{8.16}$$

$$n = (0.272\%C)(1+4.1\%Mn)(1+2.83\%P)(1-0.62\%S)(1+0.64\%Si)(1+1.16\%Cr)(1+0.52\%Ni)(1+0.27\%Cu) \quad (8.17)$$

EXERCISE 8.8

Analyze which alloying elements are more important in delaying the pearlitic and bainitic transformation periods for the following steel: 0.55% C, 0.7% Mn, 1.20% Si, 1.80% Ni, 0.7% Cr, 0.020% P, and 0.010% S.

Solution

Using Eqs. (8.16), (8.17):

$$m = (0.254\%C)(1+4.1\%Mn)(1+2.83\%P)(1-0.62\%S)(1+0.64\%Si)(1+2.33\%Cr)(1+0.52\%Ni)(1+3.14\%Mo)(1+0.27\%Cu) = (0.14)(3.87)(1.06)(0.99)(1.77)(2.63)(1.94)(1)(1) = 5.13\,s$$

$$n = (0.272\%C)(1+4.1\%Mn)(1+2.83\%P)(1-0.62\%S)(1+0.64\%Si)(1+1.16\%Cr)(1+0.52\%Ni)(1+0.27\%Cu) = (0.15)(3.87)(1.06)(0.99)(1.77)(1.81)(1.94)(1) = 3.79\,s$$

Though the effect of P and S is very strong in the formula, most steels have very low amounts of both. On the other hand, Mn, Cr, and Mo are very common alloying elements. For this steel, the amounts of Mn and Cr considerably modify the transformation periods.

Since *eutectoid steels* do not have proeutectoid constituents, the pearlitic nose is farther from the temperature axis compared to *hypereutectoid ones*, as the already formed proeutectoid cementite accelerates, by heterogeneous nucleation, the incubation of cementite nuclei promoting the pearlitic transformation.

In low-alloyed steels, the pearlitic and bainitic zones are overlapped and the same occurs with the bainitic and martensitic domains for highly alloyed. This is caused by the formation of a certain amount of martensite followed by a bainitic reaction in the residual austenite.[3]

The TTT curve of a steel, obtained with different isothermal treatments, is very valuable in industrial applications, even though most practical heat treatments are of the continuous cooling type. Superimposing the TTT curve with another curve that indicates the speed at which austenite will be cooled, is used to determine (with high accuracy) the microstructures that will be formed:

- If cooling rate is very low, pearlite.
- If cooling rate is high, bainite.
- If rate is such that the cooling curve does not cross the TTT nose at any point, martensite (*quenched steel*).

3. Deformations (strain energy) produced by the first martensite laths slightly accelerate the bainite formation mechanism.

Critical quenching rate is the minimum cooling speed required to avoid the transformation of austenite into other constituents (pearlite, bainite) before reaching the temperature M_s (Fig. 8.32). For low-alloyed Ni-Cr-Mo steels, the critical quenching rate (or continuous cooling rate, *CCR*) can be obtained by:

$$\log \mathrm{CCR} = 4.3 - 3.27(\%\mathrm{C}) - \frac{\%\mathrm{Mn} + \%\mathrm{Ni} + \%\mathrm{Cr} + \%\mathrm{Mo}}{1.6} \quad (8.18)$$

If two cylindrical bars of the same steel but with different diameters ($D_1 > D_2$) are compared when austenized and oil quenched, then the following differences may be observed:

- The larger the diameter, the lower the cooling speed and the smaller the quenched depth will be.
- The bar with D_2 will have a higher amount of retained austenite and a better resistance to cracking, because of the more uniform martensite amount in the cross-section.
- The use of a more adequate quenching medium will produce higher cooling rates and larger quenched depths (oil with a better heat removal capacity).
- The use of a more severe quenching agent increases the amount of retained austenite and decreases the susceptibility of the steel to crack.

EXERCISE 8.9

A steel shaft with a 40 mm diameter will be quenched in oil to form 99% martensite in its core. The following table shows the correlation between the continuous cooling rate (***CCR***) at the core of cylindrical bars and their diameter (***D***), after austenization and quenching in oil:

D (mm)	CCR (°C s^{-1})
500	0.17
100	2.50
25	50.00
5	667.00

Three steels will be analyzed, in order to determine their viability and their quenching abilities:

Material	% C	% Mn	% Ni	% Cr	% Mo
Steel I	0.30	0.60		1.00	0.20
Steel II	0.40	0.80	1.00	0.90	0.30
Steel III	0.35	0.80	2.20	0.90	0.30

(a) Obtain the relationship between the diameter of the cylindrical bars and the cooling rate of their core when oil quenched.
(b) Determine which steel would be ideal for this shaft.
(c) Determine which steel would suffer a higher risk of cracking during quenching and which would have the highest amount of retained austenite.

Solution

(a) Relationship between diameter and cooling rate

Sketching diameter and continuous cooling rate data (Fig. 8.41), a linear logarithmic relationship can be established:

$$\log CCR = -1.8271 \log D + 4.142$$

and thus:

$$CCR = 10^{4.142} \cdot D^{-1.8271}$$

For a cylindrical bar with 40 mm in diameter:

$$CCR = 10^{4.142} \cdot 40^{-1.8271} = 16.4°\text{C}\,\text{s}^{-1}$$

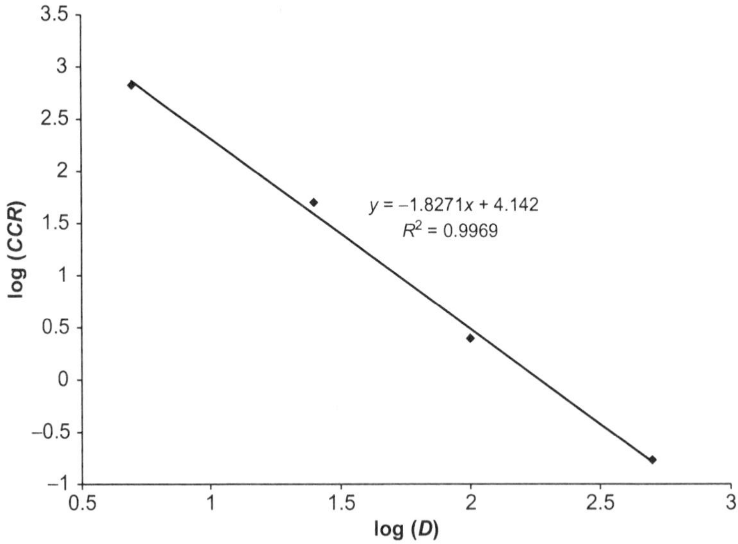

FIG. 8.41 Logarithm of diameter vs. logarithm of continuous cooling rate for steel bars.

This means that every steel bar with that diameter and $CCR > 16.4°\text{C}\,\text{s}^{-1}$ will not fully quench the core.

(b) Ideal alloy for the shaft

Using Eq. (8.18):

$$\log CCR_{\text{SteelI}} = 4.3 - 3.27 \cdot \text{C} - \frac{\text{Mn} + \text{Ni} + \text{Cr} + \text{Mo}}{1.6} = 4.3 - 0.981 - 1.125 = 2.194$$

$$CCR_{\text{SteelI}} = 10^{2.194} = 156.31°\text{C}\,\text{s}^{-1}$$

$$\log CCR_{\text{Steel II}} = 4.3 - 3.27(0.40) - \frac{0.80 + 1.0 + 0.90 + 0.30}{1.6}$$
$$= 4.3 - 1.308 - 1.875 = 1.117$$
$$CCR_{\text{Steel II}} = 10^{1.117} = 13.09°\text{C s}^{-1}$$
$$\log CCR_{\text{Steel III}} = 4.3 - 3.27(0.35) - \frac{0.80 + 2.20 + 0.90 + 0.30}{1.6}$$
$$= 4.3 - 1.1445 - 2.625 = 0.5305$$
$$CCR_{\text{Steel III}} = 10^{0.5305} = 3.39°\text{C s}^{-1}$$

This means that steels II and III can be used in a 40 mm shaft, since steel I has a *CCR* value higher than the critical one. As steel II has a lower amount of alloying elements compared to steel III, it would be a better option due to its lower price.

(c) Cracking and retained austenite

Using Eq. (8.14):

$$M_{s-\text{steel I}} = 561 - 474(\%\text{C}) - 33(\%\text{Mn}) - 17(\%\text{Ni}) - 17(\%\text{Cr}) - 21(\%\text{Mo}) - 11(\%\text{W}) - 11(\%\text{Si})$$
$$= 561 - 474(0.30) - 33(0.60) - 17(1) - 21(0.20) = 377.8°\text{C}$$
$$M_{s-\text{steel II}} = 561 - 474(0.40) - 33(0.80) - 17(1) - 17(0.90) - 21(0.30) = 306.4°\text{C}$$
$$M_{s-\text{steel III}} = 561 - 474(0.35) - 33(0.80) - 17(1) - 17(0.90) - 21(0.30) = 330.1°\text{C}$$

All three steels have $M_s > 300°\text{C}$, this means that all of them would quench adequately and with no risk of cracking.

To calculate the retained austenite, Eq. (8.15) is used:

$$V_{a-\text{steel I}} = \exp|-0.011(M_s - T_m)| = \exp|-0.011(377.8 - 25)| = 0.0206 = 2.06\%$$
$$V_{a-\text{steel II}} = \exp|-0.011(306.4 - 25)| = 0.0453 = 4.53\%$$
$$V_{a-\text{steel III}} = \exp|-0.011(330.1 - 25)| = 0.0349 = 3.49\%$$

The TTT curve depends mainly on the chemical composition of the steel, but there are *extrinsic* factors such as austenitic grain size and austenization temperature that also influence it, though to a lesser degree. For equal chemical compositions, the higher the austenitic grain size, the slower the nucleation and growth transformations of austenite will be. Cementite and ferrite nuclei are usually formed at grain boundaries (heterogeneous nucleation); therefore, the smaller the austenitic grain size, the higher the number of grain boundaries in the material, and quicker the transformations into pearlite and bainite.

On the other hand, with fine austenitic grain sizes, temperature M_s usually diminishes: grain boundaries act as obstacles to the formation of martensite as the mechanical energy stored by distortion (created by the transformation of α-shaped grains into bct) is increased, added to volume dilation and large strains in the surrounding austenite (Section 8.2.3). Furthermore, austenization temperature has the following effects:

- If the steel does not present grain growth inhibitors (Section 8.3.3), increase in temperature is accompanied by an increase in grain size, with the consequences just mentioned.
- If the austenization temperature is too high, austenite, when homogenizing, is more stable and the pearlitic and bainitic transformations are delayed: a nonhomogeneous austenite has a larger number of locations susceptible to suffering premature pearlitic and bainitic reactions.

Also, an increase in austenization temperature is translated, in general, to a decrease in M_s, which is a nondesirable effect. This is clearly evident in steels with high carbon content and highly alloyed, and seems to be caused by:

- The progressive dissolution of carbides increasing the carbon content and the amount of alloying elements solubilized in the austenite. Furthermore, austenite is more stable because of its higher homogeneity and requires a higher undercooling for the $\gamma \rightarrow \alpha$ transformation to take place, resulting in a decrease in M_s.
- The drag effect of alloying elements, which maintain a fine austenitic grain size.

When analyzing residual austenite, the increase in austenization temperature (and the related austenite stability) produces, almost always, an increase in the amount of untransformed austenite when quenching, not only for the lower M_s, but also because transforming austenite becomes more difficult. This is particularly noticeable in steels with high carbon content and high amount of alloying elements.

However, there are cases where increasing the austenization temperature (very large austenitic grains) affect M_s by moving it upwards.

EXERCISE 8.10

A steel used for springs (0.55% C, 1.20% Si, 1.80% Ni, and 0.70% Cr) is austenized at 880°C and heat treated at 650°C during nine different holding periods, and then quenched in cold water. If the Rockwell C hardness values obtained are the following:

Sample	1	2	3	4	5	6	7	8	9
Holding time (s)	10	20	40	50	60	70	80	100	1000
HRC	60	59.5	58	56	48.5	32.5	31	28.5	28

Calculate: (a) the Avrami equation and the B and k coefficients, (b) the tensile stress of pearlite, and (c) the interlamellar spacing.

Solution

(a) Avrami equation, coefficients B and k

For all transformations of the nucleation and growth type (solidification, changes in crystalline structure, precipitation, solid-state transformations with the exception of the martensitic transformation), the generalized Avrami equation can be used:

$$X = 1 - \exp\left(-B \cdot t^k\right)$$

where X is the transformed fraction, t the transformation time, B a constant that depends on temperature, and k a coefficient that depends on the growth of the phase: unidimensional ($k=2$), bidimensional ($k=3$), or tridimensional ($k=4$). The same transformed fraction can be calculated through:

$$X = \frac{H_0 - H_t}{H_0 - H_p}$$

where H_0 is the hardness of the steel in its quenched state, H_t the hardness of the steel isothermally treated after a time t, and H_p the hardness of the steel after the pearlitic transformation.

If hardness and the corresponding time for each sample (in logarithmic scale) are sketched (Fig. 8.42), $H_0 = 60\text{HRC}$ (short cooling times) and $H_p = 28\text{HRC}$ (long cooling times). These two values may be used to determine B and k through:

$$\log\left(\ln\left(\frac{1}{1-X}\right)\right) = \log B + k \log t$$

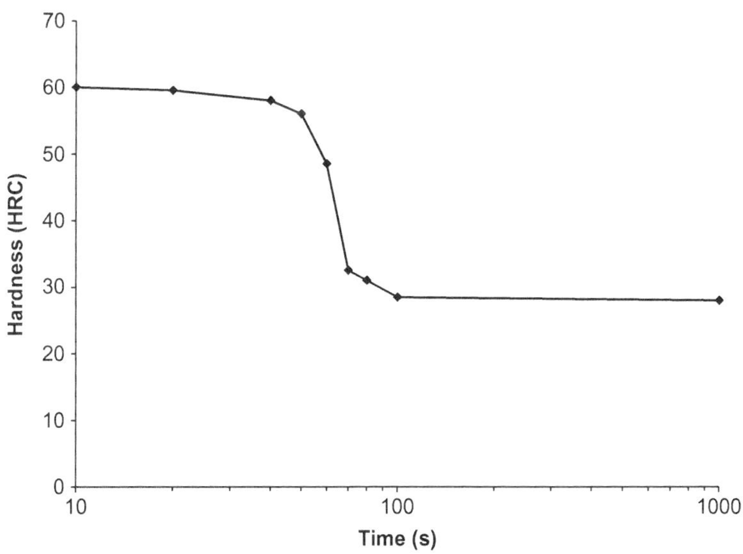

FIG. 8.42 Hardness as a function of time in logarithmic scale.

The following table shows the values for the relevant terms:

H_t	$X(\%)$	$\log(\ln 1/(1-X))$	$\log t$
60	0	–	1
59.5	0.015625	−1.80276475	1.30103
58	0.0625	−1.19018099	1.60205999
56	0.125	−0.87441662	1.69897
48.5	0.359375	−0.35133656	1.77815125
32.5	0.859375	0.29262341	1.84509804
31	0.90625	0.37422094	1.90308999
28.5	0.984375	0.61897671	2
28	1	–	3

When plotting the last two columns (Fig. 8.43), the fitting of a straight line results in:

$$y = 3.7817x - 6.9713$$

or in other words

$$\log B = -6.9713$$

$$B = 1.07 \times 10^{-7}$$

and

$$k = 3.7817$$

Finally, the Avrami law is:

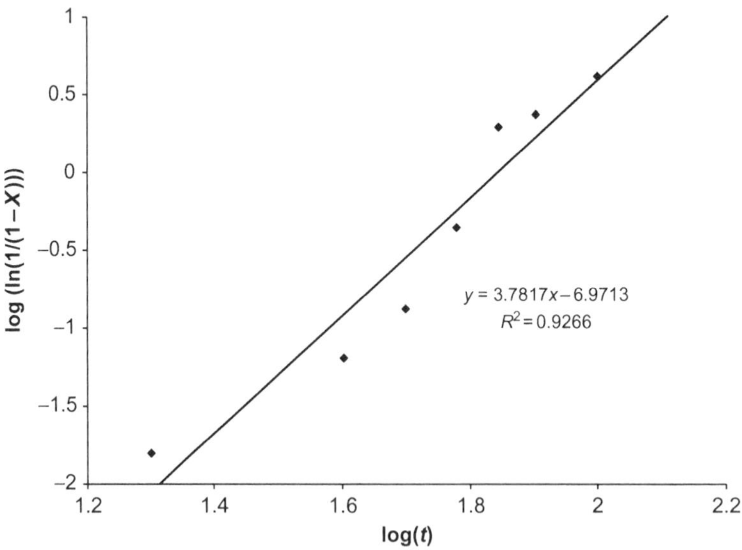

FIG. 8.43 Linear regression to estimate values of the Avrami expression.

$$X = 1 - \exp\left(-1.07 \times 10^{-7} \cdot t^{3.7817}\right)$$

(b) Tensile stresses

A good approximation can be made through the minimum hardness value:

$$\sigma_{max} = \frac{HRB}{3} \approx \frac{10HRC}{3} \approx \frac{10(28)}{3} \approx 93.3\frac{kg}{mm^2} = 915.6\text{MPa}$$

(c) Interlamellar spacing

The fraction of ferrite must be obtained (lever rule on the phase diagram):

$$f_\alpha(0.55\%C) = \frac{0.77 - 0.55}{0.77 - 0.022} = 0.2941 = 29.41\%$$

Using Eq. (8.6) and considering 0.006% N, 1.2% Si, and a "normal" grain size of ASTM 7 (or 0.0276 mm):

$$\sigma_{max} = f_\alpha^{\frac{1}{3}}\left[246.43 + 1142.81\sqrt{N} + 18.17d^{-\frac{1}{2}}\right] + \left(1 - f_\alpha^{\frac{1}{3}}\right)\left[719.26 + 3.54S_0^{-\frac{1}{2}}\right] + 97.03\text{Si}$$

$$915.6 = 0.2941^{\frac{1}{3}}\left[246.43 + 1142.81\sqrt{0.00006} + 18.17(0.0276)^{-\frac{1}{2}}\right] + \left(1 - 0.2941^{\frac{1}{3}}\right)\left[719.26 + 3.54S_0^{-\frac{1}{2}}\right] + 97.03 \quad (1.2)$$

$$S_0 = 3.96 \times 10^{-6}\text{mm} = 3.96 \times 10^{-3}\mu\text{m} = 0.004\mu\text{m}$$

which is a very fine pearlitic microstructure.

EXERCISE 8.11

If the Continuous Cooling Transformation diagram of SAE 1038 steel (Fig. 8.44) is considered: (a) analyze the percentage of ferrite and pearlite of bars with different diameters (different cooling rates) and (b) calculate the critical oil-quenching rate (Exercise 8.9).

Solution

(a) Bars with different diameters (and different cooling rates)

When cooling rate is increased (Fig. 8.18), A_3 is displaced towards lower temperatures. Thus, cylinders cooled in air would have the following properties according to the continuous cooling diagram:

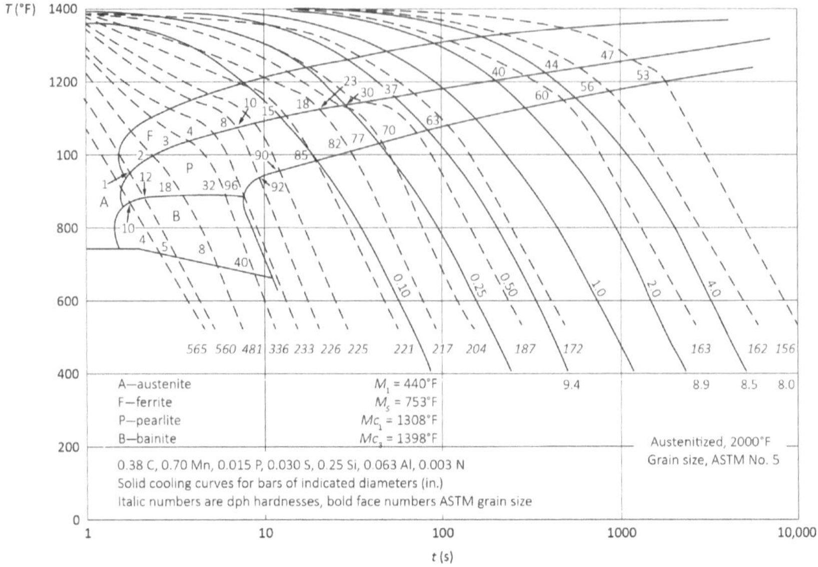

FIG. 8.44 Continuous cooling transformation diagram for SAE 1038 steel (Vander Voort, 1991).

Diameter (in.)	Constituents (approx. amounts)	Hardness (HV)
0.1	15% ferrite–85% pearlite	221–217
0.25	30% ferrite–70% pearlite	204–187
0.50	37% ferrite–63% pearlite	187–172
1.0	37%–40% ferrite–63%–60% pearlite	172–163
2.0	40% ferrite–60% pearlite	163
4.0	44% ferrite–56% pearlite	162

The amount of constituents in thick bars (4 in.) is close to those deduced from equilibrium cooling in the Fe-Fe_3C metastable phase diagram:

$$\%\text{pearlite} \approx \frac{0.38}{0.77} \approx 50\%$$

For thinner bars (faster cooling rates) the proportion of pearlite is higher than 85% (when $D_1 = 0.1$ in.), the amount of cementite in pearlite will be low (diluted pearlite) and the amount of carbon in pearlite would be:

$$(0.38)(1) = (0.02)(f_\alpha) + (C_{eut})(f_{\text{pearlite}})$$
$$(0.38)(1) = (0.02)(0.15) + (C_{eut})(0.85)$$
$$C_{eut} = 0.44\%\text{C}$$

The dilution coefficient is known as:

$$D = \frac{0.77}{C_{eut}}$$

which in this case is $D \approx 1.74$.

Increasing the amount of pearlite would also increase the tensile stress of the steel (Exercise 8.4) for equal size and interlamellar spacing.

(b) Oil-quenching rate

The microstructure with this coolant would be 10% pearlite and 90% martensite. The continuous cooling rate (Exercise 8.9) is:

$$\log CCR = 4.3 - 3.27 \cdot \mathrm{C} - \frac{\mathrm{Mn + Ni + Cr + Mo}}{1.6} = 4.3 - 3.27(0.38) - \frac{0.7}{1.6} = 2.62$$

$$CCR = 416.77\,°\mathrm{C\,s^{-1}}$$

and since

$$CCR = \frac{1\times10^4}{D^{1.8}}$$

this rate would be critical for diameters of:

$$D = \left(\frac{1\times10^4}{CCR}\right)^{1/1.8} = \left(\frac{1\times10^4}{416.77}\right)^{1/1.8} = 5.84\,\mathrm{mm} \approx 6\,\mathrm{mm}$$

EXERCISE 8.12

Obtain the critical diameter for a SAE 5140 steel part.

Solution

From Fig. 8.45, the continuous cooling rate can be calculated as:

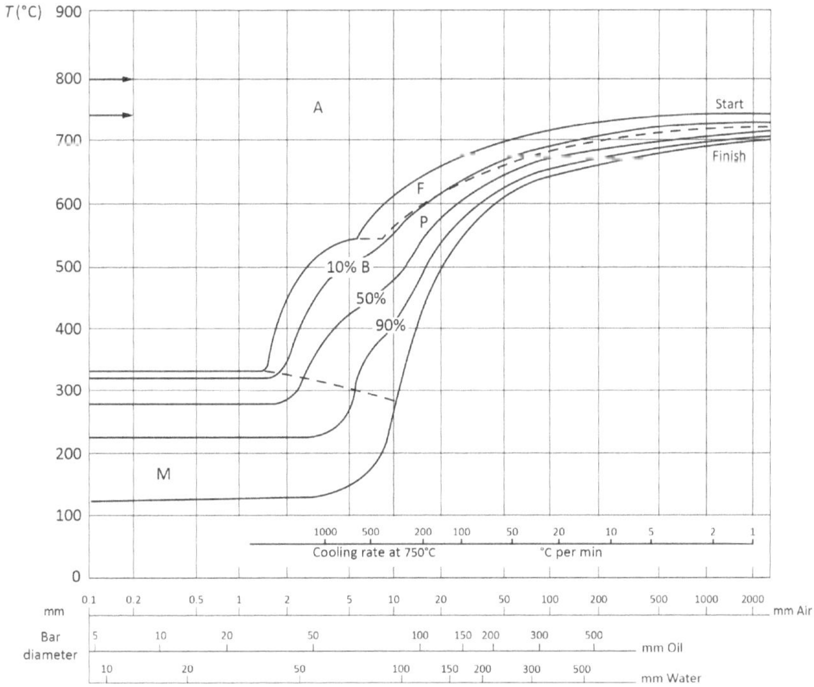

FIG. 8.45 Continuous cooling transformation diagram for SAE 5140 steel (Vander Voort, 1991).

$$\log CCR = 4.3 - 3.27 \cdot C - \frac{Mn + Ni + Cr + Mo}{1.6} = 4.3 - 3.27(0.39) - \frac{1.7}{1.6} = 1.96$$

$$CCR = 91.66°Cs^{-1}$$

And the critical diameter is:

$$D = \left(\frac{1 \times 10^4}{CCR}\right)^{1/1.8} = \left(\frac{1 \times 10^4}{91.66}\right)^{1/1.8} = 13.56\text{mm}$$

this diameter or lower would ensure a 99% martensitic structure.

Comment: a larger diameter, for example 50–100 mm would imply 99% bainitic microstructures with cooling rates of $8°Cs^{-1}$ (moderate value) reducing the risk of cracks during quenching.

8.3 HEAT TREATMENTS

The term *heat treatment* is given to the group of controlled heating and cooling processes to which a *solid* metal is subjected in order to modify its microstructure (and with it, its properties), without changing the chemical composition of the alloy.

8.3.1 Austenite Formation by Heating at $T > A_3$

A steel, irrespective of its initial structure (tempered martensite, bainite, pearlite, ferrite-pearlite, or cementite-pearlite), suffers, at temperatures above A_3, the following transformation of the initial structure:

$$\alpha + Fe_3C \rightarrow \gamma \tag{8.19}$$

In steels[4] the reaction is produced by a nucleation mechanism in the ferrite-cementite interface: at that location the formation of austenite begins with the allotropic transformation of the ferrite grains. The adjacent cementite easily disaggregates and, as a consequence, the ferrite in the interface, already transformed into austenite, absorbs the released C. The transformation of ferrite into austenite occurs faster than the decomposition of carbides.

In order for the reaction to be complete, a holding time at the austenization temperature T_1 (higher than A_{3c}) is necessary. The driving force of the reaction is proportional to the thermal interval $T_1 - A_{3c}$. This holding time is necessary, not only to transform all α into γ and disaggregate the cementite, but also to achieve, by volume diffusion of C, homogeneous austenite.

For the same temperature T_1, the larger the free ferrite/cementite interface, the shorter the required times will be. Thus, dissolution of lamellar cementite is

4. At temperatures close to A_c, the microstructure is always formed by ferrite and cementite.

faster than for globular cementite; and, similar to the morphology (lamellar or globular), austenization will require lesser times for smaller cementite sizes.

The austenite grains that are formed at T_1, by nucleation and growth, are able, with time, to be in contact with each other, and subsequently, a phenomenon of secondary growth can occur if some of them coalesce at the expense of their neighbors, which decreases the number of grain boundaries and, consequently, the energy of the system. The higher the temperature or the larger the holding period, the coarser the austenitic grain size will be; even though large holding periods are equivalent to only small increases in temperature since the relationship between them is of the semilogarithmic type (Arrhenius-type law). The higher the austenization temperatures, the lower the time to reach austenization.

Though the increase in temperature is favorable for reducing the time of both carbide redissolution and homogenization of carbon in the formed austenite, it is not desirable for temperature to be excessively higher than A_3, neither is considerably increasing the holding time, in order to avoid excessive grain growth; this is known as *overheating* and reduces the mechanical properties of the steel (Section 8.1.7). Furthermore, temperatures higher than those of overheating promote, besides grain growth, the formation of intergranular oxides in the steel, which can suffer intergranular melting when reaching temperatures close to those of *solvus*. In both cases, the steel has *burned*.

Steel overheating can be corrected by subsequent heat treatments that lead to a new austenitic grain refinement (by cooling and fast austenization at moderate temperatures), though this procedure is expensive. On the contrary, the same cannot be done for burnt steels, as regenerating the structure is impossible and the steel becomes useless.

Hypereutectoid steels are more susceptible to overheating and burning: for the same temperature increase above A_{cm} (instead of A_3), larger grain sizes are produced in hypereutectoid than in hypoeutectoid steels. It is not common to austenize hypereutectoid steels above A_{cm} and heating is usually limited to 40–60°C above A_e; hence inside the austenite mass, undissolved carbides will remain, but the risk of an excessive austenitic grain size is avoided.

If the steel has fine carbide particles or other compounds that did not dissolve during austenitic grain growth, they act as barriers to the movement of grain boundaries, hindering grain growth and avoiding risks of overheating or burning.

To inhibit austenitic grain growth of medium- or low-carbon content steels, the following elements are usually considered: Nb ($\sim$0.03%), Ti ($\sim$0.1%), V ($\sim$0.1%), and Al ($\sim$0.08%); as these elements precipitate in the shape of carbides or nitrides (e.g., Al precipitates as AlN) during cooling, following a nucleation and growth process. Afterwards, when the steel is heated again to austenization temperatures, precipitates (as long as they do not redissolve) act as effective inhibitors to grain growth by impeding the displacement of grain boundaries (*fine-grained steels*). Their inhibiting effect begins to disappear, by

TABLE 8.1 Solubility Limits of Carbides and Nitrides

Compound	Solubility Limit
AlN	$\log[\%Al][\%N] = (-7400/T) + 1.95$
MnS	$K_S = [\%Mn][\%S]\, f_s^{Mn}$
	$\log K_S = (-9020/T) + 2.929$
	$\log f_s^{Mn} = (-215/T) + 0.097(\%Mn)$
NbC	$\log[\%Nb][\%C]^{0.87} = (-7530/T) + 3.11$
NbN	$\log[\%Nb][\%N] = (-10,230) + 4.04$
Nb[CN]	$\log[\%Nb][\%C]^{0.24}[\%N]^{0.65} = (-10,400/T) + 4.09$
V_4C_3	$\log[\%V]^{4/3}[\%C] = (-10,800/T) + 7.06$
VN^4	$\log[\%V][\%N] = (-7733/T) + 2.99$
TiC	$\log[\%Ti][\%C] = (-7000/T) + 2.75$
ZrC	$\log[\%Zr][\%C] = (-8760/T) + 4.93$
TiN	$\log[\%Ti][\%N] = (-14,400/T) + 5$

redissolution of precipitates, at temperatures higher than 950°C. Some of them, for example TiN and NbC, maintain the fine austenitic grain size up to higher temperatures. Table 8.1 indicates the solubility limits of carbides and nitrides as a function of temperature.

On the other hand, and since the austenitic grain growth cannot start until ferrite has fully transformed into austenite, the alpha-forming elements are also inhibitors to grain growth. When a steel has alloying elements that are simultaneously alpha-forming and carbide-forming, as in the case of tool steels (e.g., HSS with 0.8% C, 6% W, 5% Mo, 1% V, and 4% Cr), austenitic grains are less prone to growth. Furthermore, it is important to point out that the elements that raise the *solidus* temperatures in the Fe-C diagram for high carbon contents (e.g., Cr, Mo, Co, etc.) also diminish, for the same reason, the risks of overheating.

EXERCISE 8.13

For a 0.15% C structural steel, microalloyed with 0.04% Nb, 0.1% V, and 0.02% Ti, calculate: (a) the volume fraction (f_V) of carbides of V, Nb, and Ti, and (b) the maximum precipitation temperature of the steel, during cooling.

Data: $\rho_{steel} = 7.8\,g/cm^3$, $\rho_C = 12\,g/mol$, $\rho_V = 51\,g/mol$, $\rho_{Nb} = 93\,g/mol$, $\rho_{Ti} = 48\,g/mol$, $\rho_{V_4C_3} = 5.8\,g/cm^3$, $\rho_{NbC} = 7.8\,g/cm^3$, and $\rho_{TiC} = 4.94\,g/cm^3$

Solution

(a) Volume fraction of carbides

Vanadium would form carbides with the chemical formula V_4C_3 and the following molecular weight:

$$M_{V_4C_3} = (4)\rho_V + (3)\rho_C = (4)(51) + (3)(12) = 240\,g/mol$$

The amount of carbon absorbed by 0.1% V in the steel would be:

$$\frac{(3)(12)\,gC}{(4)(51)\,gV} \cdot \frac{0.1\,gV}{100\,g\,steel} = \frac{0.017\,gC}{100\,g\,steel} \approx \frac{0.02\,gC}{100\,g\,steel}$$

That is 0.02% C would form precipitates and the remaining carbon (0.13% C) would be in solid solution in the steel. In 100 g of steel there are $(0.02 + 0.1)\,g$ of V_4C_3 or in other words:

$$\frac{0.12\,g\,V_4C_3}{100\,g\,steel}$$

Thus the volume fraction of V_4C_3 is:

$$f_{V_{V_4C_3}} = \frac{0.12\,g\,V_4C_3}{100\,g\,steel} \cdot \frac{7.8\,g\,steel}{1\,cm^3\,steel} \cdot \frac{1\,cm^3\,V_4C_3}{5.8\,g\,V_4C_3} = 0.0016 = 0.16\%$$

On the other hand, Niobium would form NbC carbides, and following the same procedure as for vanadium, the molecular weight would be:

$$M_{NbC} = (1)\rho_{Nb} + (1)\rho_C = (1)(93) + (1)(12) = 105\,g/mol$$

The amount of carbon absorbed by 0.04% Nb in the steel would be:

$$\frac{(1)(12)\,gC}{(1)(93)\,g\,Nb} \cdot \frac{0.04\,g\,Nb}{100\,g\,steel} = \frac{0.005\,gC}{100\,g\,steel}$$

Meaning 0.005% C would form precipitates and the remaining carbon (0.145% C) would be in solid solution in the steel. In 100 g of steel there are $(0.005 + 0.04)\,g$ of NbC or in other words:

$$\frac{0.045\,g\,NbC}{100\,g\,steel}$$

Thus the volume fraction of NbC is:

$$f_{V_{NbC}} = \frac{0.045\,g\,NbC}{100\,g\,steel} \cdot \frac{7.8\,g\,steel}{1\,cm^3\,steel} \cdot \frac{1\,cm^3\,NbC}{7.8\,g\,NbC} = 0.00045 = 0.045\%$$

Finally, Titanium would form TiC carbides, and once again following the same procedures, the molecular weight would be:

$$M_{TiC} = (1)\rho_{Ti} + (1)\rho_C = (1)(48) + (1)(12) = 60\,g/mol$$

The amount of carbon absorbed by 0.02% Ti in the steel would be:

$$\frac{(1)(12)\,gC}{(1)(48)\,g\,Ti} \cdot \frac{0.02\,g\,Ti}{100\,g\,steel} = \frac{0.005\,gC}{100\,g\,steel}$$

Meaning, once again that 0.005% C would form precipitates and the remaining carbon (0.145% C) would be in solid solution in the steel. In 100 g of steel there are (0.005 + 0.02)g of TiC or in other words:

$$\frac{0.025\,\mathrm{g\,TiC}}{100\,\mathrm{g\,steel}}$$

Thus the volume fraction of TiC is:

$$f_{V_{\mathrm{TiC}}} = \frac{0.025\,\mathrm{g\,TiC}}{100\,\mathrm{g\,steel}} \cdot \frac{7.8\,\mathrm{g\,steel}}{1\,\mathrm{cm^3\,steel}} \cdot \frac{1\,\mathrm{cm^3\,NbC}}{4.94\,\mathrm{g\,NbC}} = 0.00039 = 0.039\%$$

This means that 0.244% of the material is formed by carbides.

(b) Precipitation temperature

From the expressions of Table 8.1, the vanadium carbide would have the following precipitation temperature:

$$\log\,[\%\mathrm{V}]^{4/3}[\%\mathrm{C}] = \left(-\frac{10,800}{T}\right) + 7.06$$

$$\log\,[0.1]^{4/3}[0.15] = \left(-\frac{10,800}{T}\right) + 7.06$$

$$T = 1171.72\,\mathrm{K} = 898.72^\circ\mathrm{C}$$

Likewise for the niobium carbide:

$$\log\,[\%\mathrm{Nb}][\%\mathrm{C}]^{0.87} = \left(-\frac{7530}{T}\right) + 3.11$$

$$\log\,[0.04][0.15]^{0.87} = \left(-\frac{7,530}{T}\right) + 3.11$$

$$T = 1441.22\,\mathrm{K} = 1168.22^\circ\mathrm{C}$$

And finally for titanium carbide:

$$\log\,[\%\mathrm{Ti}][\%\mathrm{C}] = \left(-\frac{7000}{T}\right) + 2.75$$

$$\log\,[0.02][0.15] = \left(-\frac{7000}{T}\right) + 2.75$$

$$T = 1327.55\,\mathrm{K} = 1054.55^\circ\mathrm{C}$$

Comment: TiN is more effective in inhibiting austenitic grain growth during homogenization before forging. For example, if the steel is alloyed with 0.2% Ti and 0.01% N; using the formula of Table 8.1, the precipitation temperature would be:

$$\log\,[\%\mathrm{Ti}][\%\mathrm{N}] = \left(-\frac{14,400}{T}\right) + 5$$

$$\log\,[0.2][0.01] = \left(-\frac{14,400}{T}\right) + 5$$

$$T = 1870.38\,\mathrm{K} = 1597.38^\circ\mathrm{C}$$

which is considerably higher than the three carbides analyzed.

8.3.2 Tempering of Martensite

Transformations that occur when heating martensite for tempering purposes involve various stages, which sometimes overlap:

1. The first stage of tempering (100–200°C), involves tetragonal martensite rejecting the excess C and inducing precipitation of a carbide (ε) with compact hexagonal structure and approximate chemical formula $Fe_{2.4}C$. This precipitation can be observed by electronic microscopes and is accompanied by a carbon impoverishment of the matrix. By diffusion and precipitation of carbon, the tetragonality of the martensite is progressively reduced, and when carbon reaches 0.2% (wt), martensite has changed from the tetragonal structure to the body-centered cubic one or supersaturated ferrite (*β-martensite*). This cubic martensite can be recognized by its fast blackening when etched with Nital. The toughness, measured at room temperature, gradually increases (for most quenched steels) when increasing tempering temperature.
2. The second stage (230–300°C) is characterized by the transformation of residual austenite into lower bainite. Dilation, and sometimes, increase in hardness (when there is a large amount of residual austenite) occur. The transformation takes place by nucleation and growth.
3. The third stage (300–350°C) is accompanied by a considerable loss in toughness. The intergranular nature of cracking is evidence of embrittlement by precipitated compounds. For this temperature interval, redissolution of ε is accompanied by precipitation of rod-shaped cementite (~200 nm in length). The existence of an almost continuous network of carbides enhances brittleness, which is the reason why this stage is industrially known as *tempered martensite embrittlement*. This temperature interval is rarely used for tempering quenched steels
4. The fourth stage (>400°C) is characterized by the spheroidization of cementite particles that break the continuity of the matrix or network. Also, ferrite restoration and recrystallization occur, with an evident increase in toughness.
5. The fifth stage is typically present in steels alloyed with Ni, Cr, and Mn and impurities of Sb (800 ppm), Sn (500 ppm), P (500 ppm), and As (500 ppm). This stage is characterized by a type of embrittlement[5] known as *Krupp sickness* or *tempered embrittlement*.

   The mechanism that causes this low toughness is not fully determined, though many authors agree that it is caused by segregation of impurity elements in grain boundaries, which can reach levels of 10^4 for impurities and 10 for alloying elements (compared to the mean composition). The zone of enrichment by segregated elements is very thin as it only affects a few rows

5. Described for the first time in 1883 and extensively evident in World War I cannons.

of atoms. Mo reduces the fragility of the steel because it delays both diffusion of impurities and restoration of martensite (though this phase has a high number of dislocations, these last ones stay close to impurities instead of moving towards grain boundaries). For high purity steels, manufactured in vacuum, embrittlement is not observed.

The interaction between alloying elements (Ni, Cr, and Mn) and impurities (Sb, P, As, and Sn), and their effect on toughness, can be analyzed through the formation of intermetallic compounds (alloying element—impurity) enthalpy. If $|\Delta H^{\circ}|$ is too large, compounds precipitate in the interior of the crystals instead of at the boundary, and thus there is no embrittlement effect. Similarly, if $|\Delta H^{\circ}|$ is too small, there is little interaction between them, and insufficient driving force for segregation, thus, fragility is not increased. Moreover, this embrittlement is a diffusion-controlled reaction, with a "C-curve" and can be reduced by fast heating at 450–550°C (and fast cooling at the same zone when tempering is carried out at higher temperatures).

This embrittlement phenomenon is manifested only by loss of toughness and does not cause significant variation in other mechanical properties. Therefore the susceptibility of a steel to this type of embrittlement is usually expressed by the ratio of resilience of the steel rapidly cooled in the 450–550°C zone divided by the resilience of the steel slowly cooled, in the same zone. Both characteristics are measured at room temperature through a resilience test.

6. The sixth stage is present in some steels alloyed with carbide-forming elements (Cr, Mo, W, V, and Ti) at 600°C, known as *secondary hardness.*

These elements, dissolved in the austenite, stay in solid solution when quenched (in the martensite), and will remain in solid solution in the ferrite during tempering, provided that temperatures close to 600°C are not reached. This carbide precipitation is accompanied by an increase in hardness.

When the amount of carbide-forming elements is large, for example in high-speed steels, the hardness of martensite decreases with temperature, but close to 600°C it may increase (by precipitation of carbides) to levels of martensite without tempering.

Temperature and time combine during tempering, since all transformations are diffusion controlled. Usually tempering periods are of 30–90 min; but similar results can be obtained with higher temperatures and shorter periods. For each temperature, variation of hardness as a function of the logarithm of holding time is almost linear (Rockwell scale). According to Hollomon and Jaffe (1945), the hardness after tempering is

$$\mathrm{HRC} = H_p - M \tag{8.20}$$

where H_p is the potential hardness and M is the softening coefficient expressed as:

$$M = KT(C + \log t) \tag{8.21}$$

with K being a parameter that depends on the C content (and to a lesser degree on the alloying elements usually used in "hardened" steels), T is the absolute temperature, t is the time, and C is a constant equal to 15.9 for low-alloyed steels and 14 for alloyed ones.

EXERCISE 8.14

For two hardened steels, both with 0.3% C, one of them low-alloyed and the other alloyed, the following hardness data was obtained for a one-hour treatment and different tempering temperatures:

Tempering temperature (°C)	Hardness (HRC) low-alloyed	Hardness (HRC) alloyed
200	50	65
300	42	62
400	31	58
500	23.5	52
600	17.5	48

Estimate the potential hardness of both steels and the K coefficient of Eq. (8.21).

Solution

If hardness is plotted vs. temperature (in K) and straight lines are fitted to obtain their corresponding equation, Fig. 8.46 is obtained.

As Rockwell hardness can be expressed with Eq. (8.20), the fitted lines show that H_p is very similar for both steels (~ 87), indicating that the potential hardness of the steel is not influenced by the amount of alloying elements. Furthermore, the K

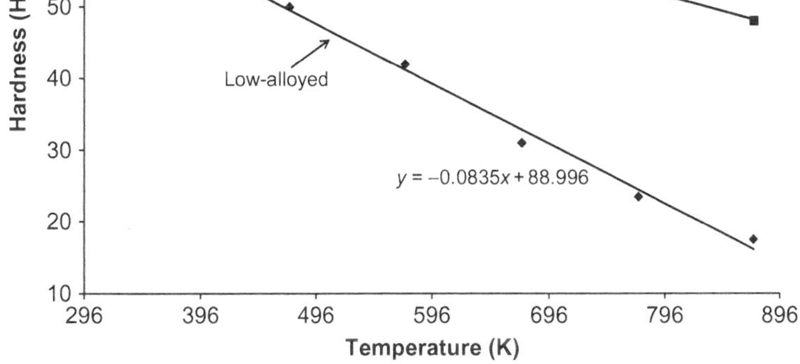

FIG. 8.46 Hardness vs. tempering temperature.

coefficient can also be obtained, since treatment time is 1 h (3600 s), $\log t = 3.556$, and considering the C values, the slope of the fitted lines and Eq. (8.21):

$$K_{low-alloy}(15.9 + 3.556) = 0.0835$$

$$K_{alloyed}(14 + 3.556) = 0.044$$

Thus:

$$K_{low-alloy} = 0.0043$$

$$K_{alloyed} = 0.0025$$

Comment: Alloying elements have little influence in the potential hardness, but a large influence on the softening coefficient (slope of the equation) due to solution and precipitation mechanisms.

The superficial color of tempered parts gives information on the tempering temperature, as an iron oxide film is formed with variable thickness (and different coloration due to the phenomenon of light interference as a function of thickness). These colors may vary from pale yellow (at 200°C), brown yellow (240°C), purple red (270°C), violet (280°C), dark blue (290°C), light blue (320°C), gray blue (340°C), etc.

If mechanical properties are analyzed, all quenched and tempered steels have a microstructure of fine and disperse carbides and restored ferrite, providing toughness and good fatigue behavior. In general, when carbides are visible in the optical microscope, its contribution to the improvement of mechanical properties is almost negligible (large sizes).

When tempering martensite, and since its hardness decreases when increasing temperature, toughness increases: the alloying elements that are found in solid solution inside martensite, stabilize it and induce a decrease in hardness. Thus, alloyed steels, in their useful zone of tempering, have higher hardness and strength compared to carbon steels with higher C content, with the advantage that this alloyed martensite will have a higher toughness (because of its lower C content).

Alloying elements stabilize both martensite hardness and strength: in steels with medium C content and low-alloyed, considering as a reference the stabilization produced by 1% Ni (as a value of 1), the indexes corresponding to 1% of other elements are: 4 for Mn, 4 for Si, 2.5 for Mo, 1.5 for Cr, and 0.5 for Co. This means that Mn and Si stabilize martensite four times more than Ni in tempering treatments. Therefore, when a martensite which softens slightly in low tempering treatments is sought, for example springs, steels alloyed with Si and Mn are used.

It must be pointed out that the transformation from martensite to tempered martensite implies contraction during the first stage, dilation during the second, and contraction during the subsequent ones. To sum up, the process results in a volume contraction of $\sim$1.4% for a 1% wt C steel.

8.3.3 Tempering of Bainite and Pearlite

Tempering bainite and pearlite takes advantage of the transformations suffered by nonmartensitic structures that, when heated, do not reach A_e, which is the $\alpha \rightarrow \gamma$ allotropic transformation temperature. In other words, heat treatments at temperatures below A_e, are of the *subcritical* type.

Bainite, constituted by needles (laths) of ferrite with carbides inside them, experiences by heating, a coalescence of the carbides, and creates at the end, a structure analogous to that of tempered martensite, but less homogeneous (because martensitic aging starts with only one constituent unlike bainite) and as a consequence, tempered bainite has lower toughness.

Subcritical heating of pearlite causes globulization of the cementite lamellae since it tends to adopt a more stable morphology: higher volume/surface ratio. The softened structure and its morphology correspond to carbides (larger than those of tempered bainite) and ferrite. The level of globulization depends on temperature and time: in order for tempering to occur it is necessary to heat the pearlite at temperatures close to those of its formation, and since the A_1 value cannot be exceeded, the temperature margin for tempering is very small. Thus, in this process, known as *subcritical tempering*, holding time is the main factor.

Heating hardly has any effect on the proeutectoid grains resulting from the decomposition of austenite, as the structure is ferritic-pearlitic and proeutectoid ferrite (since it is not saturated in C) and will not suffer modifications by heating at temperatures below A_1; on the other hand, proeutectoid cementite, in cementitic-pearlitic structures, will only undergo spheroidization, as an effect of heating, forming globular cementite.

REFERENCES

Andrews, K.W., 1965. Empirical formulae for the calculation of some transformation temperatures. J. Iron Steel Inst. 203, 721–727.

ASM International, 1992. ASM Handbook, 10th ed. Alloy Phase Diagrams, vol. 3. ASM International, Metals Park, OH.

Bain, E., Paxton, H., 1966. Alloying Elements in Steel, second ed. American Society for Metals, Metals Park, OH.

Hendricks, S., 1930. The crystal structure of cementite. Z. Kristallogr. 74, 534–545.

Hollomon, J., Jaffe, L., 1945. Time-temperature relations in tempering steel. Trans. AIME 162, 223–249.

Irvine, K., Gladman, T., Pickering, F., 1969. The strength of austenitic stainless steels. J. Iron Steel Inst. 207, 1017–1028.

Koistinen, D., Marburger, R., 1959. A general equation prescribing the extent of the austenite-martensite transformation in pure iron-carbon alloys and plain carbon steels. Acta Metall. 7 (1), 59–60.

Kurdjumov, G., Sachs, G., 1930. Über den Mechanismus der Stahlhärtung. Z. Phys. 64, 325–343.

Nehrenberg, A.E., 1946. The temperature range of Martensite formation. Trans. AIME 167, 494–498.

Pickering, F.B., 1971. Towards improved ductility and toughness. In: Clymax Molybdenum Co. Symposium, Kyoto, Japan.

Quintana, M., Gonzalez, R., 2016. Ultrafine-grain Steels: Mechanical Behavior. Amazon/CreateSpace, Lexington, KY.

Steven, W., Haynes, A., 1956. The temperature of formation of martensite and bainite in low-alloy steels. J. Iron Steel Inst. 183, 349–359.

Vander Voort, G., 1991. Atlas of Time-Temperature Diagrams for Irons and Steels. ASM International, Metals Park, OH.

BIBLIOGRAPHY

Apraiz, J., 1986. Aceros Especiales. Dossat, Madrid.

Apraiz, J., 2002. Tratamientos Térmicos de los aceros, 10th ed. Dossat, Madrid.

ASM, 1973. Metallograph Structures and Phase Diagrams, eighth ed. American Society for Metals, Metals Park, OH.

ASM, 1983. HSLA Steels Technology & Applications. American Society for Metals, Philadelphia, PA.

Bhadeshia, H., Honeycombe, R., 2006. Steels. Microstructure and Properties, third ed. Butterworth-Heinemann, London.

Calvo Rodes, R., 1956. El Acero: su elección y selección. INTA, Madrid.

Cohen, M., 1962. The strengthening of steel. Trans. Met. Soc. AIME 224, 638.

Constant, G., Delbart, A., 1956. Courbes De Transformation Des Aciers De Fabrication Francaise. IRSID, Saint Germain en Laye.

González, R., García, J., Barbés, M., 2010. Ultrafine grained HSLA steels for cold forming. J. Iron Steel Res. Int. 17 (10), 50–56.

Hansen, M., Elliot, R., 1965. Constitution of Binary Alloys, second ed. McGraw-Hill, New York.

Hebraken, L., 1967. Fundamentals of Metallography. Presses Academiques Europeennes, Brussels.

Kuo, K., 1953. The iron-carbon-molybdenum system. J. Iron Steel Inst. 173, 363.

Leslie, W., 1991. The Physical Metallurgy of Steels. McGraw-Hill, New York.

Tofaute, W., Buttinghaus, A., 1938. Die Eisenecke des Systems Eisen-Titan-Kohlenstoff. Arch. Eisenhü ttenwesen 33 (12), 33–37.

USS, 1998. The Making, Shaping and Treating of Steel, 11th ed. United States Steel, Pittsburgh, PA.

Chapter 9

Ternary Systems

9.1 INTERPRETATION OF TERNARY DIAGRAMS

9.1.1 Graphic Representation

The most common method to represent a ternary alloy, is the Roozebum method: over an equilateral triangle (Fig. 9.1) where each corner corresponds to 100% of each pure metal or compound, each side having a scale from 1% to 100% (AB, AC, and BC). An alloy M formed by $a\%$ of metal A, $b\%$ of metal B, and $c\%$ of metal C is represented by a point in Fig. 9.1. If an alloy has a constant proportion between B and C, this is:

$$\frac{b\%}{c\%} = \text{constant} \tag{9.1}$$

it will necessary be located over the AM line. Analogously, the geometric location of all alloys with percentage of A equal to $a\%$ are parallel to the side BC intersecting M. As temperature is considered an axis perpendicular to the plane ABC, different diagrams must be plotted to represent temperature variation.

Applying the phase rule for a three-component system, with a number of phases ϕ, will result in a maximum of four degrees of freedom (V).

The Roozebum representation makes the generalization of the lever rule possible when, for a certain temperature, alloy H has two phases in equilibrium, with compositions I and J. Fig. 9.2 shows, for the side AC, compositions i', h', and j' of metal C. Supposing that P_J and P_I are the respective weights of phases J and I, the total weight of alloy H (P_T) equals $P_J + P_I$. On the other hand, the total weight of C can be expressed as:

$$P_T \cdot h' = P_J \cdot j' + P_I \cdot i' \tag{9.2}$$

and since:

$$P_T \cdot h' = P_J \cdot h' + P_I \cdot h' \tag{9.3}$$

then:

$$\frac{P_J}{P_I} = \frac{h' - i'}{j' - h'} = \frac{\text{HI}}{\text{HJ}} \tag{9.4}$$

Solidification and Solid-State Transformations of Metals and Alloys. http://dx.doi.org/10.1016/B978-0-12-812607-3.00009-7

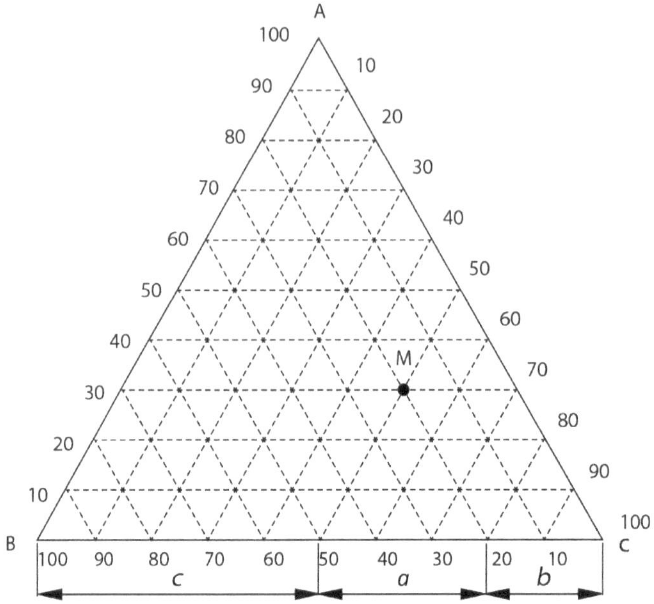

FIG. 9.1 Roozebum representation.

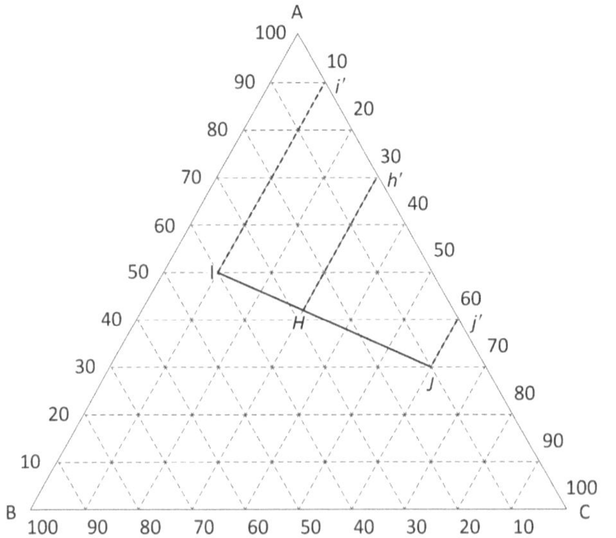

FIG. 9.2 Generalization of the lever rule for alloy H.

When alloy H, at a certain temperature, is formed by three phases S, R, and L (one of them liquid) as shown in Fig. 9.3 in the triangle SRL (sometimes known as *tie-triangle*); equilibrium is of the invariant type (with no degrees of freedom) and, at that constant temperature, proportions in weight of the phases can be determined by repetitive application of the lever rule, being P_1, P_2, and P_3

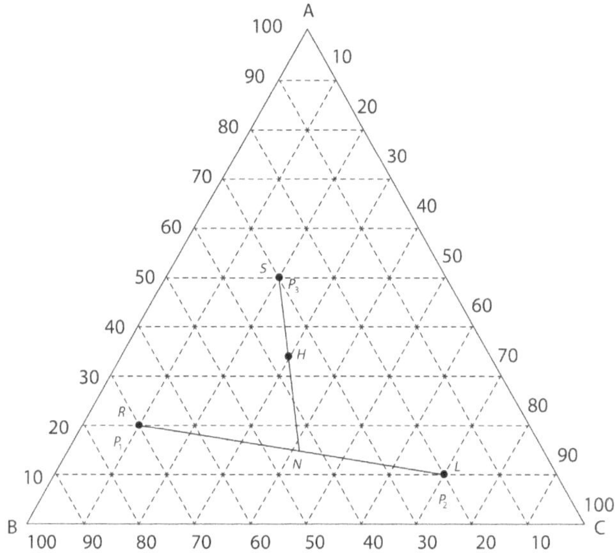

FIG. 9.3 Equilibrium between three phases S, R, and L.

the respective weights of phases R, L, and S, and the total weight of the alloy H ($P_T = P_1 + P_2 + P_3$):

$$\left.\begin{aligned} \frac{P_1}{P_1+P_2} &= \frac{\text{NL}}{\text{LR}} \\ \frac{P_2}{P_1+P_2} &= \frac{\text{NR}}{\text{RL}} \\ \frac{P_1+P_2}{P_1+P_2+P_3} &= \frac{\text{SH}}{\text{SN}} \end{aligned}\right\} \begin{aligned} \frac{P_1}{P_T} &= \frac{\text{NL}}{\text{LR}}\cdot\frac{\text{SH}}{\text{SN}} \\ \frac{P_2}{P_T} &= \frac{\text{NR}}{\text{RL}}\cdot\frac{\text{SH}}{\text{SN}} \\ P_3 &= P_T - (P_1+P_2) \end{aligned} \tag{9.5}$$

EXERCISE 9.1

A ternary alloy H (46, 31.5, 22.5) is formed by components A, B, and C, and by phases $\alpha(85,5,10)$, $\beta(5,90,5)$, and $\gamma(10,10,80)$. Calculate the amount of each phase using both, the graphic method (Fig. 9.3) and a balance of matter for the three components.

Solution

Locating the composition of H in the ternary diagram (Fig. 9.4) as well as the positions of α, β, and γ (Fig. 9.5), the amount of each phase can be calculated as:

$$f_\beta = \frac{P_1}{P_T} = \frac{\text{NL}}{\text{LR}}\cdot\frac{\text{SH}}{\text{SN}} = \frac{47}{77}\cdot\frac{35}{69} = 0.31$$

$$f_\gamma = \frac{P_2}{P_T} = \frac{\text{NR}}{\text{LR}}\cdot\frac{\text{SH}}{\text{SN}} = \frac{31}{77}\cdot\frac{35}{69} = 0.20$$

$$f_\alpha = P_3 = P_T - (P_1+P_2) = 1-(0.31+0.20) = 0.49$$

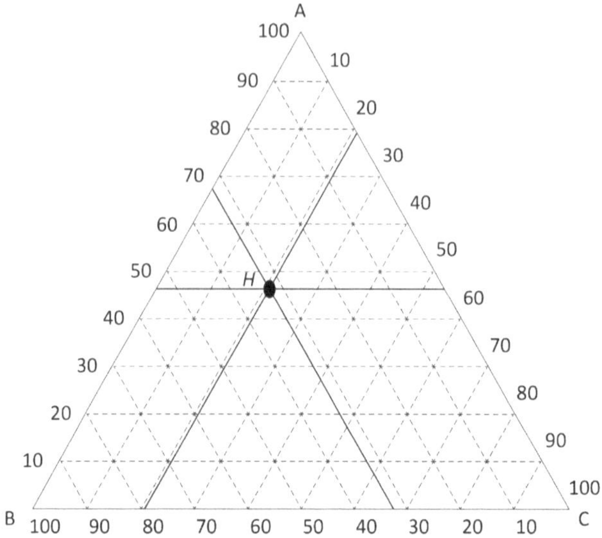

FIG. 9.4 Representation of alloy H in the ternary diagram.

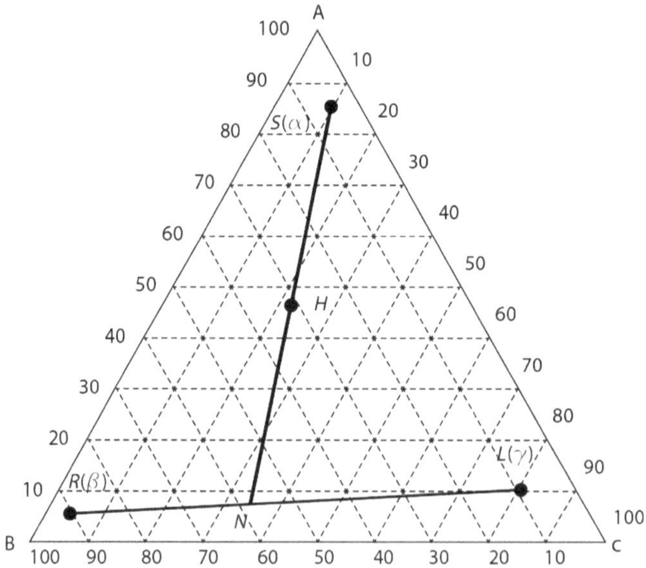

FIG. 9.5 Representation of α, β, and γ phases.

On the other hand, considering the composition of the alloy and phases, the balance of matter is:

$$46 = 85f_{\alpha} + 5f_{\beta} + 10f_{\gamma}$$

$$31.5 = 5f_{\alpha} + 90f_{\beta} + 10f_{\gamma}$$

$$22.5 = 10f_\alpha + 5f_\beta + 80f_\gamma$$

and solving the system:

$$f_\alpha = 0.5, \quad f_\beta = 0.3, \quad f_\gamma = 0.2$$

9.1.2 Equilibrium of Two Phases

If a ternary system has only a solid and a liquid phase in equilibrium, and temperature is fixed, the number of degrees of freedom is 2. If composition M of Fig. 9.6 is considered, ternary diagrams for different temperatures can be obtained (Fig. 9.7), as an example, at T_1, the solid will have a composition determined by the line S_1.

When composition J_2 of the solid is known, the one for the liquid is obtained by the intersection of the *tie-line* that joins J_2 with point 2 (the composition of

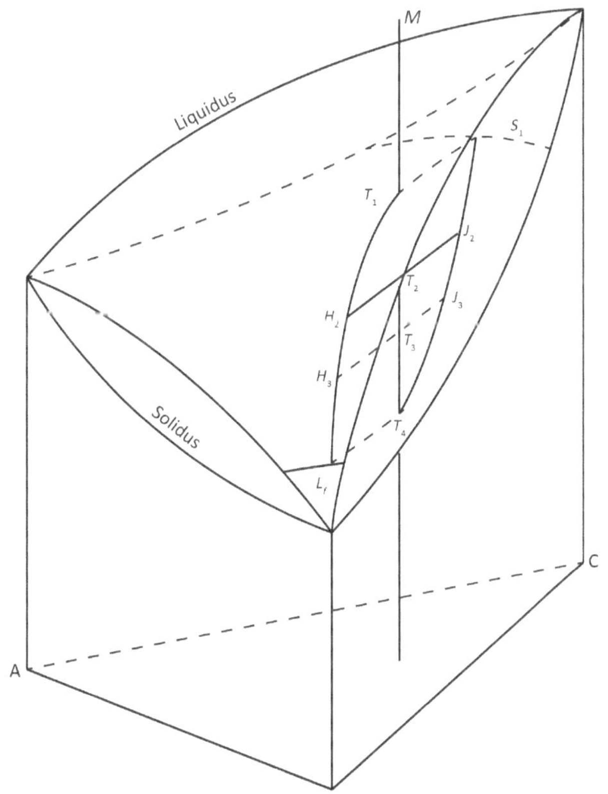

FIG. 9.6 Total solid solution in a ternary diagram.

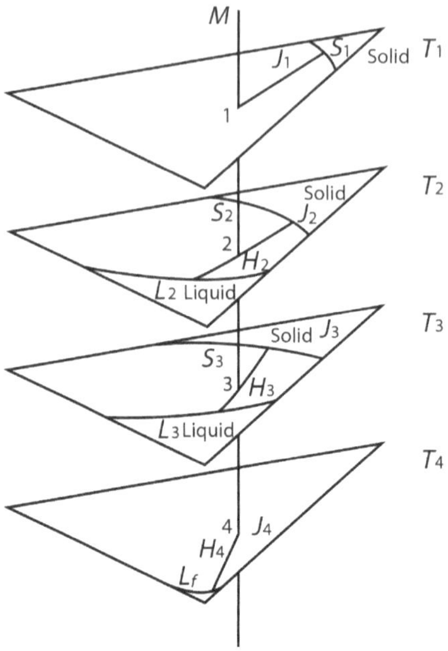

FIG. 9.7 Isothermal sections of Fig. 9.6.

alloy M in Fig. 9.7). In other words, in total solution ternary diagrams, equilibrium solidification is undetermined, presenting, compared to binary systems, various possible trajectories. For example, alloy M of Fig. 9.6 starts its solidification at T_1 and ends at T_4; though, the trajectory for the liquid can be any of the lines on the surface of *liquidus*, starting at l and ending at the curve L_f (isothermal section of the liquidus at T_4). Analogously, the *solidus* line can be any of those starting at the isotherm line S_1 (obtained at T_1) and ending at point 4. In general, the liquidus and solidus lines are not coplanar.

To determine, at each temperature, the proportions of solid and liquid, it is necessary to consider the isotherm sections of solid and liquid surfaces and the tie-line (that joins the compositions of solid and liquid); or the composition of each of the phases that allows the determination of that line.

9.1.3 Equilibrium Between Three Phases

In the case of three phases (one liquid and two solids) in equilibrium in a ternary system, the two possible reactions during solidification are:

$$\text{Liquid} \rightarrow \alpha + \beta \,(\text{binary eutectic}) \tag{9.6}$$

$$\text{Liquid} + \beta \rightarrow \alpha \,(\text{binary peritectic}) \tag{9.7}$$

The number of degrees of freedom in these cases is 1, as:

- The reaction (binary eutectic or peritectic) does not occur at constant temperature; thus, temperature can vary in an interval without breaking equilibrium.
- At each temperature of the mentioned interval, the solid and liquid compositions are perfectly defined.
- In nonequilibrium solidification, minor segregation can appear not only in proeutectic constituents but also in the binary eutectic and peritectic ones.

Fig. 9.8 is an example of a binary eutectic. To make interpretation of ternary diagrams easier, Fig. 9.9 presents the solidification of alloy M starting at point P with the formation of a solid solution β and the trajectory of liquidus following the curve PL. The binary eutectic starts upon reaching point L; from this point onwards, the trajectory of liquidus follows the valley formed by liquidus 1 and liquidus 2 profiles; meanwhile the solidus profile will be a surface (formed by the displacement of a straight line that is, joining the concentration of solids α and β, at all times, parallel to the horizontal plane).

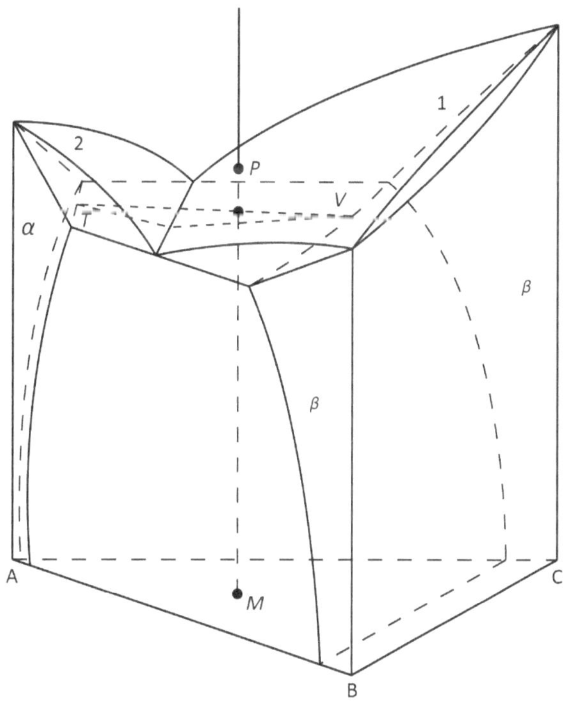

FIG. 9.8 Ternary diagram with a binary eutectic reaction.

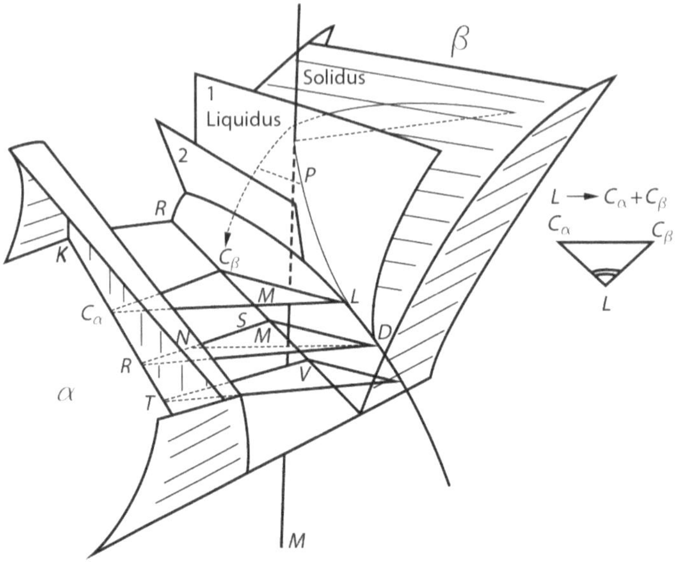

FIG. 9.9 Binary eutectic reaction in a ternary system.

To determine the weight proportions of α, β, and liquid phases, at equilibrium at T_1, it is necessary to know the tie-triangle for the mentioned isotherm T_1. The percentage of liquid will be NM/ND (Exercise 9.1), and the solidification will end at a temperature such that the vertical corresponding to the composition of alloy M intersects the line VT of the tie-triangle.

Fig. 9.10 presents an equilibrium diagram of three phases with peritectic reaction: liquid $+\beta \rightarrow \alpha$. Furthermore, Fig. 9.11 shows the start of the solidification at T_1 with the formation of β. When the separation of crystals of β from the liquid is total (when reaching the point L_1 at T_2), the reaction causing the formation of α begins (the line M will intersect, at the beginning of the peritectic reaction, the side L_1C_β of the tie-triangle). The reaction will end when all β has disappeared and only liquid and α remain: this will occur at T_f, in which the line M intersects the side L_fN of the tie-triangle.

If the side $\overline{C_\alpha}\,\overline{C_\beta}...\overline{RN}$ is extended, lines sketched from the apex $(L_1,...,L_f)$ of the different tie-triangles between the beginning and end of the binary peritectic reaction always intersect. Meanwhile, in a binary eutectic reaction, the sketched lines intersect the opposite side inside the segment joining the α and β compositions. Sometimes, in the same system, the reaction can change from binary peritectic to binary eutectic, as shown in Fig. 9.12.

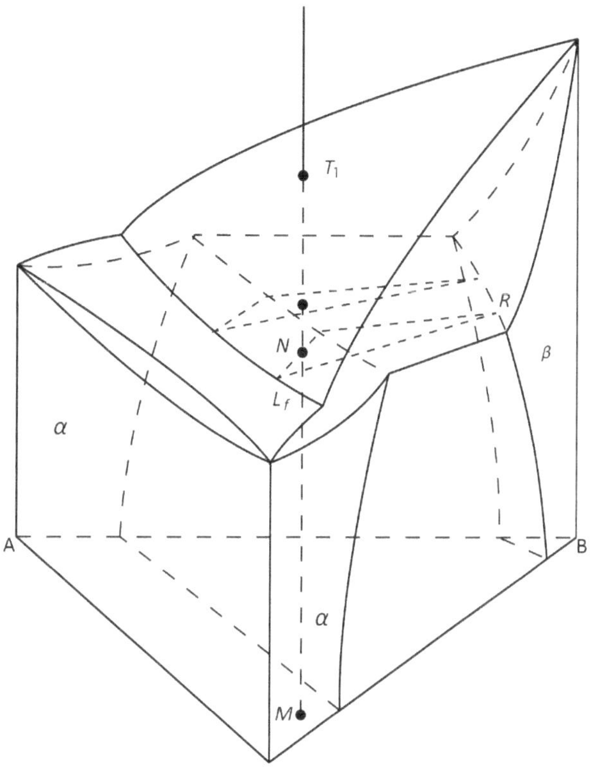

FIG. 9.10 Ternary diagram with a binary peritectic reaction.

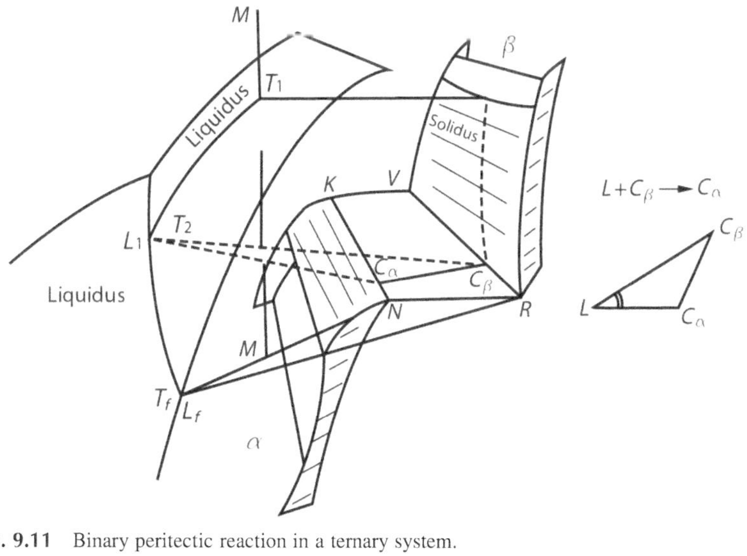

FIG. 9.11 Binary peritectic reaction in a ternary system.

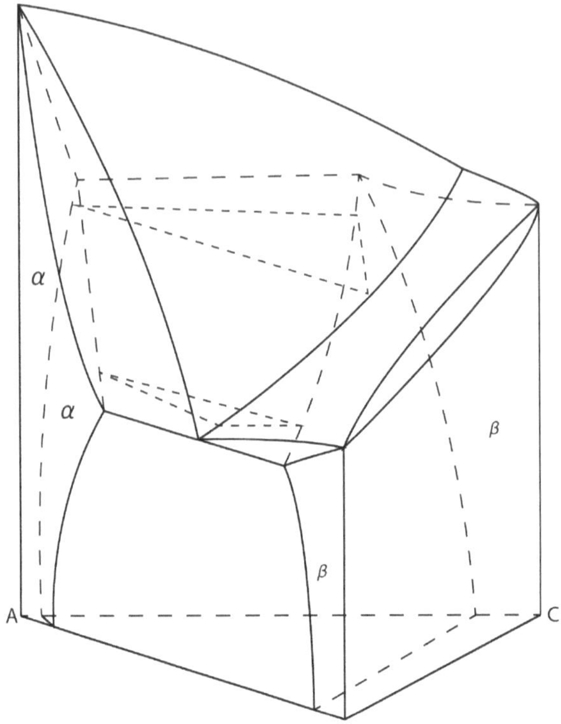

FIG. 9.12 Transformation of a binary peritectic reaction into a binary eutectic.

9.1.4 Equilibrium Between Four Phases

During solidification, the following reactions between liquid and solid phases α, β, and γ can occur:

$$\text{Liquid} \leftrightarrow \alpha + \beta + \gamma \text{ (ternary eutectic)} \tag{9.8}$$

$$\text{Liquid} + \beta \leftrightarrow \delta + \gamma \text{ (second class ternary peritectic)} \tag{9.9}$$

$$\text{Liquid} + \alpha + \beta \leftrightarrow \gamma \text{ (third class ternary peritectic)} \tag{9.10}$$

These reactions are invariant ($V=0$) and, therefore, happen at constant temperature.

In the case of the ternary eutectic (Eq. 9.8), equilibrium of the liquid with the three phases α, β, and γ (with clearly defined compositions) is shown in Figs. 9.13 and 9.14. On the other hand, Figs. 9.15 and 9.16 show the beginning of the second class peritectic reaction (Eq. 9.9) as well as the diagram with this type of reaction. Finally, Fig. 9.17 corresponds to an ideal diagram with a third class ternary peritectic reaction (Eq. 9.10).

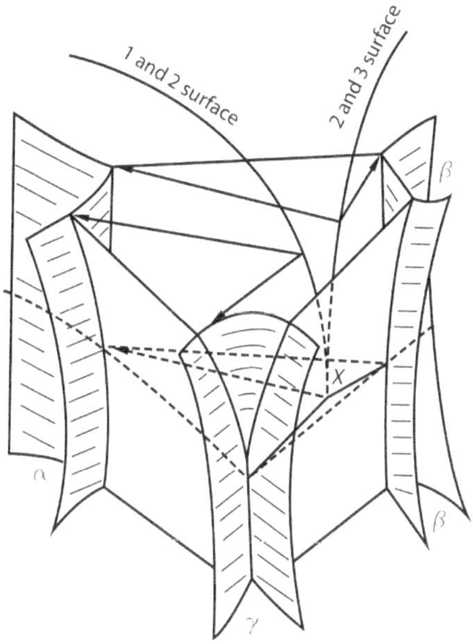

FIG. 9.13 Ternary eutectic reaction.

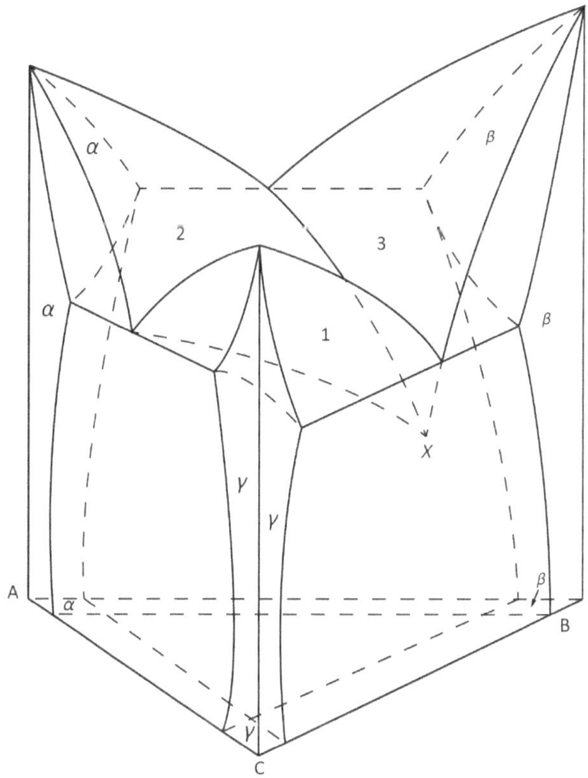

FIG. 9.14 Ternary diagram with ternary eutectic.

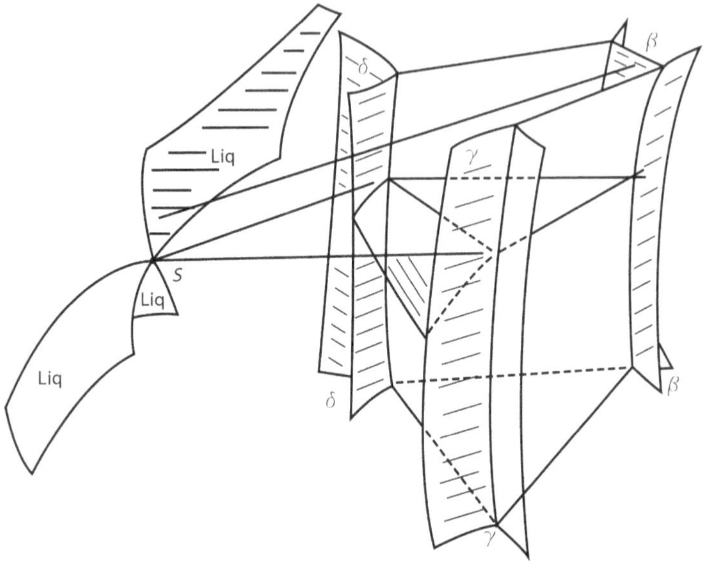

FIG. 9.15 Second class ternary peritectic reaction.

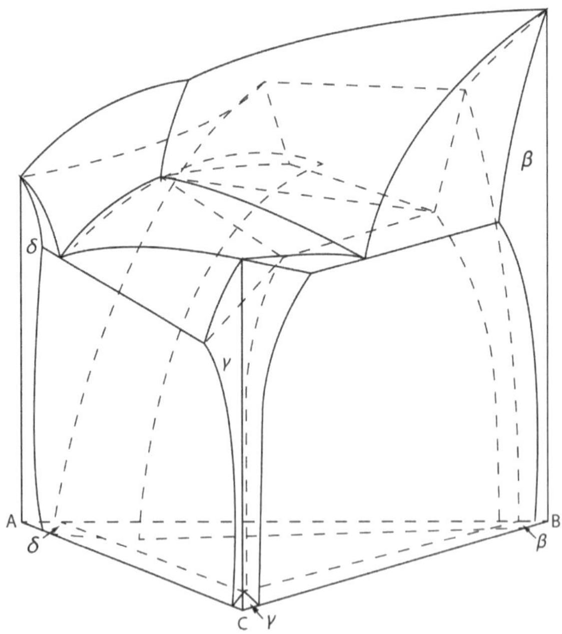

FIG. 9.16 Ternary diagram with a second class ternary peritectic reaction.

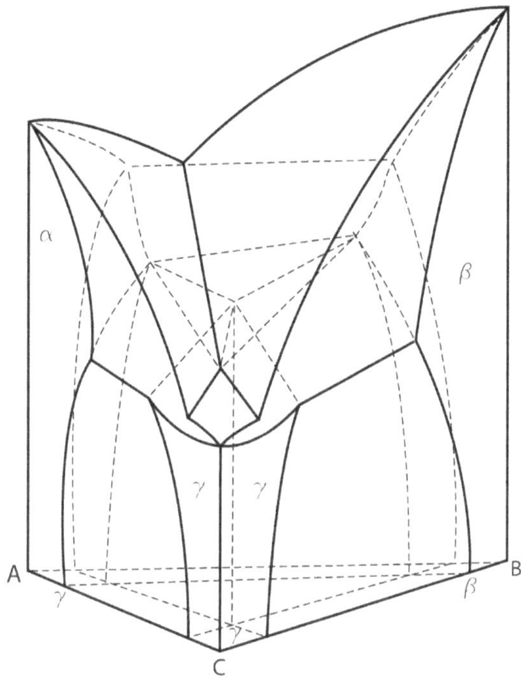

FIG. 9.17 Third class ternary peritectic reaction.

Solid-state transformations (eutectoids, binary peritectoids, ternary eutectoids, etc.) are analogous to the reactions previously explained, and other reactions (monotectic) can also be interpreted using these concepts.

EXERCISE 9.2

A ternary alloy H (40% *A*, 20% *B*, and 40% *C*) forms an eutectic solidifying into three phases:

- Solid α: 80% *A*, 5% *B*, and 15% *C*
- Solid β: 10% *A*, 70% *B*, and 20% *C*
- Solid γ: 10% *A*, 20% *B*, and 70% *C*

Calculate the fraction of α, β, and γ in the microstructure.

Solution

By plotting compositions for H, α, β, and γ in a ternary diagram, Fig. 9.18 is obtained. The amount of each phase can be calculated as:

$$f_\alpha = \frac{P_1}{P_T} = \frac{\text{NL}}{\text{LR}} \cdot \frac{\text{SH}}{\text{SN}} = \frac{31}{65} \cdot \frac{44}{50} = 42\%$$

$$f_\gamma = \frac{P_2}{P_T} = \frac{\text{NR}}{\text{LR}} \cdot \frac{\text{SH}}{\text{SN}} = \frac{34}{65} \cdot \frac{44}{50} = 46\%$$

$$f_\beta = P_3 = P_T - (P_1 + P_2) = 100 - (42 + 46) = 12\%$$

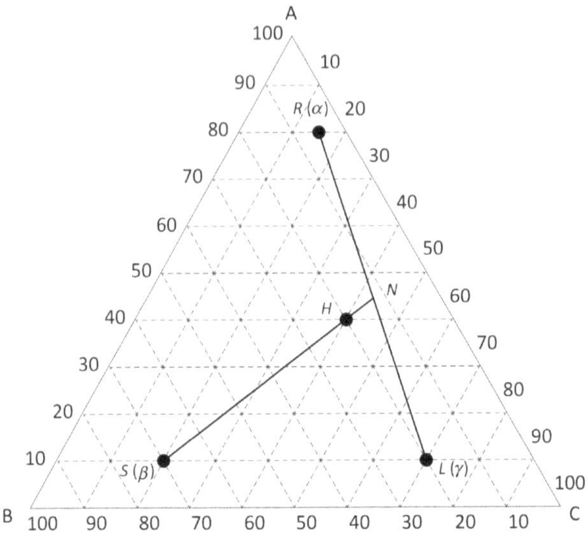

FIG. 9.18 Representation of compositions H, α, β, and γ.

9.1.4.1 Low Melting Point Alloys

Alloys of the binary or ternary types are sometimes used, due to their low melting point and constant temperature liquid formation, as safety valves which melt at a specific temperature, or to obtain high fluidity metallic liquids, among others. Table 9.1 indicates some of these alloys in an increasing melting temperature order between −48°C and 245°C, some binary eutectics, ternary eutectics, and quaternary eutectics, etc. are included as well as some hypo and hypereutectic alloys with low melting points.

9.2 TERNARY DIAGRAMS IN METALLOGRAPHY

Unlike binary systems where almost all combinations that can form any industrial metal are known, ternary systems and their diagrams have a more limited application in metallography. For the structural study of alloys, it is common to use binary diagrams and then indicate the influence which a third or fourth element has (on the equilibrium lines of the diagram).

On the other hand, unlike binary alloys, solidification structures in ternary diagrams are not completely determined: the structure obtained by equilibrium solidification depends on the chemical composition of the alloy, and also on different possible cooling trajectories over the liquidus and solidus profiles (the same occurring in solid state equilibrium transformations). Thus, for precise interpretation, the tridimensional ternary

TABLE 9.1 Low Melting Point Alloys

T_M (°C)	Denomination	Composition (%)				
−48	Binary eutectic	Cs (77.0)	K (23.0)			
−40	Binary eutectic	Cs (87.0)	Rb (13.0)			
−30	Binary eutectic	Cs (95)	Na (5.0)			
−11	Binary eutectic	K (78.0)	Na (22.0)			
−8	Binary eutectic	Rb (92.0)	Na (8.0)			
10.7	Ternary eutectic	Ga (62.5)	In (21.5)	Sn (16.0)		
10.8	Ternary eutectic	Ga (69.8)	In (17.6)	Sn (12.5)		
17	Ternary eutectic	Ga (82.0)	Sn (12.0)	Zn (6.0)		
33	Binary eutectic	Rb (68.0)	K (32.0)			
46.5	Quinary eutectic	Sn (10.65)	Bi (40.63)	Pb (22.11)	In (18.1)	Cd (8.2)
47	Quinary eutectic	Bi (44.7)	Pb (22.6)	Sn (8.3)	Cd (5.3)	In (19.1)
58.2	Quaternary eutectic	Bi (49.5)	Pb (17.6)	Sn (11.6)	In (21.3)	
60.5	Ternary eutectic	In (51.0)	Bi (32.5)	Sn (16.5)		
70	Wood's metal	Bi (50)	Pb (25)	Sn (12.5)	Sn (12.5)	
70	Lipowitz's metal	Bi (50)	Pb (26.7)	Sn (13.3)	Cd (10.0)	
70	Binary eutectic	In (67.0)	Bi (33.0)			
91.5	Ternary eutectic	Bi (51.6)	Pb (40.2)	Cd (8.2)		
95	Ternary eutectic	Bi (52.5)	Pb (32.0)	Sn (15.5)		
97	Newton's metal	Bi (50.0)	Sn (18.8)	Pb (31.2)		
98	Arcet's metal	Bi (50.0)	Sn (25.0)	Pb (25.0)		

Continued

TABLE 9.1 Low Melting Point Alloys—cont'd

T_M (°C)	Denomination	Composition (%)			
100	Onion's or Lichtenberg's metal	Bi (50.0)	Sn (20.0)	Pb (30.0)	
102.5	Ternary eutectic	Bi (54.0)	Sn (26.0)	Cd (20.0)	
109	Rose's Metal	Bi (50.0)	Pb (28.0)	Sn (22.0)	
117	Binary eutectic	In (52.0)	Sn (48.0)		
120	Binary eutectic	In (75.0)	Cd (25.0)		
123	Malotte's metal	Bi (46.1)	Sn (34.2)	Pb (19.7)	
124	Binary eutectic	Bi (55.5)	Pb (44.5)		
130	Ternary eutectic	Bi (56.0)	Sn (40.0)	Zn (4.0)	
140	Binary eutectic	Bi (58.0)	Sn (42.0)		
140	Binary eutectic	Bi (60.0)	Cd (40.0)		
183	Sn Eutectic solder	Sn (63.0)	Pb (37.0)		
185	Binary eutectic	Tl (52.0)	Bi (48.0)		
192	Soft solder	Sn (70.0)	Pb (30.0)		
198	Binary eutectic	Sn (91.0)	Zn (9.0)		
199	White metal	Sn (92.0)	Sb (8.0)		
221	Binary eutectic	Sn (96.5)	Ag (3.5)		
226	Matrix	Bi (48.0)	Pb (28.5)	Sn (14.5)	Sb (9.0)
227	Binary eutectic	Sn (99.25)	Cu (0.75)		
240	Sn/Sb Solder	Sn (95.0)	Sb (5.0)		
245	Sn/Ag Solder	Sn (95.0)	Ag (5.0)		

diagram is not enough, and the group of isothermal sections of the diagram at different temperatures as well as the position of the lines or the tie-triangles are also necessary.

Common ternary systems in metallography are: Pb-Sb-Sn which is the basis of antifriction (Babbits) and low melting point alloys, and Al-Cu-Si for light alloys.

9.2.1 Pb-Sb-Sn System

Fig. 9.19 shows the Pb-Sb, Sb-Sn, and Pb-Sn binary diagrams and the projection of the liquid profiles in a ternary fashion, while Fig. 9.20 shows these profiles in detail.

Table 9.2 shows typical alloys of this system used for several applications including wear-resistant parts, low temperature and high temperature bearings, and corrosion resistant automotive parts.

The Pb-Sb-Sn system has four phases, all of them formed by the three elements:

- Phase α is a substitutional solid solution of Pb and Sn in Sb.
- Phase δ is a substitutional solid solution of Sb and Sn in Pb.
- Phase γ is a substitutional solid solution with Sn as solvent and Pb and Sb as solutes.

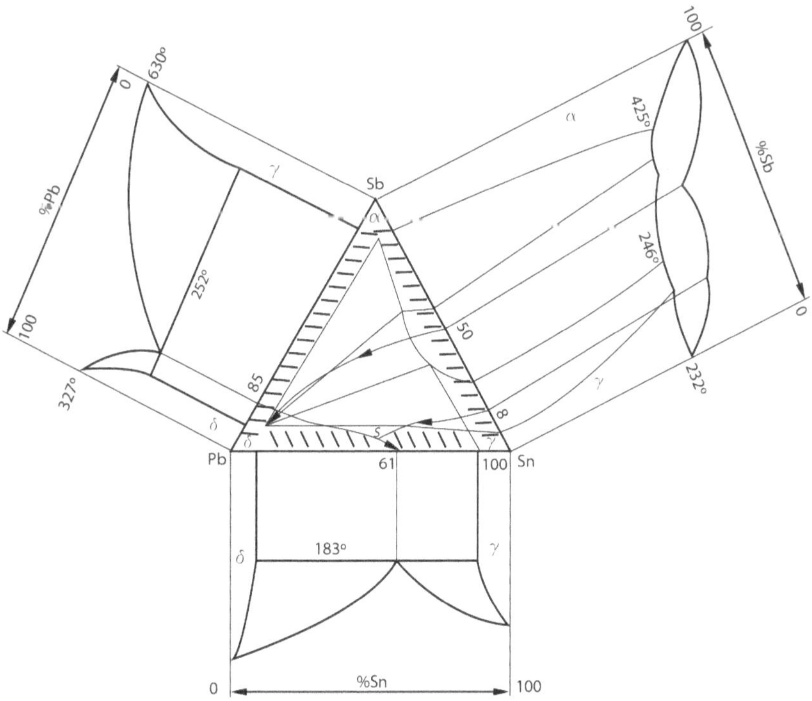

FIG. 9.19 Pb-Sb-Sn system.

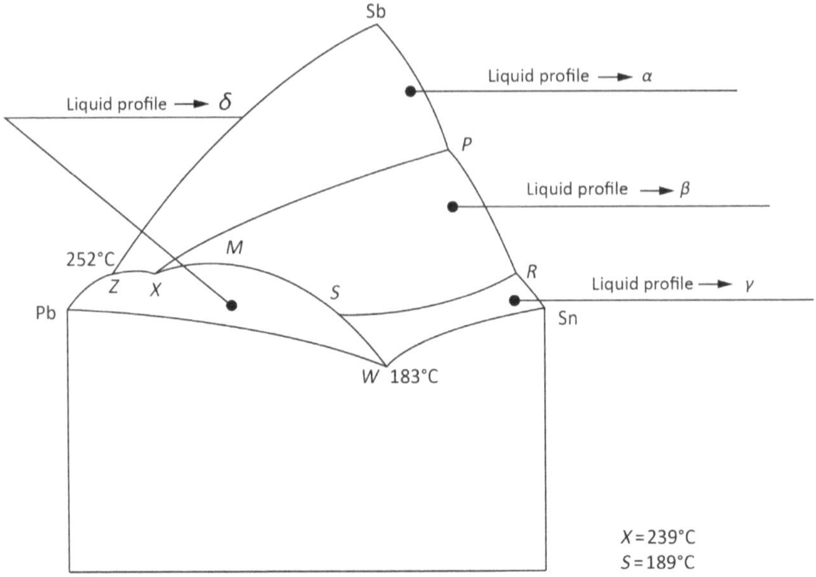

FIG. 9.20 Pb-Sb-Sn system profile.

- Phase β is the result of substituting Pb atoms in the intermetallic Sb-Sn compound of the binary Sb-Sn system.

Antifriction applications usually take advantage of a fine disperse phase β such as Pb based (SAE 14) and Sn-based (ASTM 5) Babbitt alloys (Table 9.2). Pb-based Babbitts resist wear under moderate loads and are self-lubricating materials, while Sn-based ones show a higher wear resistance as the hard intermetallic β is more abundant.

Furthermore, Sb-based Babbits can withstand a higher load if the amount of α is higher, though conformability is also a parameter to be considered when designing parts such as bearings.

Tri-phasic equilibriums (profiles in Fig. 9.20) are the following:

- Line ZX (binary eutectic): liquid $\leftrightarrow \alpha + \delta$
- Line XS (binary eutectic): liquid $\leftrightarrow \beta + \delta$
- Line SW (binary eutectic): liquid $\leftrightarrow \gamma + \delta$
- Line PX (binary peritectic): liquid $+ \alpha \leftrightarrow \beta$
- Line RS (binary peritectic): liquid $+ \beta \leftrightarrow \gamma$

In the Pb-Sb-Sn diagram, two invariant points are shown, corresponding to tetraphasic equilibriums: ternary eutexia (point X) and second class ternary peritexia (point S):

- At X, with 85% Pb-11.5% Sb-3.5% Sn, at 239°C the ternary eutectic liquid $\leftrightarrow \alpha + \beta + \delta$ reaction occurs.
- At S, with 40% Pb-2.5% Sb-57.5% Sn, at 189°C the second class ternary peritexial liquid $+ \beta \leftrightarrow \gamma + \delta$ reaction occurs.

TABLE 9.2 Some Pb-Sn-Sb Alloys

	% Pb	% Sb	% Sn	Others (%)	Application	Brinell hardness (HB)
H1	84	12	4	–	Linotype bullet casting alloy, bearings	–
H2	54	28	18	–	Linotype bullet casting alloy	–
H3	74	16	10	–	Monotype bullet casting alloy	–
Babbitt (Pb base)						
SAE 13	85	10	5	–	Light loads: automobile bearings	19
SAE 14	75	15	10	–	Medium loads: fans, pumps	22
SAE 15	83	15	1	1 As	Heavy loads: diesel engine bearings	20
G	83.5	12.75	0.75	3 As	High temperature bearings: trucks	22
Babbitt (Sn base)						
ASTM 1	–	4.5	91	4.5 Cu	Automotive industry; better resistance to corrosion and wear compared to Pb based applications	17
ASTM 2	–	7.5	89	3.5 Cu		24
ASTM 3	–	8	84	8 Cu		30
ASTM 5	–	15	65	12 Cu 18 Pb		23

Dividing the ternary diagram into four zones (α, β, γ, or δ phases), the evolution of some alloys at each of the regions can be analyzed (Fig. 9.21):

- α zone
  - Fig. 9.22 shows the differential cooling curve (T_1, dt/dT[1]) for *alloy A* (20% Pb, 70% Sb, and 10% Sn) during solidification, as well as the first solid to appear (α) at 562°C. At 332°C a new discontinuity indicates the beginning of the peritectic reaction ($L + \alpha \leftrightarrow \beta$) with the creation of β. Finally, at 239°C

1. *dt/dT* inverse of the cooling rate (*dt/dT*), decreases as a consequence of the latent heat released during solidification (phase transformation), then it increases and creates a peak (temperature arrest) in the cooling curve.

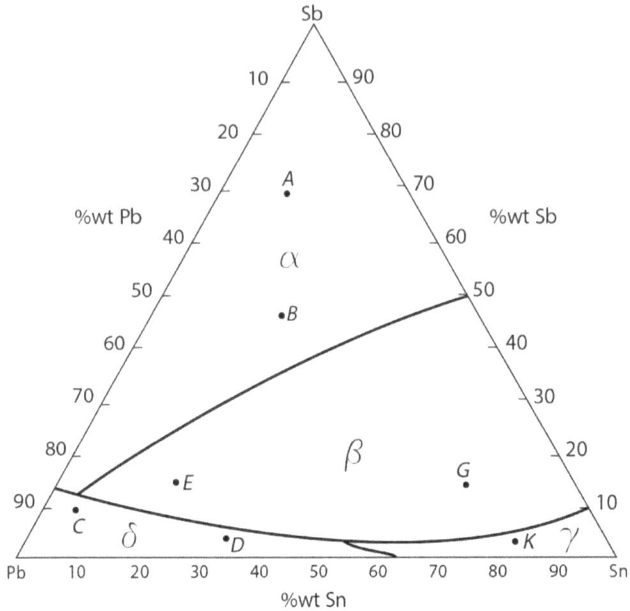

FIG. 9.21 Pb-Sb-Sn diagram. Some alloys.

an abundant amount of heat is released when the ternary eutectic solidifies, ending the solidification process of the alloy. The microstructure (Fig. 9.23) confirms the solidification stages and shows the obtained constituents.

- Fig. 9.24 shows the cooling curve of *alloy B* (35% Pb, 45% Sb, and 20% Sn), which is similar to A: the first solid (α) appears at a lower temperature (440°C) because of the lower proportion of Sb; and since the beginning of the peritectic reaction takes place at a temperature similar to A, it is expected that α is in a lower proportion. The microstructure of Fig. 9.25 presents a large proportion of β, with straight boundaries that surround the remains of primitive α crystals.

- δ zone
  - Fig. 9.26 corresponds to the cooling curve of *alloy C* (86% Pb, 11% Sb, and 3% Sn) with two peak points: one at 255°C (start of solidification of dendrites δ) and another at 239°C which is the final temperature of solidification by ternary eutectic reaction. The microstructure (Fig. 9.27) shows dendrites with rounded shapes (disperse constituent δ) and an eutectic (matrix constituent) formed by α, β, and δ.
  - Fig. 9.28 shows the cooling behavior of *alloy D* (65% Pb, 5% Sb, and 30% Sn) with three critical points: the first at 232°C corresponds to the formation of the first solid (δ); the second at 220°C marks the start of the binary eutectic reaction (liquid $\leftrightarrow \delta + \beta$), following the trajectory of the line XS of Fig. 9.19; and finally, the third corresponds to the peritectic point *S*. It can be observed that for this alloy, as well as alloy C, the

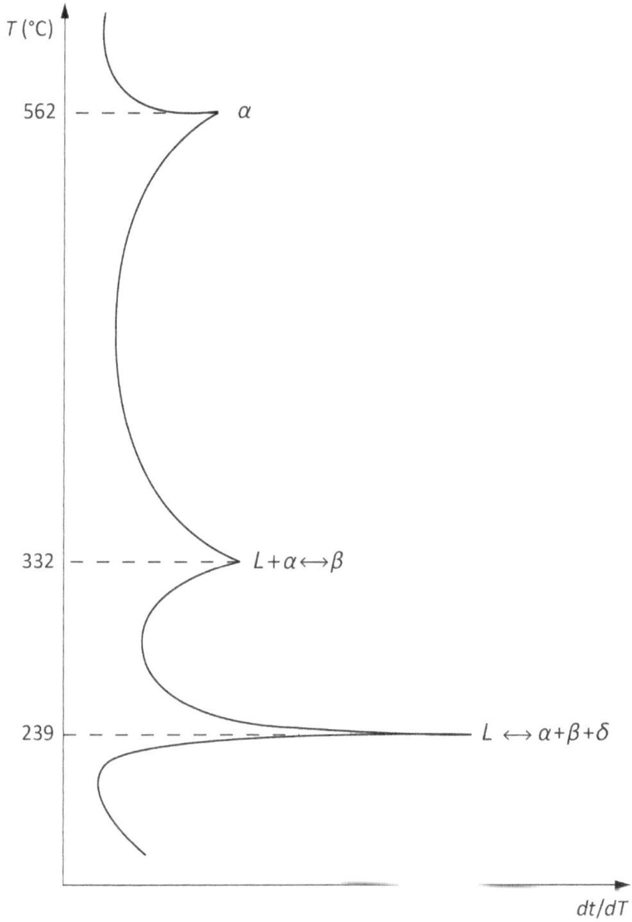

FIG. 9.22 Differential cooling curve of alloy A (20% Pb, 70% Sb, and 10% Sn).

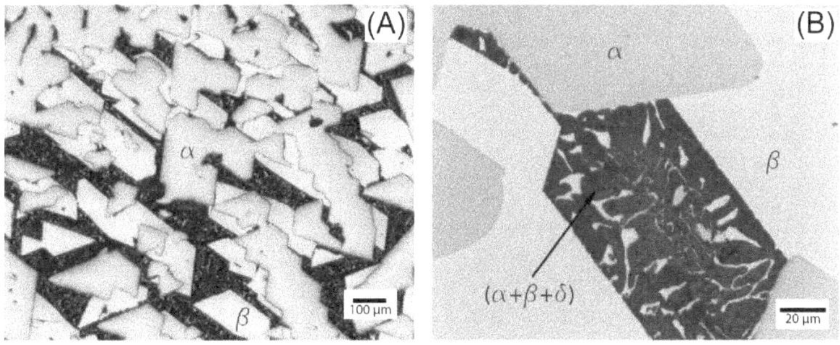

FIG. 9.23 Microstructure of alloy A showing α and β constituents (A) and detailed ternary eutectic (B).

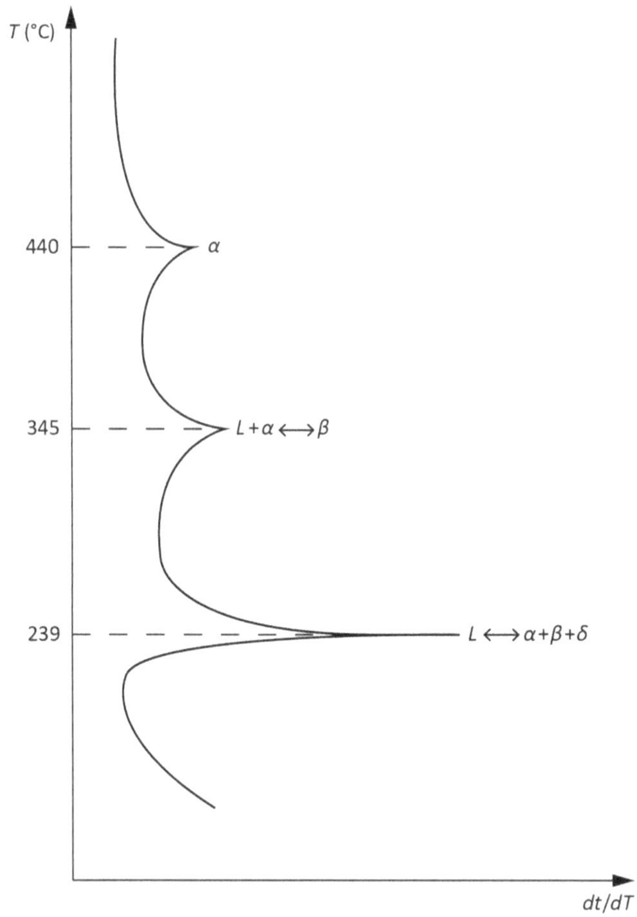

FIG. 9.24 Differential cooling curve of alloy B (35% Pb, 45% Sb, and 20% Sn).

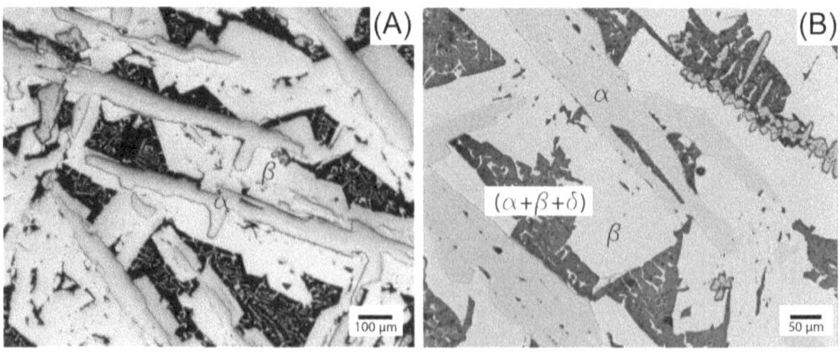

FIG. 9.25 Microstructure of alloy B showing α and β constituents (A) and detailed ternary eutectic (B).

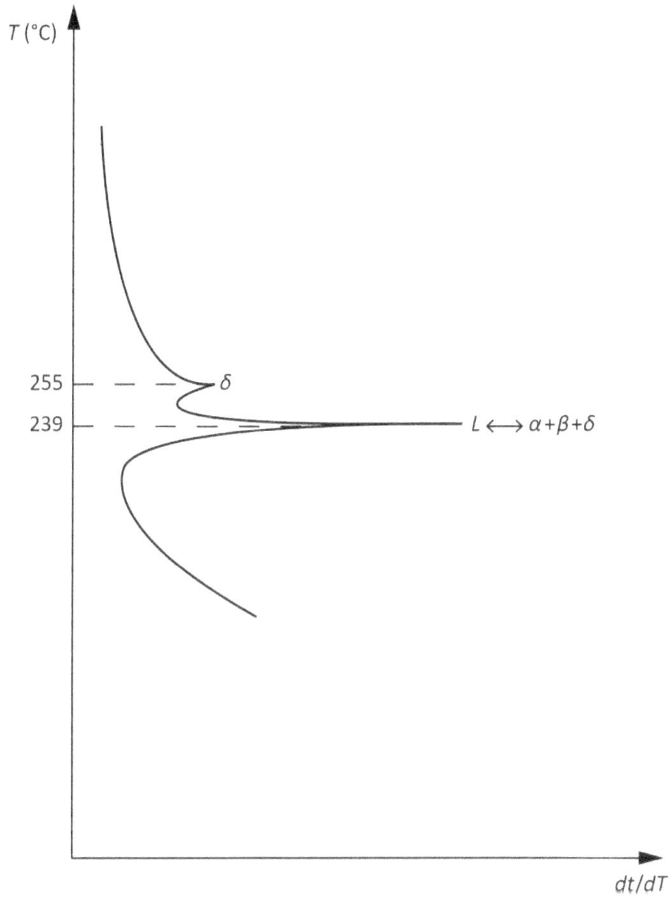

FIG. 9.26 Differential cooling curve of alloy C (86% Pb, 11% Sb, and 3% Sn).

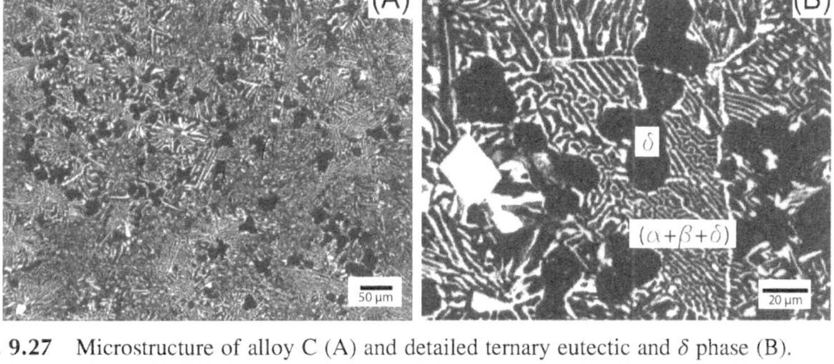

FIG. 9.27 Microstructure of alloy C (A) and detailed ternary eutectic and δ phase (B).

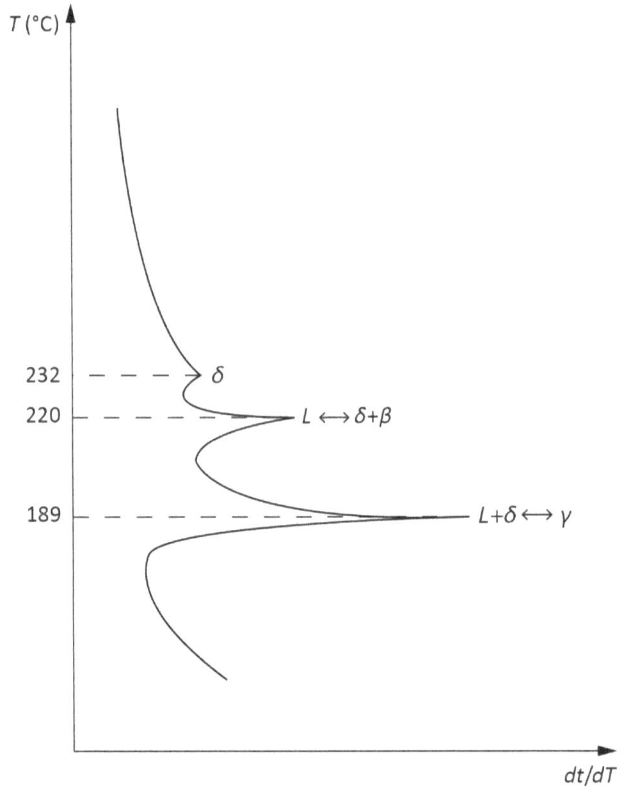

FIG. 9.28 Differential cooling curve of alloy D (65% Pb, 5% Sb, and 30% Sn).

first phase to appear is δ, rich in Pb. The microstructure of Fig. 9.29 shows large dendrites δ surrounded by a binary eutectic (lighter color) and a dark aggregate formed by γ and δ, product of the peritectic reaction (very fine grained and with "filigree" shape).

- β zone
  - Fig. 9.30 indicates the cooling curve of *alloy E* (75% Pb, 15% Sb, and 10% Sn) with two peak points: the first at 270°C corresponds to the formation of the first β crystals and the second, at 239°C corresponds to the ternary eutectic. The microstructure (Fig. 9.31) shows crystals of β surrounded by the ternary eutectic.
  - Fig. 9.32 shows the cooling curve of *alloy G* (20% Pb, 15% Sb, and 65% Sn) following a descending direction towards the binary peritectic *RS*: solidification starts by forming β at 280°C, and at 210°C the binary peritectic reaction (liquid + $\beta \leftrightarrow \gamma$) occurs, while at 189°C solidification ends with the ternary peritectic reaction (liquid + $\beta \leftrightarrow \gamma + \delta$). The microstructure of this alloy is shown in Fig. 9.33.

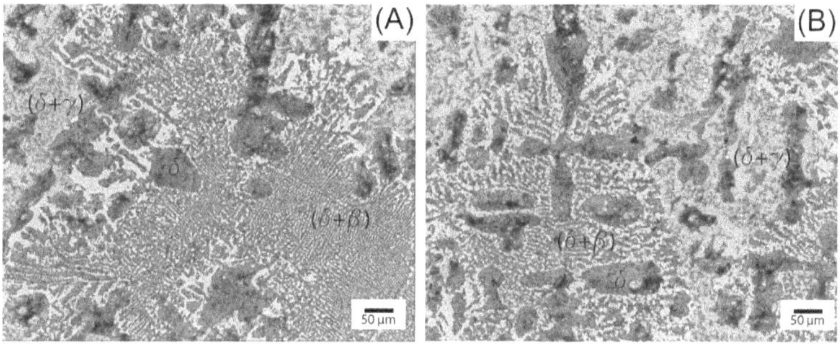

FIG. 9.29 Microstructure of alloy D showing δ, $\delta+\beta$, and $\delta+\gamma$ zones (A) and evidence of dendrite morphology (B).

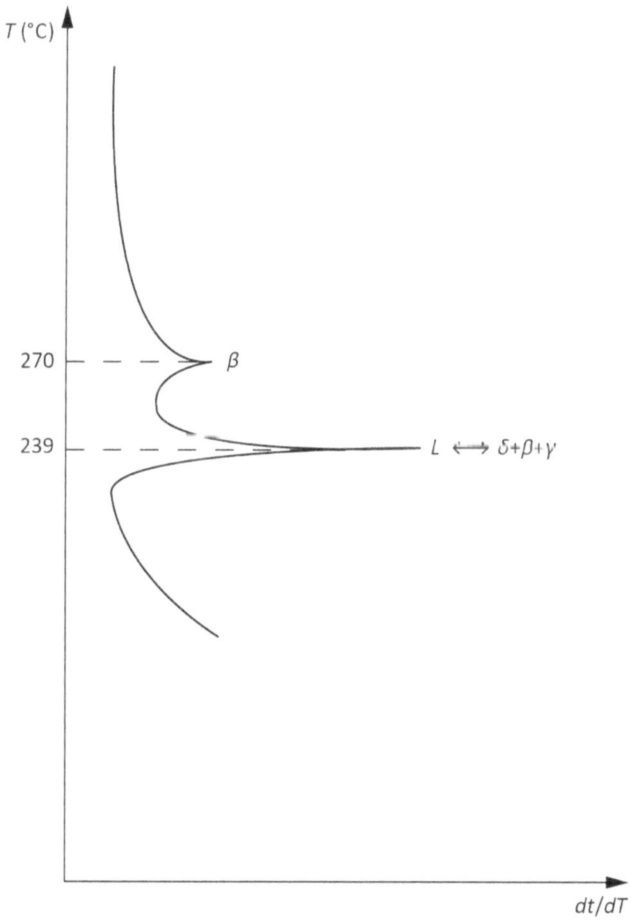

FIG. 9.30 Differential cooling curve of alloy E (75% Pb, 15% Sb, and 10% Sn).

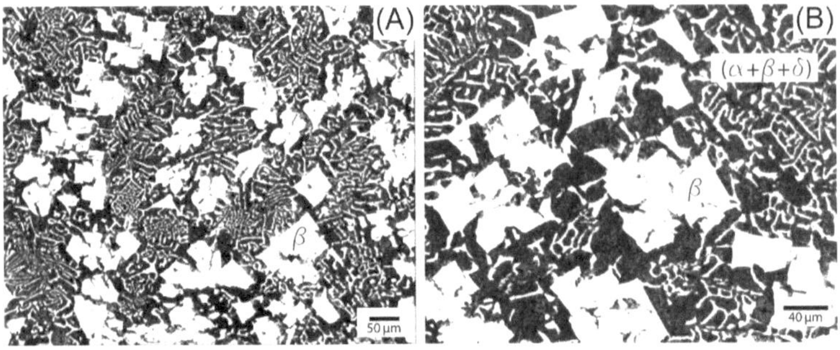

FIG. 9.31 Microstructure of alloy E showing β constituent (A) and detailed ternary eutectic (B).

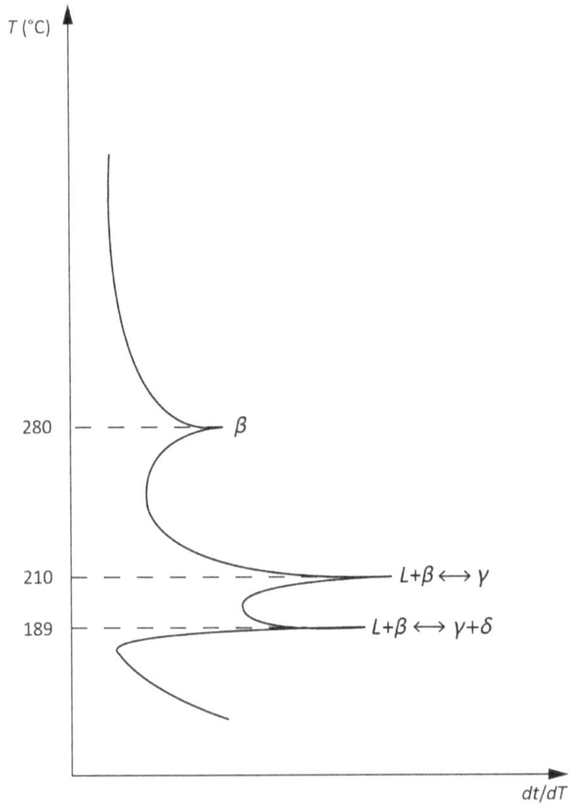

FIG. 9.32 Differential cooling curve of alloy G (20% Pb, 15% Sb, and 65% Sn).

- γ zone
  - Fig. 9.34 illustrates the cooling behavior of alloy K (17% Pb, 3% Sb, and 80% Sn) with a first peak at 205°C due to the formation of dendrites γ: the solidification trajectory in the liquidus profile follows a curve ending at

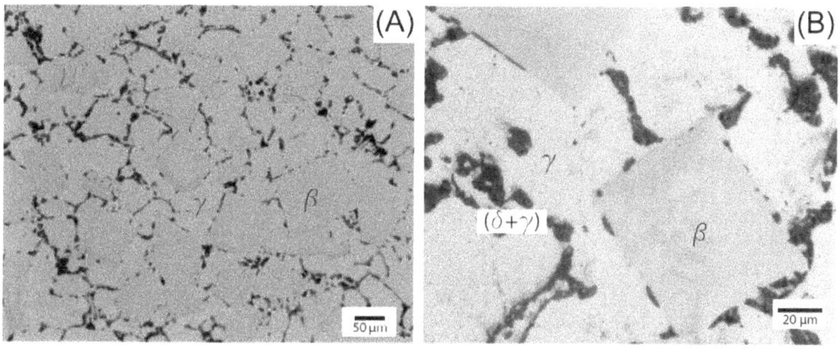

FIG. 9.33 Microstructure of alloy G showing β constituent (A) and detailed peritectic (B).

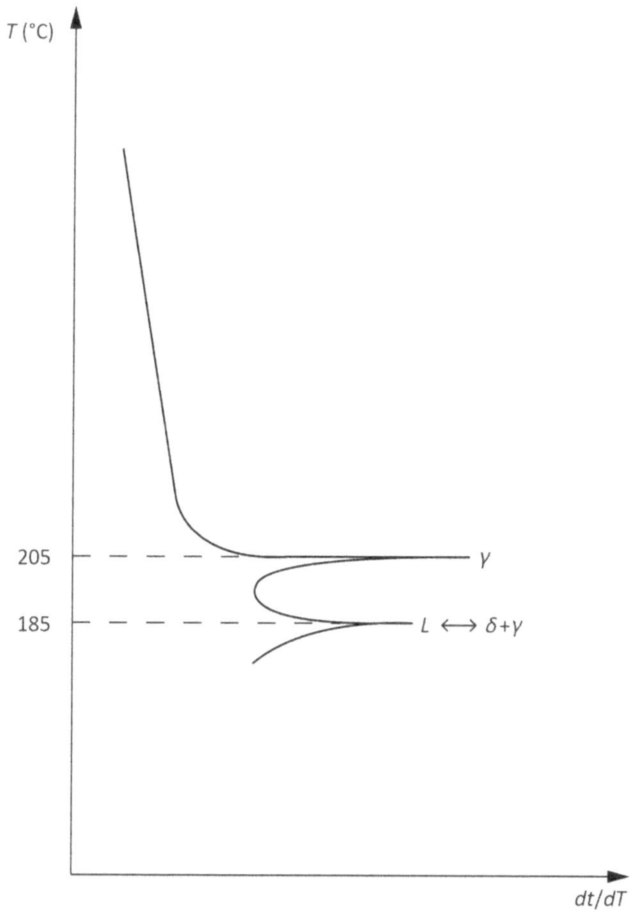

FIG. 9.34 Differential cooling curve of alloy K (17% Pb, 3% Sb, and 80% Sn).

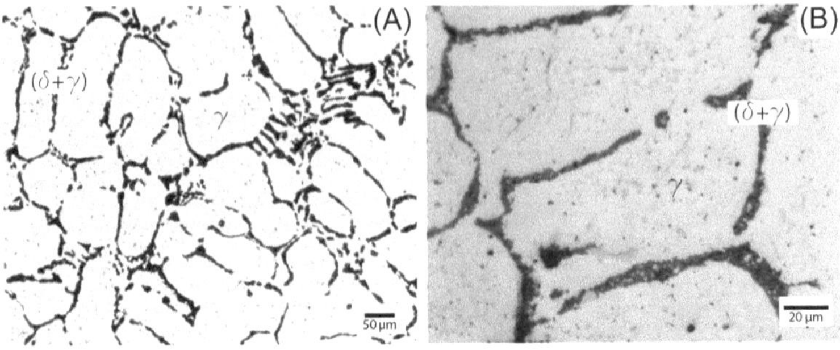

FIG. 9.35 Microstructure of alloy K showing γ constituent (A) and detailed binary eutectic (B).

W (Fig. 9.20) of the binary eutectic reaction (liquid $\leftrightarrow \delta + \gamma$). The microstructure of Fig. 9.35 shows primary dendrites γ (with rounded shapes) surrounded by binary eutectic. This microstructure is similar to those of hypereutectic binary alloys used in soldering.

9.2.2 *Al-Cu-Si System*

One of the most common ternary systems used as light alloy is Al-Cu-Si (Fig. 9.36) due to its applicability in the casting of parts, as Al-Cu alloys considerably improve their fluidity by adding Si. The selection of an alloy adequate for this type of parts must consider moldability, mechanical requirements, oxidation-corrosion resistance, and economic factors. For example Al-4% Cu alloys increase their fluidity by 20% by adding 5% Si, however this addition lowers the machinability properties.

The most common Al alloys for casting can be divided in: Al-Si, Al-Cu, Al-Cu-Si, Al-Mg, Al-Mg-Si, and Al-Zn-Mg (some of them previously explained). The Al-Cu-Si system forms a combination of α (Al lattice with Si and Cu in solid solution), Si particles and Al_2Cu.

Low amounts of Si ($<1\%$) practically maintain the maximum solubility of Cu in Al (5.7% at 572°C). Furthermore, Si forms a quasi-binary system with Al_2Cu (Fig. 9.37), with an eutectic at 571°C and 4.5% wt Si, with no intermediate phases. Alloys with this eutectic constituent are not recommended for high-toughness applications.

Solubility of both Cu and Si in Al_2Cu - Si is practically null, which leads to a weight ratio of Al_2Cu/Si of ~21, almost 11 times more Al_2Cu (black constituent) than Si in Fig. 9.38.

A liquid with a composition of 5.5% Si-27% Cu-67.5% Al suffers the following ternary eutectic at 525°C:

$$L \leftrightarrow Al + Al_2Cu + Si \tag{9.11}$$

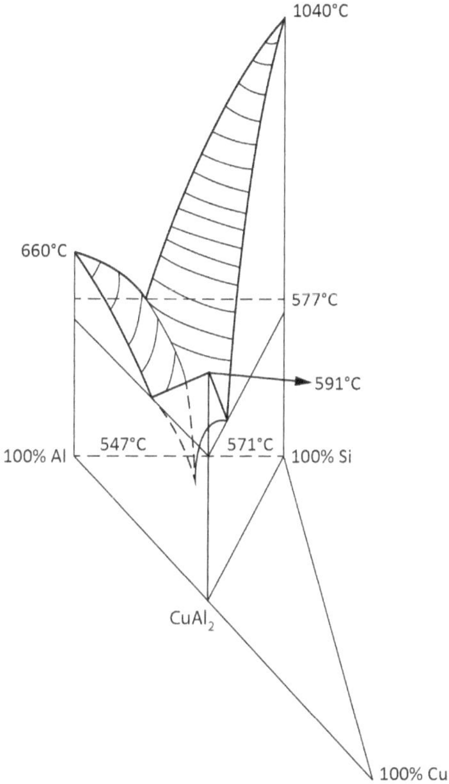

FIG. 9.36 Al-Cu-Si system (Plaza et al., 1995).

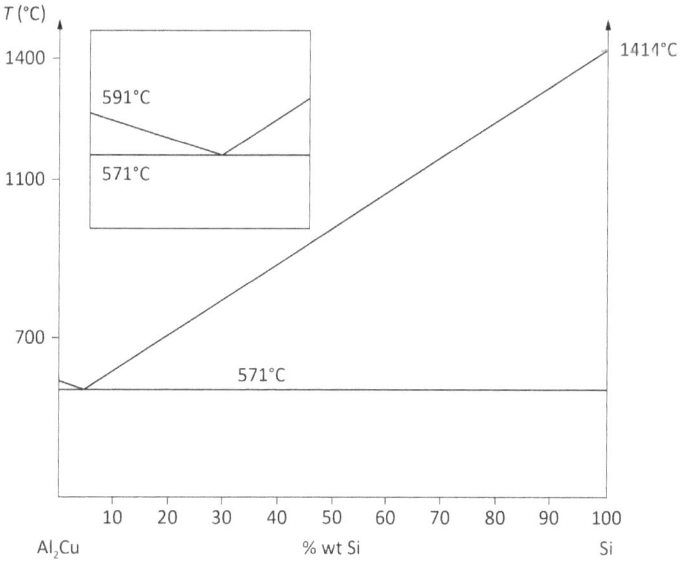

FIG. 9.37 Al_2Cu-Si phase diagram.

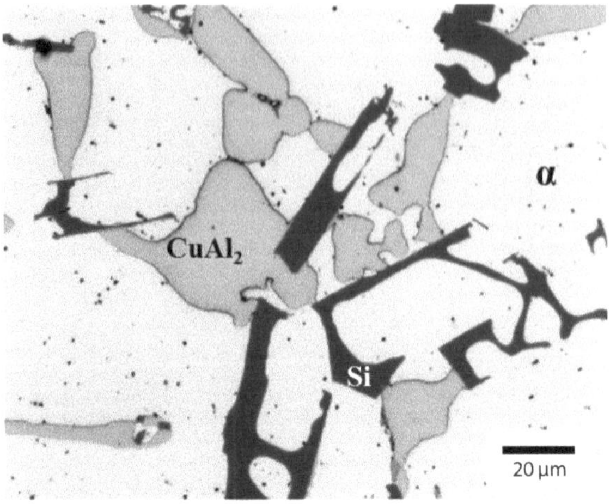

FIG. 9.38 Micrograph of alloy 63.5% Al-30% Cu-6.5% Si.

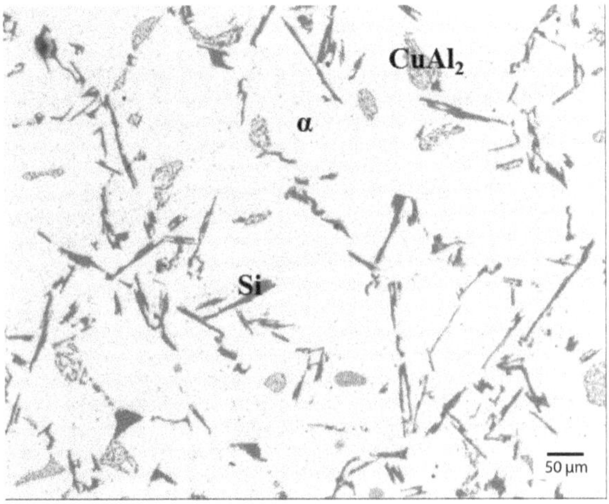

FIG. 9.39 Micrograph of alloy 69.2% Al-25.5% Cu-5.3% Si.

with weight proportions of 44.7% Al (α), 50.8% Al_2Cu, and 4.5% Si, which are equal to volumetric proportions of six times more Al_2Cu than Si, and nine times more Al than Si (Figs. 9.39–9.41). This eutectic also has low toughness.

For ternary alloys with low amount of Cu, binary eutectic reactions are influenced by Si as a nucleating agent, and thus, the morphology they present is similar to unmodified siluminum (Chapter 4). The addition of Cu reduces the size of α-Si clusters, which results in small polyhedra Si zones (Fig. 9.42).

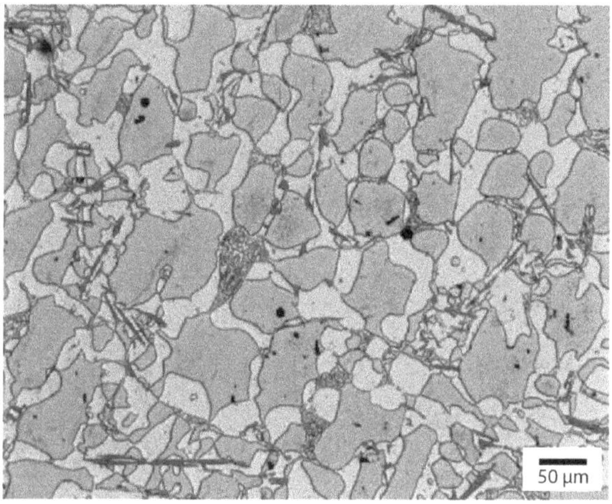

FIG. 9.40 Micrograph of alloy 61% Al-35% Cu-4% Si.

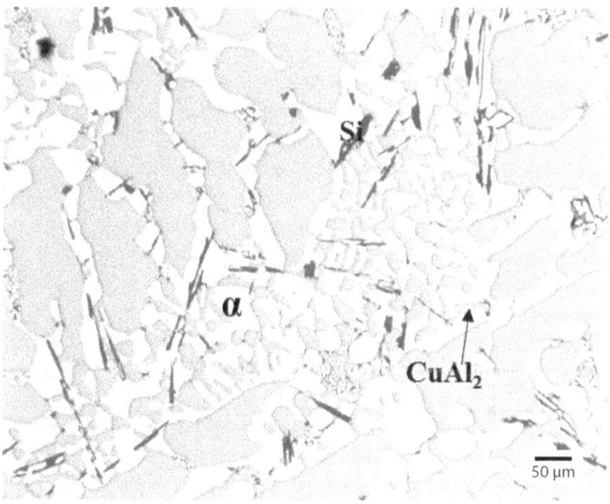

FIG. 9.41 Micrograph of alloy 84% Al-10% Cu-6% Si.

The mechanical properties of common industrial Al-Cu-Si alloys, including its heat treatment state are presented in Table 9.3.

Though increasing the amount of Cu and Si may result in a constant solidifying temperature as low as 525°C for the 67.5% Al-27% Cu-5.5% Si alloy, the industrial application is limited by the amount of hard particles producing very low toughness, therefore common casting technologies always use alloys rich in Al. Nevertheless, techniques such as rheocasting, which take advantage of the semisolid state and produce more rounded particles, result in higher mechanical properties, along wear resistance capacity (Fig. 9.43).

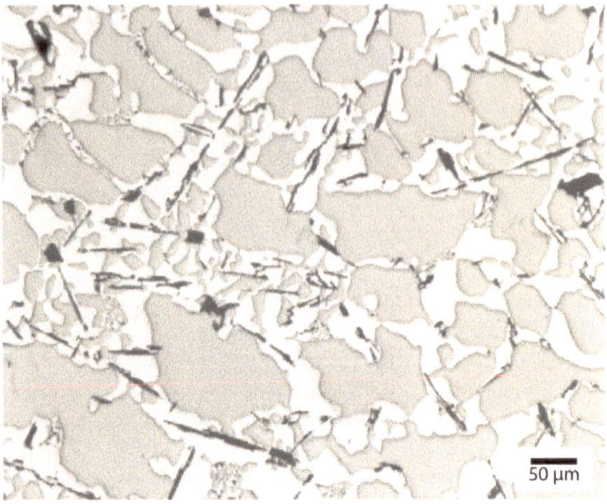

FIG. 9.42 Micrograph of alloy 78% Al-10% Cu-12% Si.

TABLE 9.3 Mechanical Properties of Industrial Al-Cu-Si Alloys

Alloy	T_M (°C)	σ_t (MPa)	σ_{max} (MPa)	Elongation (%)	Heat-Treatment State
Al-4.5 Cu-0.25 Si	206	347	436	12	T7
Al-4 Cu-3 Si	208	95	145	2.5	As-cast
Al-4.5 Cu-1.1 Si	295	110	220	8.5	T4
Al-4.5 Cu-2.5 Si	296	130	255	2.0	T4
Al-4.5 Cu-5.5 Si	308	110	195	2.0	As-cast
Al-3.5 Cu-6 Si	319	125	185	2.0	As-cast
Al-3.5 Cu-8.5 Si	380	165	330	3.0	As-cast
Al-2.5 Cu-10.5 Si	383	150	310	3.5	As-cast
Al-3.8 Cu-11.2 Si	384	165	330	2.5	As-cast
Al-4.5 Cu-17 Si	390	180	180	1	As-cast
Al-12 Si	413	145	300	2.5	As-cast
Al-5.2 Si	443	55	130	8	As-cast

T4—solubility, quenching, and natural aging treatment; T7—solubility, quenching, and stabilization treatment.

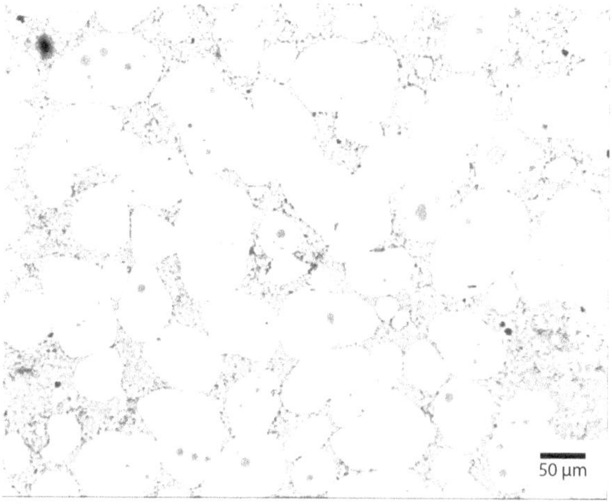

FIG. 9.43 Semisolid rheocasting microstructure.

EXERCISE 9.3

Determine the amounts of constituents α, β, and γ for the ternary eutectic H 67.5% Al-27% Cu-5.5% Si.

Data: α (98.35, 0, 1.65), β (46.4, 53.2, 0), and γ (0, 0, 100)

Solution

Locating the composition of H, α, β, and γ in the ternary diagram (Fig. 9.44), the amount of each phase can be calculated as:

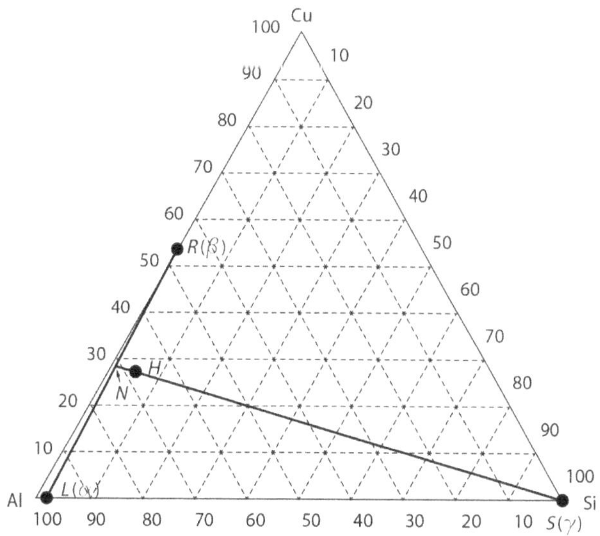

FIG. 9.44 Representation of H, α, β, and γ compositions.

$$f_\beta = \frac{P_1}{P_T} = \frac{NL}{LR} \cdot \frac{SH}{SN} = \frac{28}{53} \cdot \frac{84}{88} = 0.504 = 50.4\%$$

$$f_\alpha = \frac{P_2}{P_T} = \frac{NR}{LR} \cdot \frac{SH}{SN} = \frac{25}{53} \cdot \frac{84}{88} = 0.45 = 45\%$$

$$f_\gamma = P_3 = P_T - (P_1 + P_2) = 1 - (0.504 + 0.45) = 0.046 = 4.6\%$$

meaning 45% α (Al), 50.4% β (Al_2Cu), and 4.6% γ (Si).

REFERENCE

Plaza, D., Perosanz, J.A., Verdeja, J.I., 1995. La Estructura micrográfica, determinante para los límites en composición de aleaciones ligeras. El sistema Al-Cu-Si. Revista de Minas (11–12), 99–107.

BIBLIOGRAPHY

Apraiz, J., 1978. Fabricación de Hierros, Aceros y Fundiciones. Urmo, Bilbao.

Belzunce, J., Menendez, F., Verdeja, J.I., Pero-Sanz, J.A., 1977. Estructuras de solidificación: aleaciones Pb-Sb-Sn. Rev. Tec. Met. 222, 33–40.

Brewer, L., Chang, S.G., 1973. Metals hand. Am. Soc. Met. 8, 422–425.

Flinn, R.A., Trojan, P.K., 2006. Engineering Materials and Their Applications. Hougton Mifflin, Boston, MA.

Pero-Sanz, J., 1967. Diagramas Ternarios. CITG, San Sebastián.

Rhines, F., 1956. Phase Diagrams in Metallurgy: Their Development and Application. McGraw-Hill, New York.

Trosenqvist, T., 2004. Principles of Extractive Metallurgy, second ed. McGraw-Hill, New York.

Weast, R., 2014. Handbook of Chemistry and Physics, 95th ed. The Chemical Rubber Co., Cleveland, OH.

Index

Note: Page numbers followed by *f* indicate figures, *t* indicate tables, *b* indicate boxes, and *np* indicate footnote.

A

Activity coefficient, 127–128
Affinity, 35–37, 43, 67–78
  eutectic, 67, 74–78
  intermetallic, 67, 71–74
  solid solution, 67, 78–91
Age hardening, 221–230
  overaging, 222
Alloys
  Babbits, 342, 343*t*
  hardened, 222–223
  Heussler alloy, 251–252
  low melting point alloys, 338
Alpha-stabilizing elements, 280–285
Andrews equation, 261–262
Annealing, 162, 163*f*, 221
  homogenization, 221
Anti-phase, 250–251, 250*f*
  boundary, 251–252
Atomic
  Barrett diameter, 14–16*t*
  bond, 1, 5–6, 26
  diameter, 14–16*t*
  Goldschmidt diameter, 14–16*t*
  insertion, 79–82
  interatomic distance, 26
  interatomic forces, 5–6
  kinetic energy, 4–6
  mean speed, 4–6
  number, 10–13*t*, 13–14
  octahedral, 79–82
  packing factor, 17–18, 17–18*b*
  radius, 14, 18
  stiffness of bond, 25–28
  tetrahedral, 27
  weight, 10–13*t*, 13–14
Austenite, 212–214*t*, 230, 233, 244, 258–268
  austenite-stabilizing elements, 259, 261–262, 278–280, 283, 285, 293
  Koistinen-Marburger equation, 302–303
  transformation kinetics, 258–262
Avrami equation, 309–310

B

Bainite
  lower bainite, 294–295, 295*f*
  tempering, 323
  transformation, 293–296
  upper bainite, 294, 294*f*
Blowholes, 196, 196–197*f*, 201
Bravais lattice, 6–7, 13–14
  density, 10–13*t*, 17*b*
Brazing, 36, 36–37*b*

C

Carbides, 221–222, 232–234
  carbide-stabilizing elements, 283*np*, 303, 316–320
Casting
  ability, 202–204
  ferrous, 272, 283–284
  hot tearing, 206–208
  soundness, 204–205
  system, 187, 187*f*
  vacuum, 180, 184, 185*f*, 189, 194
Cementite, 255, 256*f*, 258, 268, 270, 272–273
  primary, 272
  tertiary, 262, 267–270, 269*f*
  vermicular, 262, 269–270
Centerline feeding resistance, 203
Chemical corrector, 194
Chiller, 182–183, 186
Chvorinov equation, 59–65, 175
Cold-work, 20–21, 25, 25*f*
Compounds
  abnormal valence, 236
  affinity, 241–244
  electronic, 235–238
  Hume-Rothery, 236
  intermetallic, 230–238
  interstitial, 232, 248–249
Conductivity
  electric, 1
  thermal, 20

Constituent, 69–70, 73–78
  disperse, 69, 70–71*b*, 75–76
  matrix, 69, 70–71*b*, 71, 73–78
  primary, 73–74, 76–78
Continuous cooling rate (CCR), 305
Continuous cooling transformation diagram
  1038 steel, 312*f*
  5140 steel, 313*f*
Contraction/shrinkage, 6, 14, 176–187
Controlled deformation process, 286
Controlled rolling process, 286
Cooling
  curves, 68–69, 68*f*, 70*f*, 71–73, 72*f*, 75*f*, 78*f*, 85–87, 86*f*, 210, 211*f*
    differential, 343, 345–351*f*
  rates, 255, 269, 275–278, 289, 291, 301, 304
Copper
  deoxidized phosphorous, 111
  oxygen-free, 111–112
  oxygen-free high conductivity, 111
  tough-pitch, 111
Coring, 86, 89, 143, 156–157, 161, 163–164*b*
Creep, 20
Crystalline system, 4–18
  body-centered cubic/bcc, 6–18
  body-centered tetragonal/bct, 6–18
  compact hexagonal/hcp, 6–18
  face-centered cubic/fcc, 6–18
Crystals
  heterodesmic, 230
  homodesmic, 230

D

Davenport's X constituent, 293
Debye-Scherrer method, 14
Deformation, 1, 20–21, 25–26, 28
  cold-work, 20–21
  hot-work, 20–21
Degrees of freedom, 94–95, 112
Dendrite, 34, 50–51, 171, 178, 203–204
Dendritic grain growth, 34, 51*f*, 62*f*, 63
Density, 10–13*t*, 13–14, 17, 175–176, 177*t*, 178, 205
Diffusion, 46–48, 51–52, 52*f*, 54*f*, 55, 65
  constant, 51, 54
  variable, 52, 54
Dilatometry, 259, 261
Disaggregation, 37, 45–46
Dislocation, 43
  tangles, 43
Driving force, 39, 51
Ductile-brittle transition temperature (DBTT), 285
Dulong-Petit law, 58

E

Electronegativity, 1–4, 2–3*t*
Energy
  activation, 42
  excess of free energy, 128
  grain boundary free energy, 34, 37, 38*f*
  interaction, 128
  interfacial free energy, 36–37*t*, 39
  partial molar free energy, 127
  surface free energy, 31–39
Enthalpy, 18–20, 127–128
Entropy, 18–20, 21*t*, 42, 128
Equilibrium, 93–96
  bivariant, 94–95
  distance, 6, 14–16*t*, 26–28
  factors, 94
  invariant, 93–130, 326–329, 334, 342–343
  monovariant, 94–95
  perfect, 133
  phase rule, 87, 325
  tetra-phasic equilibriums, 342–343
  tri-phasic equilibriums, 342
Eutectic, 67, 74–78, 93–130
  abnormal, 104–108
  acicular, 77
  affinity, 67, 74–78
  binary, 94, 96–112, 210
  globular, 77
  hypereutectic, 96–97
  hypoeutectic, 96–97, 103–107, 103*f*, 130
  lamellar, 77
  needle, 77
  reaction, 95–112
  rod-like, 77
  sandwich, 77
Eutectoid, 95, 126, 126*f*, 128
  binary austenite, 259, 278
  point, 276, 278, 280
  reaction, 261, 278
  temperature, 261, 276, 278, 280, 281*f*

Expansion, 14, 17, 20
linear expansion coefficient, 17, 24*t*
Extensometry, 26
Exudation, 153–154

F
Ferrite, 255–323
allotriomorphic, 269
ferrite-stabilizing elements, 280–285
proeutectoid ferrite, 269, 274–275, 278, 291–292, 323
Filler metal, 35–37, 96
Fine-grained steels, 315–316
Force
attraction, 5–6
repulsion, 6

G
Gamma-stabilizing elements, 259, 261–262, 278–280, 283, 285, 293
Gibbs
free energy, 18–28, 31–39, 126–130
free energy curve, 31
law, 94–95
Grain, 1, 6–7
chill, 62–63, 171, 172*f*
cluster, 39–42, 40*f*, 45–46, 58–59
columnar, 62–63
elongated, 25
equiaxed, 62–63
growth rate, 50–56
size, 43
Graphite stabilizing elements, 284–285
Guinier-Preston zones, 223*np*

H
Hooke's law, 43
Horizontal rule, 87
Hot shortness, 37
Hot tearing, 206–208
Hot-work temperature, 25
Hume-Rothery rules, 78–79
Hydrogen embrittlement, 111–112
Hypereutectic, 272–273
white castings, 272–273
Hypereutectoid
annealed steels, 273
steels, 259, 263, 268, 273, 291, 315
Hypoeutectic, 270–271
white castings, 270–271
Hypoeutectoid
annealed steels, 273–274
steels, 259, 263, 269, 273, 278, 280–281, 291, 303, 315

I
Ingot skin, 196, 201
Insolubility, 67–78
Interface, 31, 32*f*, 34, 35*f*, 39–40, 46, 59–62
Intergranular
corrosion, 37, 39
melting, 165–169
wetting, 38–39, 38*f*
Intermetallic, 67, 71–73, 72*f*
compounds, 230–238
Interphase, 113–114, 124, 145
Invariant reaction, 95, 124–126

J
Johnson-Mehl-Avrami law, 53, 217–218

K
Kurdjumov expression, 294–295

L
Latent heat, 31, 60–62
fusion/melting, 18–20, 39, 51, 59–60, 63–65
Lattice parameter, 13–14
Le Chatelier rule, 95, 210
Ledeburite, 263–265, 270, 271–272*f*, 272–273, 283–284
Lever rule, 88, 96–97, 100–103, 114–119
Liquid contraction, 176–187

M
Maier and Kelley function, 128
Martensite, 290, 291*f*, 294–301, 296*f*, 299*f*, 304, 307–308, 314, 319–322
aged steels, 229
Hollomon-Jaffe equation, 301
invariant plane strain transformation, 299
Irvine equation, 301–302
Krupp sickness, 319
Nerenberg equation, 301
secondary hardness, 320
shearing mechanisms, 299
tempered embrittlement, 319
tempering, 319–322
transformation, 296–303

Matrix constituent, 70, 73–74, 77–78, 96–97
Maxwell-Boltzmann law, 4–6
Mechanical resistance, 43
Meniscus, 34–35
Metallic bond, 1, 5–6
Metallic glass, 46, 50, 58–59
Microalloying, 232, 285–289
Microstructure
  acicular, 245
  banded, 25
  cluster, 39–42, 42*f*, 45–46, 58–59
  cold worked, 20
  columnar, 151–153, 171–176
  equiaxed, 7*f*, 25, 62–63, 152–153, 171, 204–205
  lamellar, 37
  partially recrystallized, 25*f*
  recrystallized, 7*f*, 20–21
  solidification texture, 62
  strain-free, 25
Monotectic, 95, 119–123

N

Nucleation, 39–42, 210–222, 212*f*, 234, 239–242, 244, 251
  barriers, 216–217
  heterogeneous, 43, 46–48, 47*f*, 49*f*, 50, 55, 55*f*, 60
  homogeneous, 39–42, 46, 47*f*, 48–50, 49*f*, 54, 55*f*
  nucleating agent, 43–45
  rate, 45–50, 52–53, 56–58
Nucleus, 39–42, 50–51
  critical radius, 40–41, 43–44, 52–53
  supercritic, 42, 45–46

O

Order/disorder transformations, 248–253
  long-range order parameter, 249–250
Overaging, 222
Overheating, 279–280, 315–319

P

Parker's technique, 123
Partition coefficient, 88, 133, 135*f*, 136–137, 143–145, 148–149, 150*t*, 151–153
Pauling criteria, 2–3*t*
Pearlite
  coarse, 290–291, 291*f*
  fine, 290–292, 311
  tempering, 323
  transformation, 289–293
Penetration angle, 37
Peritectic, 95, 114, 118, 130
  binary reaction, 112–118, 331–332, 331*f*, 333*f*, 342, 344, 348
  2nd class reaction, 334–338
  3rd class reaction, 334–338
Peritectoid, 95, 126
Phase, 93
  final, 230–238
  intermediate, 230–238, 244
Phase diagram
  Al-Cu, 225*f*
  Al-Mg, 166*f*
  Al-Si, 106*f*
  Al-Zn, 219–221, 220*f*
  Au-Si, 77, 78*f*
  Bi-Cu, 70, 70*f*
  Cu-Al, 102*f*
  Cu-Be, 228*f*
  Cu-O, 110*f*
  Cu-Sn, 166*f*, 239*f*
  Cu-Zn, 235*f*
  Fe-C-Mn (gamma section), 280*f*
  Fe-C-Mo (gamma section), 284*f*
  Fe-C-Si (gamma section), 283*f*
  Fe-C-Ti (gamma section), 285*f*
  Fe-Cr, 89*f*
  Fe-Fe_3C, 262–268, 265*f*
  Fe-Mn, 278, 279*f*
  Fe-S, 125*f*
  Fe-Si, 282*f*
  Ga-As, 73*f*
  Ge-In, 70*f*
  Ice-Salt, 77*f*
  Na-Zn, 124*f*
  Ni-Cu, 87*f*, 139*f*
  Pb-Sb, 220*f*
  Pb-Sn, 96*f*
  Pt-Ag, 118*f*
  Zn-Pb, 122*f*
Phase diagram computational calculation methods, 126–130
Phase rule, 87
Phase transformation kinetics, 68–69
Pickering equation, 273–275
Pipe, 178, 179*f*, 184, 194–195
Plating, 35
Precipitates, 261, 315–316
  Precipitate-forming elements, 285–289
Precipitation, 210–221
  discontinuous, 211–219

hardening heat treatment, 221–230
localized, 211–219, 221–222

Q

Quenching, 221–223, 225, 227, 230, 240, 285–286, 296, 302, 308
rate, 305

R

Radius
critical, 40–43, 40*f*, 45, 52–53
Recrystallization, 7*f*, 20–21, 25, 25*f*
dynamic, 20–21
static, 20
Refractory metals, 20
Rheocasting, 355–358
Rim-core junction, 196
Riser, 178–184, 186–187, 199–200, 204
Roozebum representation, 325, 326*f*

S

Scheil equation, 136–139
Segregation, 143, 155, 157, 202
band, 143
dendritic, 161–162
gravity, 154–156
in gasses, 156–157
indexes, 158–160
inverse, 153–154
local, 156–158
macrosegregation, 142–160
microsegregation, 143, 161–164
minor segregation, 161–162, 164
negative, 144–145, 152*f*, 157*f*
normal, 143–153
positive, 144–145, 152*f*, 157*f*
vertical, 155–156
Shrinkage/Contraction, 6, 14, 176–187
macroshrinkage, 178–180, 183–186, 199–200
microshrinkage, 178–180, 185–186, 197–199, 204–205
Sievert's Law, 188–191
Sintectic, 95, 123–124
Sintering, 156
Skin bubbles, 196, 200–202
Softening coefficient, 320–322
Soldering, 35
Solidification
contour, 150
curve, 56–59, 69, 71–72
equilibrium, 69, 86
guided, 186
invariant, 93–130
real, 59
temperature, 18–28, 31, 43, 74
texture, 62
unidirectional, 133–142
zone melting, 133–142
Solid solution, 67, 78–79, 89–91
binary, 83–85*t*
interstitial, 67, 79–82
substitutional, 67, 78–79
Solubility, 39, 56, 67–78
curve, 216
partial, 83–85
relative, 85
total, 85–91
transformations, 210–238, 244–248
Solute distribution law, 136, 140–142
Solution affinity, 67, 85
Specific heat, 18–20, 127–128
Spinodals, 219–221
Sprue, 187
Steels
dual phase (DP), 286
high-strength low-alloy (HSLA), 285–286
hypereutectoid, 259, 263, 268, 273, 291, 315
hypereutectoid annealed, 273
hypoeutectoid, 259, 263, 269, 273, 278, 280–281, 291–292, 303, 315
hypoeutectoid annealed, 273
maraging, 229, 252–254
stainless steels, 221–222, 230, 231*t*, 234, 283
Steven-Haynes bainite equation, 295–296
Steven-Haynes martensite equation, 301–302
Stiffness of bond, 25–28
Stirring, 44–45, 146*f*, 151
Stoichiometric composition, 71–74
Structural hardening, 222–227, 224*t*, 227*f*, 229–230
Superlattices, 248–250
Surface free energy, 31–39
Syphoning, 184
System component, 93–94, 104–105, 127–128

T

Tammann method, 97–100, 98*f*
Temperature
critical, 210, 250
gradient, 59–60
melting, 18–20, 21–24*t*, 25, 32–35
solidification, 18–28, 31, 43

Tempering, 222, 229, 285, 297
  bainite, 323
  martensite, 319–322
  pearlite, 323
  subcritical, 323
Ternary alloys, 325–358
  Al-Cu-Si, 352–358
  Pb-Sb-Sn, 341–358
Thermal
  capacity, 58, 58*t*
  hysteresis, 263
  stress, 20
Tie-line, 87, 329–330
Tie-triangle, 326–329, 332, 341
Transformation
  allotropic, 17, 209, 238–243, 255, 258–259, 261, 263, 269, 289, 293, 314, 323
  rate, 126–127, 277, 289
Troostite, 290, 291*f*
TTT curve, 53–54, 54*f*, 58–59, 290*f*, 295–296, 303–305, 307

U

Undercooling, 31–44, 41*f*, 44*f*, 46–48, 47*f*, 49*f*, 50–51, 52*f*, 53, 55*f*, 60–62, 75, 151–153, 171, 204, 215, 240–241, 265*f*, 266, 293, 295, 301, 308
  constitutional, 151–153
Unit cell, 6–8, 8*f*, 13–14, 17

V

Vacancy, 43
Vein, 157–158

W

Wetting, 34–39, 38*f*, 43, 46–48, 49*f*, 50
  angle θ, 34–35, 38–39, 46–48, 49*f*, 50
Widmastätten
  plates, 269
  structure, 245, 270*f*, 277–278, 293
Work hardening, 20–21
Work temperatures, 25

X

X-ray diffractometry, 6–7, 14

Y

Yield stress, 1, 20, 192*np*, 207–209, 230
Young's modulus, 26–28, 27*t*

Kolibri 360 GmbH
Ferdinand-Jühlke-Straße 7
99095 Erfurt, Deutschland
produktsicherheit@kolibri360.de